AF785510

Emergence, Complexity and Computation

Volume 5

Series Editors

Ivan Zelinka, Ostrava-Poruba, Czech Republic
Andrew Adamatzky, Bristol, UK
Guanrong Chen, Kowloon Tong, Hong Kong, People's Republic of China

Editorial Board

Otto Rössler, Tübingen, Germany
Edward Ott, College Park, USA
Ajith Abraham, MirLabs, USA
Vaclav Snasel, Czech Republic
Jouni Lampinen, Finland
Emilio Corchado, Spain
Hendrik Richter, Germany
Martin Middendorf, Germany
Mohammed Chadli, France
Juan C. Burguillo, Spain
Sergej Čelikovský, Czech Republic
Donald Davendra, Czech Republic
Juan A. Rodriguez-Aguilar, Spain
Ana Lucia C. Bazzan, Porto Alegre, Brazil
Gheorghe Păun, Bucharest, Romania
Linqiang Pan, Wuhan, China
Ivo Vondrák, Czech Republic
Andrew Ilachinski, USA

For further volumes:
http://www.springer.com/series/10624

Ivan Zelinka · Ali Sanayei
Hector Zenil · Otto E. Rössler
Editors

How Nature Works

Complexity in Interdisciplinary Research and Applications

Springer

Editors
Ivan Zelinka
Faculty of Electrical Engineering
and Computer Science
Department of Computer Science
VŠB-TUO
Ostrava-Poruba
Czech Republic

Ali Sanayei
Institute for Theoretical Physics
University of Tübingen
Tübingen
Germany

Hector Zenil
Department of Computer Science
Kroto Research Institute
University of Sheffield
Portobello
UK

Otto E. Rössler
Institute for Physical and Theoretical
Chemistry
University of Tübingen
Tübingen
Germany

ISSN 2194-7287 ISSN 2194-7295 (electronic)
ISBN 978-3-319-00253-8 ISBN 978-3-319-00254-5 (eBook)
DOI 10.1007/978-3-319-00254-5
Springer Cham Heidelberg New York Dordrecht London

Library of Congress Control Number: 2013942658

© Springer International Publishing Switzerland 2014
This work is subject to copyright. All rights are reserved by the Publisher, whether the whole or part of the material is concerned, specifically the rights of translation, reprinting, reuse of illustrations, recitation, broadcasting, reproduction on microfilms or in any other physical way, and transmission or information storage and retrieval, electronic adaptation, computer software, or by similar or dissimilar methodology now known or hereafter developed. Exempted from this legal reservation are brief excerpts in connection with reviews or scholarly analysis or material supplied specifically for the purpose of being entered and executed on a computer system, for exclusive use by the purchaser of the work. Duplication of this publication or parts thereof is permitted only under the provisions of the Copyright Law of the Publisher's location, in its current version, and permission for use must always be obtained from Springer. Permissions for use may be obtained through Rights Link at the Copyright Clearance Center. Violations are liable to prosecution under the respective Copyright Law.
The use of general descriptive names, registered names, trademarks, service marks, etc. in this publication does not imply, even in the absence of a specific statement, that such names are exempt from the relevant protective laws and regulations and therefore free for general use.
While the advice and information in this book are believed to be true and accurate at the date of publication, neither the authors nor the editors nor the publisher can accept any legal responsibility for any errors or omissions that may be made. The publisher makes no warranty, express or implied, with respect to the material contained herein.

Printed on acid-free paper

Springer is part of Springer Science+Business Media (www.springer.com)

Preface

The book you are holding in your hands is the outcome of the "2012 Interdisciplinary Symposium on Complex Systems" held at the beautiful Mediterranean island of Kos of pre-and post-Socratic fame. The event was conceived as a continuation of our series of symposia in the science of complex systems. Only through bringing scientists and philosophers alike from different areas of modern research together, as the organizers felt, the necessary dialogue and heat can be generated in which a new paradigm can take a more definitive shape. The paradigm itself—a "science of complexity" in a both overarching and sharply delineated sense—has different and often not convergent trends because a "single" definition of complexity does not exist. Perhaps, one of the main reasons for not yet having a unique definition is due to the lack of agreement whether complexity is an inherent property of Nature or it only appears when one builds a model for a given phenomenon. Accordingly, prestigious scientists and philosophers with different points of view were brought together along with youthful budding researchers, who jointly contributed previously unheard examples to form an efficient engine to produce new work. Sense of wonder as to what would happen in the next talk was palpable several times. The panel discussion—not reproduced here—did its own part in fostering a spirit of friendship and progress in spite of different viewpoints. The finished papers reproduced here reflect this unique spirit of mental and physical cooperation across disciplines and continents.

The motivation to prepare this book was based on a few facts. The main one is that the research field on complexity is an interesting area that is under intensive investigation from the viewpoint of many branches of science today. Complexity theory with its applications can be found in biology, physics, economy, chemical technologies, aircraft industry, job scheduling, urban planning, and others. Complex systems and their behavior are very important in engineering, because such behavior can be used in many interesting applications as well as in the interdisciplinary combinations forming on a theoretical level. This book is written to present simplified versions of experiments and thought experiments to show how, in principle, complexity can be put to use. Collecting different reasoned opinions with likely different ideas and then assembling a book comprising theoretical physics, mathematics, engineering, and philosophy might appear odd; nevertheless, we found that our endeavor achieved a high efficiency at arriving at a synthesis in this manner. Furthermore, the history of science has demonstrated that pure mathematics and physics can combine beautifully with practical engineering, as the example of "the strangest man" shows.[1]

The book consists of 12 selected papers of the symposium starting with a comprehensive overview and classification of complexity problems. Readers can also find other interesting papers about complexity, its observation, modeling, and its applications in solving various problems. More concretely, readers will have an encounter with the structural complexity of vortex flows, the use of chaotic dynamics within evolutionary algorithms, complexity in synthetic biology, types of complexity hidden inside evolutionary dynamics and possible controlling methods, complexity of rugged landscapes, and more. All selected papers represent innovative ideas, philosophical overviews, and state-of-the-art discussions on the aspects of complexity.

The book can be useful as an instructional material for senior undergraduate and entry-level graduate students in computer science, physics, applied mathematics, and engineering-type work in the area of complexity. Researchers, who are interested how complexity and evolutionary algorithms are merged together as well as researchers interested in the ramifications of complexity in various fields of science and its applications, will find this book very useful as a stepping stone. The book can also be valuable as a resource of material for practitioners who want to apply complexity to solve real-life problems in their own challenging applications. It goes without saying that this book does not encompass all aspects of complexity types and fields of research due to its limited space. Only the main ideas and results of selected papers are reported here. The authors and editors hope that readers will be inspired to do their own experiments and simulations, based on information reported in this book, thereby moving beyond the scope of the book.

As a token of the participants' affection for genuine progress in the decomplexification of the universe, we dedicate this book to Fabiola Gianotti who bears much responsibility for the discovery in 2012 of the Higgs Boson in fundamental physics. At the same time, the question still remains open whether knowing more about Nature can decomplexify it, or else makes it even nimbler to the fishing mind.

January 2013

Ivan Zelinka
Ali Sanayei
Hector Zenil
Otto E. Rössler

---

[1] Four years before his death, Niels Bohr told a colleague that of all the people who had ever visited this institute, Dirac—an engineer by training—was "the strangest man" (cf. Graham Farmelo's colorful biography of Paul Dirac under this title).

Contents

Complexity Decomplexified: A List of 200+ Results Encountered Over 55 Years

Otto E. Rössler

Abstract The present list was compiled by a "specialist for non-specialization" who owes his scientific identity to the masters of three disciplines: physicist Carl-Friedrich von Weizsäcker, biologist Konrad Lorenz and mathematician Bob Rosen. With the in retrospect best findings compressed into a line or two, the synopsis brings hidden patterns to the fore. Simultaneously, the individual results become maximally vulnerable—so as to facilitate improvement or falsification.

1 Philosophical Preface

Descartes re-invented the rational world of Heraclitus. Specifically he asked the following question (paraphrased): "Do the 'assignment conditions' that we find ourselves glued to (the body, the Now, the qualia including color and joy) represent an acceptable state of affairs?" The answer is "yes," Descartes proposed: if and only if the other two conditions that hold us in their grip—the "laws" and the "initial conditions" that momentarily apply within the laws (to use Newton's later terms)—are mathematically consistent. As long as this "machine conjecture" is fulfilled empirically, an infinite privilege separates the conscious observer from all other inhabitants of the world: The others become "mere machines" in the experience of the first (so that he may, for example, do a brain operation on one of them to save this life). Lévinas called this state of one's being totally outside the other's interior side, "exteriority". The subject has the option of not misusing the infinite power of exteriority, by acting fairly towards the poor "machine" of the other so as if it possessed a subjective side of its own—even though this cannot be proved and indeed is absurd to assume (were there not the miracle of the first's

O. E. Rössler (✉)
Faculty of Science, University of Tubingen, Auf der Morgenstelle 8, 72076 Tubingen, Germany
e-mail: oeross00@yahoo.com

I. Zelinka et al. (eds.), *How Nature Works*, Emergence, Complexity and Computation 5, DOI: 10.1007/978-3-319-00254-5_1,
© Springer International Publishing Switzerland 2014

own consciousness). A single act of not misusing the infinite power of exteriority, performed by the inmate of the dream of consciousness on a fellow machine, would then put the Dream-Giving Instance to shame—unless it is benevolent itself. The fact that this risk is being taken by the DGI is living proof—according to Descartes—that the chain of colorful subjective Nows, imposed on the victim of consciousness, is not a "bad dream". But this insurance applies only as long as the "steel fibers" of the Cartesian coordinates, which Descartes had proposed to search for to mathematically fit the colorless sub-portion of experience (its "Hades part"), prove to be consistent—a machine. This empirical question endows the study of the steel fibers with maximal dignity. In the Greek Hades, all quantitative relations valid in our own upper world were preserved—except for the "blood" that endows them with color and substance. Therefore the merely relational (or "shadow") part of experience becomes an instrument by which to do good to one's fellow inhabitants of the dream who by their being machines are totally given into the dreamer's hand, as hostages. This "exteriority theory" (Lévinas) endows science with an infinite dignity—as long as it is empirically consistent. The modern part of this Cartesian checking task—to include quantum mechanics with its indeterminism and nonlocality as being explicable by a "micro assignment" in a transfinitely exact deterministic universe—was taken up by Everett in the footsteps of Einstein. This rational "endo" approach is likely to succeed in an experiment first conceived by Susan J. Feingold and later proposed to ESA by Anton Zeilinger more than a decade ago. (I thank Ali Sanayei and Ivan Zelinka for discussions and Stephen Wolfram for encouragement.)

2 The List

Introductory sentence: This is a list of scientific findings, almost all published, accrued over time. They can be referred-to via the Roman number added at the end of each "twitter." Later related work is filed-in so that the list is not entirely chronological. Sometimes several points are made out of one paper to facilitate continuation and/or criticism by the reader.

Energy-saving voice-signal proportional amplitude-modulation (made distortion-free by negative feedback between rectified high-frequency output and low-frequency input) I

Z-incision (a non-mutilating circumcision method) II

"Invisible machines," consisting of virtually infinitely many non-negative chemical variables that are almost all zero initially, (with arbitrarily long delays at very low concentrations) III

Chemical evolution is a special case: It forms an Erdös-type growing automaton (similarly Stu Kauffman, Joel Cohen and Koichiro Matsuno) IV

Far-from-equilibrium statistical mechanics and chemical kinetics predict the emergence of life with C–C–C– backbones in liquid water on earth and Europa and Enceladus (and with B–N–B–N– backbones in liquid ammonia inside Jupiter; with Artur P. Schmidt) V

Teilhard's "second arrow" in statistical thermodynamics is a valid description of the implied asymptotic approach towards "point Omega" VI

"Recursive evolution": Evolution improves evolution in the first place (with Michael Conrad in the footsteps of John von Neumann and John Holland) VII

Unlike "metabolic adaptation" (Darwin) which is non-predictive in its history-dependent details, "positional adaptation" (discovered in a discussion with Konrad Lorenz as being of equal rank) is predictive VIII

"What are brains for?" is a well-posed scientific question (in that new science of deductive biology) IX

"The Rossler task" (Michael Conrad) or the "decision-type traveling salesman problem" (as its re-discoverers Garey and Johnson called it 5 years later in their great book "Computers and Intractability") X

Ric Charnov's "optimal foraging theory" is closely related (finding things "just in time" is what brains are for) XI

Gödel's "incompleteness theorem" can be seen as a limiting solution to the NP-complete "traveling salesman-with-alarmclocks problem" (so that incompleteness suddenly becomes intuitive mathematically) XII

"The bacterial brain" (residing in the cell membrane along with sensors and motors) implements a local solution to the traveling-salesman-with-alarmclocks problem (with Hans Bremermann and Dan Koshland) XIII

"The brain equation" yields a maximally efficient local solution to the decision-type traveling-salesman-with-alarmclocks problem XIV

The brain equation attaches a positive or negative weight to all closest sources of different types in a lawfully smeared-out, distance- and time-dependent fashion, so that an optimum "sum direction" results (since all directions are attached an either finite or infinite, positive or negative sum weight) XV

Nonexistence of a "eusocial brain equation" (with Wilfried Musterle and Oswald Berthold) XVI

"A universal brain": The brain equation combined with a powerful "universal simulator" (or synonymously "cognitive map system" or "VR"–virtual-reality–machine in the sense of William Gibson) XVII

The combined system (brain equation plus artificial cognitive map system with overlap buffer and long-term storage device) is what Bill Seaman calls a "Neo-sentient" XVIII

The "sinc algorithm" (real-space equivalent to a Fourier window in frequency space) can be approximated by a multi-level, multi-resolution, both ascending and descending, Reichardt-von-Foerster type multi-layer neural net (with Bernhard Uehleke) XIX

"Tolerance attractors" form under recurrence in such a neural net (implementing Poincaré-Zeeman-Poston-DalCin "tolerance theory" by realizing Heinz von Foerster's 1960 early prediction of "Platonic ideation") XX

The technical problem of "fast picture-shifting" within such multi-resolution-level neural nets or wavelets, while solved by nature, still eludes science (with Michael Klein) XXI

"Pandaka-pygmaea Institute" proposed in the Biophysical Journal to solve the Platonic and other brain problems by investigating this 0.9 cm smallest fish's brain (along with that of its normal-sized close relative of 20-cm, Gobius niger) XXII

The attractive positive sum-potential in the brain equation—"happiness"—is predictably displayed by the young of social animals XXIII

One of the sub-potentials in the brain equation—"bonding"—is predictably displayed by all social animals XXIV

Two distinct displays (like happiness and bonding) can—through an evolutionary accident called "Huxley evolutionary ritualization"—acquire a mutual functional overlap XXV

This Julian-Huxleian accident happened independently in the evolution of two mammalian species: "Tail-wagging" signals both bonding and happiness in wolves, and the Smiley face signals both bonding and happiness in humans (similarly Jan van Hoof and Frans de Waal) XXVI

"All Animals Are Autistic" (AAAA): because the brain equation, being an autonomous optimizer, is autistic by definition XXVII

Every brain-equation-carrier is "alive" independently of hardware because it solves the universal positional-adaptation problem, solving which is as vital as solving the metabolic-adaptation problem (hence "chemical life" and "brain life" have the same rank) XXVIII

Universal brains are "mirror-competent" by virtue of their high simulational capability XXIX

Unlike humans and some other vertebrate and perhaps invertebrate species, wolves do not have a universal brain (their VR component is too weak for mirror-competence: their mirror neurons do not suffice) XXX

Smile-laughter overlap + strong bonding + mirror-competence = sufficient condition for an "epigenetic function change" in the sense of Robert Rosen: The overwhelming "personogenetic function change" (PFC) XXXI

The PFC consists in the invention of the "suspicion of benevolence shown by the other" (which leads to a state of "being moved" in a positive feed-back, engulfing both sides in a maximal bonding bout) XXXII

The PFC represents an example of "creation out of nothing" (the suspicion of, and in response production of, benevolence) XXXIII

"Was Mom totally moved [stirred] like scrambled eggs?" [the German word "gerührt" means both being stirred and being moved], asked 3-year-old Jonas after having shown an unexpected favor (in "Jonas' World—The Thinking of a Child" [in German], edited by Reimara and Otto E. Rossler, p. 23) XXXIV

"Person attractor" (Detlev Linke): The new stable mode of functioning that arises in the PFC XXXV

The PFC can be seen to be nothing but a misunderstanding (a haphazard convergence concocted in the universal simulator): were it not interactively confirmed XXXVI

The fact that the PFC represents a joint functional trap allows one to speak of "Nature's Shadchen trick" (with permission of Roger Malina) XXXVII

The person attractor resembles a "folie à deux" (a form of "animal schizophrenia" or "AI schizophrenia"), compared to the physiological autistic functioning of the two "autonomous optimizers with cognition" involved XXXVIII

The PFC constitutes a miracle, worked by the toddler (the only scientifically proved worked miracle) XXXIX

Watching this "creation-out-of-nothing" being achieved by the toddler is a maximally moving event (there appears to exist no recorded documentation of this "holy of holies" of humankind) XL

The mutually confirmed suspicion of benevolence acquires the character of an "objective truth" (there is no older objective truth) XLI

The miracle goes still further: A third fictitious person is involved in the persogenesis (called "god" or "Buddha" etc. in different cultures): the Dream-Giving Instance DGI, or synonymously the "non-I" (or even the "palpable emptiness behind the dream," cf. http://www.youtube.com/watch?v=YWwe8HzUJu8) XLII

The "non-I" arises concomitantly with the "I" and the "you" (the two other persons created in the PFC) XLIII

Mathematical proof that the orangutan brain is functionally superior to the human brain (with Michael Langer;independently Willie Smits in the book "Thinkers of the Jungle") XLIV

Friendly teasing jokes ("humor") are implicit in the PFC (as discovered by Jesse Bering in interaction with his gorilla, scientificamerican 2010) XLV

Being able to ask a factual question is a new behavioral trait that is made possible by the PFC XLVI

"Nonautistic languaging" automatically develops as a consequence of the PFC (similarly C. Andy Hilgartner) XLVII

Human society in all its essential aspects is formed as a consequence of the PFC: Society is based on asking questions and giving answers, enabled by the mutual trust between persons XLVIII

"Personology" = "Adamology" = "person made out of soil" (with Jürgen Jonas and Michael Langer) IL

The "physiological autism" of every autonomous optimizer with cognition persists in human beings endowed with an innate "smile blindness" (when the latter is strong enough to prevent the epigenetic PFC from occurring) L

Most alleged autism in humans is "pseudo-autism": A lesser-than-average fluency in some social conventions LI

The causal explanation of autism enables a causal therapy: The caretaker deliberately produces an "acoustic smile" when momentarily happy (the acoustic smile consists in making a tender bonding noise); a proof is the healed haircutter, featured in: 29.01.08, 22.15 VOX, Stern TV Reportage, Autisten) LII

The fact that the caretaker must be the essential bonding partner proves that modern child cribs are a collective tragedy (their uninformed use explains the current global rise in autism) LIII

The offered "causal therapy of autism" was shunned by the profession for 45 years (only Gregory Bateson approved of it, Konrad Lorenz said he believed it but it was "too hard to understand" for him) LIV

The likely reason for the professional silence is the fact that the person attractor is "too easy to elicit": Young mirror-competent bonding animals can predictably be lured into the personogenetic function change (which prospect, first expressed in the 1975 San Diego Biomedical Symposium., violates a deep subconscious taboo—the "Chewbacca taboo" as it can be called in honor of George Lucas' 1978 enchanting character) LV

"Galactic export" is the technical term for exporting the personogenetic bifurcation to non-human mirror-competent bonding animals or brain-equation-controlled robots (the "small step" of recruiting a second life form into personhood is "a giant leap for mankind") LVI

Evolutionarily speaking, the epigenetic PFC is a "lethal factor" since it replaces natural selection by person-controlled fairness LVII

The PFC nevertheless is the opposite of being "evolutionarily lethal": It represents a "jump up towards point Omega" (which thereby ceases to be asymptotic, i.e., only reachable after an infinite cosmic time) LVIII

The planet-wide shying-away from galactic export is an example of a collective subconscious "speciesism" LIX

The fear is palpable ever since Margaret Howe, Gregory Bateson and John C. Lilly's student, Margaret Howe, tried to adopt a male dolphin 48 years ago; also Koko (Francine Patterson's gorilla life partner) and Kanzi (Susan Savage-Rumbaugh's grown-up adopted bonobo child) are both tragically underrated (compare the photos in Bill Seaman's book "Neosentience") LX

Stephen Spielberg played on the same taboo in his movie "AI" which brings-in the added feature that this non-biochemical person is potentially immortal (a fact played down tactfully) LXI

Leo Szilard introduced non-human persons in his 1948 sci-fi story "The Voice of the Dolphins" written in the aftermath of his failure to prevent his other brain child (the bomb) from being dropped LXII

The "Rosette phenomenon" of sperm whales (the carriers of the most sophisticated brains on earth) deserves to be taken seriously: What function serves their daily meeting? (Cf. the sci-fi story "The Tale of the Whale" abriged in the book "Neosentience") LXIII

Nowhere-differentiable attractors(with John L. Hudson, Carsten knudsen and IchiroTsuda LXIV

"Vertical exteriority": Matching term, in the theological sense of Edmond Jabès (with Nils Röller, Klaus Sander and Kai Grehn) LXV

"A program can force the programmer to reply" (with Christa Sommerer and Adolf Muschg) LXVI

"Simulacron Three" (story by Daniel F. Galouye 1964) and "A Puppeteer's World"["Welt am Draht"] (movie by Rainer Werner Fassbinder 1973) are two anticipations of the same insight, followed by the "Matrix" movie, Ray Kurzweil's "Singularity Theory" and Eric Klien's "Lifeboat project" LXVII

The "Turing test"—a test for personhood—got passed for the first time in ancient Rome: By the Cretan slave and subsequent stoic philosopher Epictetus (as I learned from Bob Rosen) LXVIII

Limitology (joint work with Yukio-Pegio Gunji, John Casti and Joseph Traub) LXIX

An equation for a universal immune system (with Robert A. Lutz) LXX

A chemical universal circuit (with Dietrich Hoffmann) LXXI

Differentiable automata exist mathematically (because certain ordinary differential equations can approximately-if-consistently be described by automata theory) LXXII

“Well-stirred automata” (fluid; liquid): Exist physically LXXIII

Reaction scheme and simulation of a temperature-compensated chemical clock (with Wolfgang Engelmann and Marc Lefranc) LXXIV

“Ultra-long-term, continuous-stirred-tank-reactor version” of the Belousov-Zhabotinsky reaction: To check for a “late explosion” in the number of variables produced as onset of a chemical evolution (with Michael and Debbie Conrad) LXXV

“Traffic-light version of the Belousov-Zhabotinsky reaction” (with Wolfgang Engelmann and Reimara Rossler) LXXVI

“Slinky attractor” (with Okan Gurel and Eberhard Hopf) LXXVII

“Reinjection principle” is valid in more than two-dimensional phase spaces (independently Floris Takens and Christian Mira) LXXVIII

A chaotic electronic multivibrator, built with Hartmut Waible in 1975 LXXIX

“Rossler attractor” (Norman Packard and Ralph Abraham) LXXX

“Spiral chaos” LXXXI

“Screw-type chaos” LXXXII

“The sound of chaos” is known to everyone (idling motor; hoarse voice; baby cries [H. Herzel]) LXXXIII

Chaos (a stereoscopic sound movie, made with Reimara Rossler and Thomas Wiehr 1976), now on YouTube LXXXIV

“Chaos = disciplined tangle” (term due to Alfred Klemm who turned 100 on February 15, 2013) and chaos in a piecewise linear system (with Igor Gumowski) LXXXV

“Hyperchaos” (name courtesy Paul Rapp) LXXXVI

“The sound of hyperchaos” (like raindrops falling on a car’s roof): in the “chaos” movie LXXXVII

“Running electric fan suspended from a long rope”: Olafur Eliasson’s independent hyperchaos LXXXVIII

“X-attractor” in 4 dimensional flows (is still elusive) LXXXIX

“Playdough task”: To be given to thousands of toddlers to find the X-attractor XC

“Atrio-ventricular heart chaos” (with Reimara Rossler and Herbert D. Landahl) XCI

“Endocrinological chaos” (with Reimara Rossler and Peter Sadowski, independently Colin Sparrow and Christophe Letellier) XCII

Chaos in the Zhabotinsky reaction (with Klaus Wegmann, in parallel to John L. Hudson) XCIII

"Cloud attractor" (with James A. Yorke and John L. Hudson) XCIV

"Folded-towel map" (in parallel Masaya Yamaguti's "folded handkerchief map") XCV

"Punctured hyperchaos" is source of every transfinitely exact self-similarity or self-affineness in maps (with Michael Klein) XCVI

"The chaotic hierarchy" (the simplest equation was subsequently to be found by Gerold Baier and Sven Sahle) XCVII

Explicit Smale-Urysohn "solenoid attractor" (with Pal Fischer and W.R. Smith) XCVIII

"Transfinitely-invertible attractors" (being almost everywhere so, with György Targonski) XCIX

Explicit Poincaré recurrence in a 2-D invertible map (with Georg C. Hartmann) C

Generic Milnor (-like) attractor (with Francisco Doria and Georg C. Hartmann) CI

"Flare attractors" (with Georg C. Hartmann, and with Vela Vilupillai in late homage to Richard Goodwin) CII

"Society of flare attractors" as a model of the economy (with Georg C. Hartmann) CIII

"Hyperfat attractors" (with John L. Hudson) CIV

"The fat hierarchy" (with Erik Mosekilde) CV

Particle indistinguishability is transfinitely exact (with Hans Primas, Martin Hoffmann and Joe Ford) CVI

"Deterministic entropy" (with Hans H. Diebner) CVII

"Gibbs-Sackur cell" in phase space CVIII

"Classical unit action" (in the system-specific Sackur cell) CIX

"Micro time reversals": In the Sackur cell of the observer (with Richard Wages) CX

"An estimate of Planck's constant" (based on the Sackur cell) CXI

Causal (exo) explanation of quantum mechanics (with Peter Weibel) CXII

Endophysics (with David Finkelstein and John Casti) CXIII

"Boscovich covariance" (with Werner Kutzelnigg, Edgar Heilbronner, Jens Meier and Matthias Schramm) CXIV

"Causal explanation of spin" on exo-level (with Michael Conrad) CXV

"All-alpha electron chemistry" or "single-spin chemistry" in ultra-strong magnetic fields (with Dieter Fröhlich, Gerald Caris, Günter Häfelinger and Frank Kuske) CXVI

Hyperchaos generated by a single feedback circuit (with René Thomas, Markus Eiswirth and Vasileios Basios) CXVII

"Cession twin of action": Cession has h/c for its own quantum (with Claudia Giannetti) CXVIII

Everett's global Psi-function: Replaced by Boltzmann's global H-function on the exo-level (with Siegfried Zielinski) CXIX

Everett's observer-centered explanation of nonlocality (his 1957 paper, p. 149, left column): Confirmed CXX

"The momentarily consciousness-bearing Sackur cell in the brain": Determines both h and c (a conjecture, with Reimara Rossler and Peter Weibel) CXXI

"VX-diagram" (correlated photons measured in two mutually receding spaceships): "The completed Einstein–Podolsky–Rosen paradox" (with John S. Bell; in belated parallelism to Susan J. Feingold and in parallelism to Roger Penrose) CXXII

"Partially satellite-based VX-experiment": It will prove that more than one quantum world exists (independently conceived, and actually proposed to ESA, by Anton Zeilinger in 2001) CXXIII

"Counterfactual superluminal telegraph" (with Uwe Niedersen and Jürgen Parisi) CXXIV

"Everett immortality" (with Markus Fix and Bryce DeWitt) CXXV

"Aging equation" (with Reimara Rossler and Peter Kloeden) CXXVI

Evolutionary explanation of the higher female longevity (with Reimara Rossler, Peter Kloeden and Bob May) CXXVII

"Constant-temperature physico-chemical time-of-life clock": Predicted present in the body (with Reimara Rossler) CXXVIII

Melatonin as the likely "handle" of the time-of-life clock (with Reimara Rossler and Peter Kloeden) CXXIX

"Lampsacus hometown of all persons on the Internet" (with Valentino Braitenberg and Gerhard J. Lischka 1994) CXXX

Attempt made to found Lampsacus in homage to Anaxagoras (with Ezer Weizmann and Mohamed ElNaschie) [a verbal quotation from Beer Sheva: "This is what Israel was meant for"] CXXXI

"Earth-Moon University" in Lampsacus (with Wilfried Kriese, Artur P. Schmidt and George E. Lasker) CXXXII

"Pyramid of knowledge" of 16 levels in Lampsacus: A method to create an explosion of knowledge CXXXIII

"WM-diagram": Simultaneously-emitted signals, sent up and down across different levels in gravity, plotted along the upper and lower time axes drawn in parallel (with Dieter Fröhlich) CXXXIV

Gravitational-redshift proportional size increase is implicit in WM diagram if c is globally constant (with Dieter Fröhlich, Heinrich Kuypers and Jürgen Parisi) CXXXV

The most energetic photon possible has the Planck mass(courtesy Heinrich Kuypers) CXXXVI

All black holes are "almost-black holes": For they are never finished in finite outer time (with Dieter Fröhlich, Heinrich Kuypers, Hans Diebner and Mohamed El-Naschie) CXXXVII

Non-uniqueness of simultaneity: Present on rotating cylinder (with Dieter Fröhlich, Normann Kleiner and Francisco J. Muller) CXXXVII

Corrected proof of angular-momentum conservation in gravity (with Heinrich Kuypers and Martin Pfaff) CXXXIX

Seeming invariance of transversal size in the locally isotropic gravitational size increase (is a parallel to the better-known seeming invariance of transversal size in the likewise locally isotropic Lorentz contraction) CXL

Einstein's "gravitational time dilation" possesses three new corollaries: Length, Mass and Charge all suffer a proportional—or anti-proportional—change (T-L-M-Ch theorem or "Telemach" for short) CXLI

General relativity is in for a far-reaching mathematical and physical re-interpretation CXLII

Speed of light c is globally constant in gravitation (Max Abraham rehabilitated) CXLIII

Nonexistence of longitudinal gravitational waves (as a corollary) CXLIV

Nonexistence of gravitons (as a corollary) CXLV

The famous "indirect evidence for gravitational waves" (Hulse-Taylor): explained by tidal friction in the invisible white dwarf companion (with Dieter Fröhlich and René Stettler) CXLVI

A "Reeb foliation in spacetime": exists around every rotating black hole (with Dieter Fröhlich, following stimulation by Art Winfree) CXLVII

Kerr metric: Disproved (as a corollary to the new Reeb foliation) CXLVIII

Ur-meter: Disproved (via Telemach theorem) CIL

Ur-kilogram: Disproved (via Telemach theorem) CL

Charge conservation in physics: Disproved (via Telemach theorem) CLI

Black holes: Shaved of one of their 3 main "hairs": Charge (only total mass and angular momentum remain) CLII

Reissner-Nordström metric: Disproved (via Telemach theorem) CLIII

Eddington-Finkelstein and Kruskal-Szekeres transformation: Shown to be unphysical via Telemach CLIV

Bekenstein theory: Disproved via Telemach CLV

Hawking radiation: Disproved via Telemach (with apology to a world hero) CLVI

"Coordinate singularity at the horizon": Rehabilitated as a physical singularity via Telemach CLVII

"Interior Schwarzschild solution": Disproved via Telemach for finite outer time CLVIII

"Singularity theorem" inside black hole horizon: Disproved via Telemach for finite outer times CLIX

"Wormholes": disproved via Telemach for finite outer times CLX

Upper half of "Flamm's paraboloid": replaced by a generic 2-pseudosphere (the lower half disappears for finite outer time) CLXI

Sackur-cell explanation of Planck's constant h as an endo phenomenon: Implies non-existence on the exo level of all field particles CLXII

The exo-nonexistence of the field particles implies that supersymmetry is non-existent in physics CLXIII

The human Lorenz matrix of facial expressions: A universal natural facial-expressions simulator (with Wilfried Musterle) CLXIV

Equation for a one-dimensional—i.e. purely temporal—brain (with Michael Conrad and Behram Kursunoglu) CLXV

Evil is a contagious disease (unlike the good, evil cannot arise spontaneously) CLXVI

Children and adults form two different species ethologically speaking (with Konrad Lorenz) CLXVII

"Pongo goneotrophicus" (meaning "the parent-feeding ape") is the correct biological name for Homo sapiens CLXVIII

Biochemical life (including Forward's nuclear-chemical life on neutron stars), and "brain life": Two independent forms of life (Hanns Ruder kindly introduced me to Robert Forward's scientific sci-fi book "The Dragon's Egg") CLXIX

Electrons have finite volume owing to Telemach (in confirmation of Dirac) CLXX

Corollary: String theory is qualitatively (if not quantitatively) confirmed at last CLXXI

The new empirical confirmation of string theory implies that a successful generation of black holes at particle colliders has become much more likely CLXXII

Freshly generated black holes are undetectable by the detectors of particle colliders because they are uncharged and non-evaporating CLXXIII

The empirical, ten-orders-of-magnitude wide "quasar scaling law" (from quasars to microquasars) extends downwards by some 50 more orders of magnitude, owing to the new properties of black holes CLXXIV

There exists no more unstoppable and voracious parasite in the universe than a black hole CLXXV

Miniature black holes grow exponentially inside solid matter (once they got stuck) CLXXVI

"Clifford conjecture": Finite-universe solutions to the Einstein equation are unphysical (with Walter Ratjen); if so, there exists no "Gödel solution" in General Relativity and no time travel CLXXVII

The fractal dimensionality of the cosmos is close to unity, not only empirically over a large range as is well known, but also theoretically ("Fournier-Mandelbrot solution" to the Einstein equation) CLXXVIII

A first consistent history of galaxy formation is taking shape (with dark ellipticals as the end stage) CLXXIX

The newly discovered, very far-away, mature old galaxy BX442 (more than ten billion light years old) is only the first example of its kind (besides still older quasars) CLXXX

Low-surface-brightness large galaxies ("black galaxies") are at least 50 billion years old (with Henry Gebhardt and Boris Hagel) CLXXXI

Giacconi's ultra-faint equidistributed X-ray sources most likely are ultra-distant ultra-high-redshift quasars ("blazars") so that redshift measurements are highly desirable (with Dieter Fröhlich) CLXXXII

The microwave background radiation predictably merges smoothly with equal-temperature galactic-halo objects (hence the raw data of the Planck mission deserve to be published) CLXXXIII

There exist differentiable dynamical systems that are made up, not of locally parallel lines as customary, but rather of 2-D locally parallel surfaces (Bouligand-Winfree theory, with Joachim Peinke) CLXXXIV

Re-discovery of Zwicky-Chandrasekhar "dynamical friction" (with Dieter Fröhlich and Normann Kleiner in contact with Ilya Prigogine, Alfred Klemm, Joachim Peinke and Jürgen Parisi) CLXXXV

The new "little hook," at the current end of the Hubble-(Perlmutter-Schmitt-Riess) line, holds true in a non-expanding Fournier-Mandelbrot cosmos (with Dieter Fröhlich, Ramis Movassagh and Anthony Moore) CLXXXVI

Dynamical friction of low-mass particles: Numerically confirmed (with Klaus Sonnleitner) and independently by Ramis Movassagh CLXXXVII

"Deterministic statistical thermodynamics" (with Hans Diebner) CLXXXVIII

"Deterministic statistical cryodynamics": as a new fundamental science besides "deterministic statistical thermodynamics" (with Klaus Sonnleitner, Frank Kuske and Christophe Letellier) CLXXXIX

"Deterministic ectropy": Exists in statistical cryodynamics (with Ali Sanayei) CXC

The smaller (almost-) black hole in a pair-interaction gets, via formation of a blue-sky catastrophe in the sense of Ralph Abraham, re-circulated with all its in-falling particles jointly making it up, since it never got finished over finite external time (with Dieter Fröhlich) CXCI

Black hole mergers are a source of both charged and uncharged cosmic rays CXCII

Conjecture: 50 percent of all matter in the cosmos is (almost-) black holes (with Dieter Fröhlich) CXCIII

"Metabállon anapaúetai" (metabolizing it remains at rest): Heraclitus' transfinitely recycling metabolic cosmology at last proven valid after 2 ½ millennia CXCIV

Abramowicz's "topology inversion" near a black-hole's horizon: Confirmed (with Dieter Fröhlich) CXCV

Lawful "identity jumps across space" between 3 indistinguishable classical particles on a ring (with Peter Weibel and Richard Wages) CXCVI

In a classical atom containing two indistinguishable electrons, two spherical shells are formed much as this is the case in nature with ortho-helium (with Dietrich Hoffmann and George Kampis) CXCVII

The "flotor" (Ralph Hollis): A transluminally fast measuring device? (with Peter Plath) CXCVIII

The "counterfactual superluminal telegraph" is subluminally confirmable (with Uwe Niedersen) CXCIX

Counterfactual world-change machine (with Jürgen Parisi and Koichiro Matsuno) CC

History of the transfinitely exact indistinguishability: Anaxagoras, Gregorius of Naziance, the Mutakallimún, Bruno, Spinoza, Leibniz, Gibbs, Pauli, Weyl, Primas (with Martin Hoffmann, Joe Ford, Peter Weibel, Alexandre Ganoczy, Richard Wages, Rudolf Matzka Elisabeth von Samsonow, Jurgen Heiter, Anna-Sophie Mahler) CCI

"Everett-Schrödinger Russian Roulette" (with Markus Fix) CCII

Unit "el-action": A new universal conserved quantity (like the unit action) CCIII
Unit "el-cession": A new universal conserved quantity (like the unit cession) CCIV

"G-zero" is a new fundamental constant replacing the universal gravitational constant G and the universal vacuum permeability constant mu-zero, with both remaining locally constant (similarly Richard J. Cook and György Darvas) CCV

The nonlinear "simultaneity generator" in the brain forms a qualitative analog to general relativity (in dialogue with Eva Ruhnau) CCVI

Cryodynamics and thermodynamics, combined with black-hole theory, allow for an eternal cosmos in the footsteps of Heraclitus and Boltzmann CCVII

Nonexistence of WIMPs, since cold dark matter got disproved CCVIII

Nonexistence of dark energy, since accelerated expansion got disproved CCIX

Nonexistence of Big Bang and space expansion and a cosmic origin of galactic background radiation since cryodynamics explains the Lemaitre-Hubble-Perlmutter-Schmidt-Riess-law in a stationary fractal cosmos CCX

Nonexistence of the Big Bang: Follows also from the new global constancy of the speed of light c that is implicit in Telemach CCXI

Nonexistence of "inflation": Follows from the demonstrated absence of space expansion CCXII

Nonexistence of "primordial" nucleosynthesis: Follows from the absence of space expansion CCXIII

The observed absence of the Greisen-Zatsepin-Kuzmin cutoff for high cosmic ray energies confirms the absence of a very distant origin of the microwave background radiation CCXIV

The many decades-old problem of the "survival conditions of the scientific-technological world" (C.F. von Weizsäcker): Confirmed as a pressing problem for humankind CCXV

The new results on black holes (facilitated production; non-evaporation; unchargedness; exponential growth inside matter) upset the safety equation for any attempt at producing black holes on earth CCXVI

The LHC experiment designed to produce black holes besides searching for the Higgs was shown to be unsafe if run at history-making energies and luminosities CCXVII

CERN's refusal to update its "LHC Safety Assessment Group Report" over 4 ½ years entails that its stop may come too late to prevent a black-hole-induced Armageddon after a silent period of a few years CCXVIII

An attempt to convene an "LHC safety conference" made by Markus Goritschnig and many other scientists fizzled: This even though a court humbly endorsed it and although a whole country briefly left CERN (and although the United Nations' Security Council was called upon) CCXIX

Leo Szilard's 1948 proposal to slow-down scientific progress by installing the anonymous peer-review system (in his sci-fi story "The Mark Gable Foundation") is co-responsible for the dogmatic refusal by CERN to check a published scientific proof CCXX

Sunyaev-Zeldovich inverted Compton scattering predictably cannot take place between the Milky-Way's microwave background radiation's photons and the electrons of distant galaxy clusters (with Boris Hagel, Ali Sanayei and Frank Kuske) CCXXI

The new science of "Cryodynamics" is likely to be able to stabilize Tokamak-type fusion reactors so as to generate unlimited free energy for humankind (I thank Ivan Zelinka and Eric Klien for encouragement) CCXXII

Remark: Friedrich Valjavek kindly compiled an annotated bibliography up to 2002, so that the majority of references can be found there: http://www.wissensnavigator.com/documents/RosslerBibliography.pdf.

3 Discussion

It was an unexpected and undeserved chance to be allowed to try and sum up in a maximally brief form some of the topics that in retrospect would come to my mind as having been dealt with over time.

It is not easy to say something in excuse or explanation at the end. What is really important of the things achieved or tried? Some topics enjoyed a lot of resonance—especially the **chaos** animation and sounds (http://www.youtube.com/watch?v=Tmmdg2P1RIM)—, others that appeared to be equally exciting produced little or none up until now. And still others triggered a lot of counter resonance, as I had hoped, although the requested disproof—so that the danger seen could be called-off—did not materialize.

What are the important positive points in retrospect? **Evolution theory**—extending across planets and different liquid media and even across the boundaries of chemistry into nuclear chemistry (in confirmation of Robert Forward) is the oldest melody. The second-oldest is **person theory**. Or love theory (which is the same thing). Here the recruitment of not yet kissed-awake promising mirror-competent bonding individuals of other species, some provably more intelligent hardware wise (like Willie Smits' orangutans), is close to my heart. The **brain equation** makes it possible to include artificial analogs in the embrace. Wilfried Musterle gave the most moving contribution, with Javier Movellan's recent video showing what I have in mind (http://www.hansonrobotics.com/). Love theory and Emmanuel Lévinas belong together.

Darwin's principle of metabolic adaptation was complemented by the principle of **positional adaptation**, which can be seen as the biggest formal advance, with the implied new traveling salesman problem leading to the Brain Equation about which Bill Seaman and I wrote a book. Transfinite **indistinguishability** (Pauli-Primas) and the **Sackur cell**, as an explanation of Planck's constant h, form another important strand. The Einstein of EPR fame and his youthful pupil Everett acquire a new shining. This strand is on its way toward a new comprehensive theory in the footsteps of Boscovich called **endophysics**.

Lately, deterministic evolution à la Teilhard—based on far-from-equilibrium thermodynamics—got complemented by thermodynamics'new sister discipline, **cryodynamics**, with its second (anti-entropic) arrow. **Cosmology** becomes an element of a larger theory. Here the circle to Teilhard closes. A new view of Einstein's happiest thought resulted in the discovery of the global constancy of the speed of light, c (gothic-R and Telemach theorem). Cosmology thereby ceases to be globally time-dependent—no big bang any more. Heraclitus' "metabállon anapaúetai"—metabolizing it rests unchanged—got unexpectedly confirmed. At the same time, cryodynamics predictably can bring the sun's fire down on earth because it offers a technological means for stabilizing the "Iter." So a lot of money is implicit in the further investigation of cryodynamics—in case a government or private institution gets interested (http://www.youtube.com/watch?v= s73D0VIofFo).

Finally, the mystery of color and sweetness and the Now, the assignment of being called-up by name at this very moment, cannot go unmentioned. Buber called it the "light that never gets extinguished." There is nothing, not even death, that can stop one's being allowed to see color and the good intention of another soul and the Consciousness Giver. "Pánta de oiakízei keraunós"—everything is joysticked by the lightning-thrower. The rare word "oiax" (root of oiakízo) refers to a little lever by which the helm of a ship was controlled (a precursor of the joystick). Heraclitus was as sweet as he was fearless—like Jacob. Thank you, my readers and friends of all age groups for your infinite tolerance. (For J.O.R.)

The Cause of Complexity in Nature: An Analytical and Computational Approach

Klaus Mainzer

Abstract This work is going to present the cause of complexity in nature from an analytical and computational point of view. The cause of complex pattern formation is explained by the local activity of cells in complex systems which are analytically modeled by nonlinear reaction-diffusion equations in physics, chemistry, biology, and brain research. There are not only rigorous analytical criteria of local activity and the edge of chaos, but also constructive procedures to visualize them by computer simulations. In technology, the question arises whether these criteria and procedures can be used to construct artificial life and artificial minds.

Keywords Complexity · Computation · Nonlinearity · reaction-diffusion equations · Nature · Edge of chaos

1 Introduction

According to several prominent authors, a main part of twenty-first century science will be on complexity research. The intuitive idea is that global patterns and structures emerge from locally interacting elements like atoms in laser beams, molecules in chemical reactions, proteins in cells, cells in organs, neurons in brains, agents in markets etc. by self-organization [1]. But what is the cause of self-organization? Complexity phenomena have been reported from many disciplines (e.g., biology, chemistry, ecology, physics, sociology, economy etc.) and analyzed from various perspectives such as Schrödinger's order from disorder [2], Prigogine's dissipative structure [3], Haken's synergetics [4], Langton's edge of

K. Mainzer (✉)
Chair of Complexity Research/Philosophy of Science, Munich Center for Technology in Society, Technische Universität München, Munich, Germany
e-mail: mainzer@tum.de

I. Zelinka et al. (eds.), *How Nature Works*, Emergence, Complexity and Computation 5, DOI: 10.1007/978-3-319-00254-5_2,
© Springer International Publishing Switzerland 2014

chaos [5] etc. Steven Wolfram declared computer experiments with pattern formation of cellular automata as "new kind of science" [6]. But concepts of complexity are often based on experiments, examples or metaphors only. We argue for a mathematically precise and rigorous definition and analytical theory of local activity as the cause of self-organizing complexity which can be tested in an explicit and constructive manner [7, 8].

Boltzmann's struggle in understanding the physical principles distinguishing between living and non-living matter, Schrödinger's negative entropy in metabolisms, Turing's basis of morphogenesis [9], Prigogine's intuition of the instability of the inhomogeneous, and Haken's synergetics are in fact all direct manifestations of a fundamental principle of locality. It can be considered the complement of the second law of thermodynamics explaining the emergence of order from disorder instead of disorder from order, in a quantitative way, at least for reaction diffusion systems.

The principle of local activity is precisely the missing concept to explain the emergence of complex patterns in a homogeneous medium. Cellular automata are an illustrative model of local active cells with pattern formation. They can be characterized as complex dynamical systems in the strictly mathematical sense with corresponding equations and proofs. In short, there are analytical models of cellular automata, in order to find precise answers and predictions of pattern formation [10].

The local principle can be generalized and proven for the class of nonlinear reaction-diffusion systems in physics, chemistry, biology, and brain research. The principle of local activity is the cause of symmetry breaking in homogeneous media. The Brusselator model was one of the first systems of equations used to explain self-organizing chemical reactions of the reaction-diffusion type. Based on this model, a theory of dissipative structures operating far from thermodynamic equilibrium was developed [11]. Under stability theory techniques, Prigogine and his group derived a critical bifurcation boundary for the uncoupled cell. They studied stationary and dynamic patterns emerging in the neighborhood of this boundary. But, except for the stability boundaries, the far from thermodynamic-equilibrium theory is too coarse to predict sharper and more precise domain of emergent behavior. Especially, they ignored the relatively small subset of the edge of chaos where the emergence of complexity is most likely.

The local activity theory is applied, again, to the Gierer-Meinhardt-equations to illustrate the emergence of complexity. On the basis of autocatalysis and lateral inhibition, Gierer and Meinhardt proposed a mathematical model [12] to explain pattern formation (morphogenesis) in living systems. Using numerical integration, they were able to produce a number of patterns relevant to the formation of biological structures. An analytical treatment of the Gierer-Meinhardt model from the synergetics perspective was presented in Ref. [13] using the order parameter concept combined with linear stability analysis. While the later approach offers significant contributions in understanding the dynamics of the Gierer-Meinhardt model, it is still too coarse to predict the precise domain in the cell parameter space where emergent behavior may occur. The local activity theory offers a rigorous and effective tool for sharpening existing results in the sense that it can identify more precisely those regions in the cell parameter space which are capable of emergent behaviors, and also in fine tuning such regions into a relatively small subset called the edge of chaos where the emergence of complex phenomena is most likely [14].

The long-lasting action and pace-maker potentials of the Purkinje fiber of the heart were first described by the Hodgkin-Huxley equations of the cardiac Purkinje fiber model of morphogenesis in [11]. The bifurcation diagrams of the corresponding computer simulations supply a possible explanation for why a heart with a normal heart-rate may stop beating suddenly: The cell parameter of a normal heart is located in a locally active unstable domain and just nearby an edge of chaos. The membrane potential along a fiber is simulated in a Hodgkin-Huxley model by a computer [15]. Computer simulations show that oscillatory patterns, chaotic patterns, or divergent patterns may emerge if the selected cell parameters are located in locally active domains but nearby the edge of chaos. This research demonstrates once again the effectiveness of the local activity theory in choosing the parameters for the emergence of complex (static and dynamic) patterns in a homogeneous lattice formed by coupled locally active cells.

One of the marvels of the HH (Hodgkin-Huxley) equations is its ability to generate an action potential (spikes) in response to an external current excitation emulating the net synaptic current excitation. Although the spikes were consistently generated numerically, no one knew the physical and mathematical origin of the action potential (spikes). Local activity can be demonstrated to be the origin of spikes. In particular, neurons can be shown to be poised near a tiny subset of the local activity domain, which we call the edge of chaos. The domain of the edge of chaos is determined by deriving an explicit scalar complexity function of a small-signal Hodgkin-Huxley circuit model [16].

A completely new application of local activity is the so-called edge of chaos where most complex phenomena emerge. In this particular case, a hidden excitability allows a unit to be destabilized when interacting with dissipative environments. Although a diffusion process has a tendency to equalize differences, an originally dead or inactive cell becomes alive or active upon coupling with other cells by diffusion [17]. This phenomenon seems to be counterintuitive, but can mathematically rigorously be proven and confirmed with different applications in reality.

In the parameter spaces of reaction diffusion systems, the domains of local activity can be visualized in computer simulations. The edges of chaos are very small regions with the ability of creation. They seem to be hidden in the domains of activity like pearls in shells on the sea ground. Actually, in research, the "edge of chaos" was only used as metaphor, but not as mathematically precise concept. Its discovery in the parameter spaces of dynamical systems is, to our best knowledge, completely new in complexity research. The local activity principle and its pearl, the edge of chaos, are couched in rigorous mathematics. Above all, they are characterized by constructive procedures to compute and visualize their complexity. Therefore, researchers from other disciplines that can describe their dynamical systems via differential equations, such as reaction-diffusion equations, can actually calculate easily with a computer the parameter values where complexity and creativity can occur.

We argue that the principle of local activity and edge of chaos are really fundamental in science. Twenty-first-century engineering sciences are more and more inspired by the life sciences. In systems biology, cells, organs, and organisms are considered complex systems which can be modeled by complex networks with great similarity to electronic circuits. In the tradition of engineering sciences, synthetic biology uses schemes of systems biology to construct new artificial organisms for practical applications (e.g., new bacteria for cleaning polluted water). Robots become more and more autonomous and self-organizing systems. The question arises whether the principle of local activity can be applied in these technical systems.

2 Mathematical Definition of Local Activity and Edge of Chaos

The principle of local activity had originated from electronic circuits, but can easily be translated into other non-electrical homogeneous media [8]. The transistor is an example of a locally-active device, whereby a “small” (low-power) input signal can be converted into a “large” (high power) output signal at the expense of an energy supply (namely a battery). No radios, televisions, and computers can function without using locally-active devices such as transistors. For the formation of complex biological and chemical patterns, Schrödinger and Prigogine demanded nonlinear dynamics and an energy source as necessary conditions. But, for the exhibition of patterns in an electronic circuit (i.e., non-uniform voltage distributions), the demand for nonlinearity and energy source is too crude. In fact, no patterns can emerge from circuits with cells made of only batteries and nonlinear circuit elements which are not locally active.

In general, a spatially continuous or discrete medium made of identical cells interacting with all cells located within a neighborhood is said to manifest complexity if the homogeneous medium can exhibit a non-homogeneous static or spatio-temporal pattern under homogeneous initial and boundary conditions. The principle of local activity can be formulated mathematically in an axiomatic way without mentioning any circuit models. Moreover, any proposed unified theory on complexity should not be based on observations from a particular collection of examples and explained in terms that make sense only for a particular discipline, say chemistry. Rather it must be couched in discipline-free concepts, which means mathematics, being the only universal scientific language.

However, in order to keep physical intuition and motivation behind this concept, we start with a special class of spatially-extended dynamical systems, namely the reaction-diffusion equations which are familiar in physics and chemistry.

Our first definition of local activity refers to a discretized spatial model which can easily be illustrated by cellular nonlinear networks. All results apply to the associated systems of continuous reaction-diffusion partial differential equations which will be later analyzed e.g., in fluid dynamics of physics and chemistry. Let us consider a spatial lattice of identical cells located at grid points and changing their states by local reaction-diffusion. In general, the change of a local cellular state depends on all the other cellular states in the spatial lattice and (at least in some cases) on the local diffusion of the cell.

In mathematical terms, the dynamics of the whole system is defined by a system of discrete reaction-diffusion equations describing the changes of the local cellular states in the spatial lattice [7, 18] or corresponding continuous reaction-diffusion partial differential equations. In the discretized as well as continuous case, the state variables pertain to reaction cells lumped at lattice points $r \triangleq (j,k,l)$. This dynamical system implements kinetic equations in chemistry and circuit models in network theories. The equations can be represented in the following compact vector form with

$$\mathbf{V}_a \triangleq [V_1, V_2, \ldots, V_m]^T \text{ and } \mathbf{V}_b \triangleq [V_{m+1}, V_{m+2}, \ldots, V_n]^T:$$

$$\dot{\mathbf{V}}_a = \mathbf{f}_a(\mathbf{V}_a, \mathbf{V}_b) + \mathbf{D}\nabla^2\mathbf{V}_a \tag{1}$$

$$\dot{\mathbf{V}}_b = \mathbf{f}_b(\mathbf{V}_a, \mathbf{V}_b) \tag{2}$$

where

$\mathbf{f}_a(\mathbf{V}_a, \mathbf{V}_b) \in \mathbb{R}^m$ denotes the first m components $f_1(\mathbf{V}_a, \mathbf{V}_b)$, $f_2(\mathbf{V}_a, \mathbf{V}_b), \ldots f_m(\mathbf{V}_a, \mathbf{V}_b)$ of the kinetic term,

$\mathbf{f}_b(\mathbf{V}_a, \mathbf{V}_b) \in \mathbb{R}^m$ denotes the remaining (n–m) components $f_{m+1}(\mathbf{V}_a, \mathbf{V}_b), f_{m+2}(\mathbf{V}_a, \mathbf{V}_b), \ldots f_n(\mathbf{V}_a, \mathbf{V}_b)$,

D denotes an m × m diagonal matrix defined by $\mathbf{D}_{\sigma\sigma} = D_\sigma$, and

$\nabla^2\mathbf{V}_a \in \mathbb{R}^m$ denotes a m × 1 vector defined by the m discrete Laplacian Operators $\nabla^2 V_\sigma$, $\sigma = 1, 2, \ldots$, m.

The state variables $\mathbf{V}_a$ and $\mathbf{V}_b$ pertain to only one isolated cell located at the lattice coordinate r = (j,k,l) or to a point r $\in \mathbb{R}^3$ in the continuous case. Any dynamical system has some tunable control parameters $\boldsymbol{\mu} = [\mu_1\mu_1\ldots\mu_\rho]^T$ associated with changing conditions of the system. Hence, the kinetic term in the reaction-diffusion equation for each cell at location r = (j,k,l) is described by the following cell kinetic equations:

$$\dot{\mathbf{V}}_a = \mathbf{f}_a(\mathbf{V}_a(\mathbf{r}), \mathbf{V}_b(\mathbf{r}); \boldsymbol{\mu}) \tag{3}$$

$$\dot{\mathbf{V}}_b = \mathbf{f}_b(\mathbf{V}_a(\mathbf{r}), \mathbf{V}_b(\mathbf{r}); \boldsymbol{\mu}) \tag{4}$$

For an $N \times N \times N$ cubic lattice, there are N^3 identical cells, each one described these cell kinetic equations. Since only the first m state variables $\mathbf{V}_a(\mathrm{r}) = [V_1(\mathrm{r}), V_2(\mathrm{r}), \ldots V_m(\mathrm{r})]^T$ of each cell can interact with the neighboring cells via the diffusion term $D\nabla^2\mathbf{V}_a(\mathrm{r})$, only the energy and matter associated with the first m state variables can flow into a neighbor cell. Henceforth, the state variables $V_1(\mathrm{r}), V_2(\mathrm{r}), \ldots V_m(\mathrm{r})$ in $\mathbf{V}_a(\mathrm{r})$ are called port variables in analogy with the transfer of goods between islands. The remaining state variables $V_{m+1}(\mathrm{r}), V_{m+2}(\mathrm{r}), \ldots V_n(\mathrm{r})$ in $\mathbf{V}_b(\mathrm{r})$ are called non-port variables. The concept of local activity will be defined in terms of only port variables.

Since the diffusion term can play only a dissipative and hence stabilizing role with $D_i > 0$ in the reaction-diffusion equations, the origin of any complex phenomenon exhibited by these equations can only come from the cell kinetic equations. It can rigorously be proved that if the cell kinetic equations are not locally active for a fixed control parameter $\boldsymbol{\mu} = \boldsymbol{\mu}_0$, the reaction-diffusion equations cannot exhibit any complexity regardless of the choices of the diffusion coefficients $D_i > 0$. Moreover, explicit mathematical criteria can be given for testing any cell kinetic equation for local activity. With these criteria, one can identify the active parameter domain A of the parameter space $\boldsymbol{\mu} \in \mathbb{R}^\rho$ where a cell kinetic equation is locally active. Since the complement $\mathcal{P} = \mathbb{R}^\rho \backslash \mathcal{A}$ can only lead to homogeneous solution of the reaction-diffusion equations, it is called the passive parameter domain.

Since complexity can occur only for parameters located in the active parameter region $\mathcal{A}$, it follows that local activity is indeed the origin of complexity. A locally-active cell kinetic equation can exhibit complex dynamics such as limit cycles or chaos, even if the cells are uncoupled from each other by setting all diffusion coefficients to zero. It is not surprising that coupling such cells could give rise to complex spatio-temporal phenomena. What is surprising and counter-intuitive is that there exists a proper subset ε of the active parameter domain $\mathcal{A}$, called the edge of chaos where the uncoupled cell kinetic equation is asymptotically stable.

To illustrate this state biologically, a cell is inert or dead in the sense that the concentrations of its enzymes achieved a constant equilibrium. In interaction, however, the cellular system pulses or become alive in the sense that the concentrations of the enzymes in each cell will oscillate indefinitely. By coupling these dead cells via a dissipative diffusion environment, it may be possible for the reaction-diffusion equation to exhibit non-homogeneous patterns and other spatio-temporal phenomena for appropriate diffusion coefficients. The criteria for the edge of chaos correspond mathematically to Prigogine's instability of the homogeneous and Turing's properties of instability as origin of morphogenesis.

Since local activity is defined only with respect to m port variables in $\mathbf{V}_a = [V_1, V_2, \ldots V_m]^{\mathrm{T}}$, and since it does not involve the diffusion coefficients $D_1, D_2, \ldots, D_m$ in the reaction-diffusion equation, an interaction term $\mathbf{I}_a \triangleq D\nabla^2\mathbf{V}_a$ can be defined as port input vector at the diffusion-driven ports. By substituting this input term into the compact vector form of the reaction-diffusion equations, we get the cell kinetic equations

$$\dot{\mathbf{V}}_a(\mathbf{r}) = \mathbf{f}_a(\mathbf{V}_a(\mathbf{r}),\ \mathbf{V}_b(\mathbf{r}); \boldsymbol{\mu}) + \mathbf{I}_a \tag{5}$$

$$\dot{\mathbf{V}}_b(\mathbf{r}) = \mathbf{f}_b(\mathbf{V}_a(\mathbf{r}), \mathbf{V}_b(\mathbf{r}); \boldsymbol{\mu}) \tag{6}$$

with control parameter $\boldsymbol{\mu} \in \mathbb{R}^\rho$. They are called forced cell kinetic equations, because $\mathbf{I}_a \in \mathbb{R}^m$ can be physically interpreted as an external force applied at the m diffusion ports. Since all the cells are identical, this equation represents the state dynamics of an isolated cell driven at its diffusion ports by an external force $\mathbf{I}_a \in \mathbb{R}^m$ representing the external world.

The equilibrium states of an isolated cell can be obtained by setting the state change to zero with $\dot{\mathbf{V}}_\mathbf{a} = 0$ and $\dot{\mathbf{V}}_b = 0$, namely

$$0 = \mathbf{f}_a(\mathbf{V}_a, \mathbf{V}_b; \boldsymbol{\mu}) + \mathbf{I}_a \tag{7}$$

$$0 = \mathbf{f}_b(\mathbf{V}_a, \mathbf{V}_b; \boldsymbol{\mu}), \tag{8}$$

and solving these equations for $\mathbf{V}_a \in \mathbb{R}^m$ and $\mathbf{V}_b \in \mathbb{R}^{n-m}$ for each fixed parameter $\boldsymbol{\mu} \in \mathbb{R}^\rho$. In general, there are multiple equilibrium points for each input $\mathbf{I}_a \in \mathbb{R}^m$.

Let $\overline{\mathbf{V}}_a$ and $\overline{\mathbf{V}}_b$ denote the coordinates of any cell equilibrium point $\mathcal{Q}$, where $\mathcal{Q}$ depends on the constant input $\overline{\mathbf{I}}_a \in \mathbb{R}^m$ and control parameter $\boldsymbol{\mu} \in \mathbb{R}^\rho$. In the case of an infinitesimal change $\mathbf{i}_a(\mathrm{t})$ of the constant input $\mathbf{I}_a$, we consider infinitesimal deviations $\mathbf{v}_a(\mathrm{t})$ and $\mathbf{v}_b(\mathrm{t})$ in the neighborhood of the equilibrium point $\mathcal{Q}$ with coordinates $\overline{\mathbf{V}}_a$ and $\overline{\mathbf{V}}_b$, namely

$$\mathbf{V}_a(\mathrm{t}) \triangleq \overline{\mathbf{V}}_a + \mathbf{v}_a(\mathrm{t}) \tag{9}$$

$$\mathbf{V}_b(\mathrm{t}) \triangleq \overline{\mathbf{V}}_b + \mathbf{v}_b(\mathrm{t}) \tag{10}$$

$$\mathbf{I}_a(\mathrm{t}) \triangleq \overline{\mathbf{I}}_b + \mathbf{i}_a(\mathrm{t}). \tag{11}$$

Linearized Cell State Equations. In order to approximate the forced cell kinetic dynamics at the cell equilibrium point with constant input $\mathbf{I}_a = \overline{\mathbf{I}}_a$, we use the Taylor series expansion of $\mathbf{f}_a(\mathbf{V}_a, \mathbf{V}_b; \boldsymbol{\mu})$ and $\mathbf{f}_b(\mathbf{V}_a, \mathbf{V}_b; \boldsymbol{\mu})$ about the cell equilibrium point $\mathcal{Q}(\mathbf{V}_a = \overline{\mathbf{V}}_a, \mathbf{V}_b = \overline{\mathbf{V}}_b)$. In general, a Taylor series is a representation of a function as an infinite sum of terms that are calculated from the values of the function's derivatives at a single point. It is usual to approximate a function by using a finite number of terms of its Taylor series. Taylor's theorem gives quantitative estimates on the error in this approximation. Any finite number of initial terms of the Taylor series of a function is called a Taylor polynomial. The Taylor series of a function is the limit of that function's Taylor polynomials, provided that the limit exists. If we delete the higher-order terms of the Taylor series, we obtain linearized cell state equations. They can be interpreted as the cell dynamics along a tangent plane at the cell equilibrium point $\mathcal{Q}$ which depends on the input $\mathbf{I}_a$ and the control parameter value $\boldsymbol{\mu}$, namely

$$\frac{d\mathbf{v}_a(t)}{dt} = \mathbf{A}_{11}(\mathcal{Q})\mathbf{v}_a(\mathrm{t}) + \mathbf{A}_{12}(\mathcal{Q})\mathbf{v}_b(\mathrm{t}) + \mathbf{i}_a(\mathrm{t}) \tag{12}$$

$$\frac{d\mathbf{v}_b(t)}{dt} = \mathbf{A}_{21}(\mathcal{Q})\mathbf{v}_a(\mathrm{t}) + \mathbf{A}_{22}(\mathcal{Q})\mathbf{v}_b(\mathrm{t}) \tag{13}$$

where

$$\mathbf{A}_{11}(\mathcal{Q}) \triangleq \frac{\partial \mathbf{f}_a(\mathbf{V}_a, \mathbf{V}_b; \boldsymbol{\mu})}{\partial \mathbf{V}_a} | \mathbf{V}_a = \overline{\mathbf{V}}_a, \mathbf{V}_b = \overline{\mathbf{V}}_b, \tag{14}$$

$$\mathbf{A}_{12}(\mathcal{Q}) \triangleq \frac{\partial \mathbf{f}_a(\mathbf{V}_a, \mathbf{V}_b; \boldsymbol{\mu})}{\partial \mathbf{V}_b} | \mathbf{V}_a = \overline{\mathbf{V}}_a, \mathbf{V}_b = \overline{\mathbf{V}}_b \tag{15}$$

$$\mathbf{A}_{21}(\mathcal{Q}) \triangleq \frac{\partial \mathbf{f}_b(\mathbf{V}_a, \mathbf{V}_b; \boldsymbol{\mu})}{\partial \mathbf{V}_a} | \mathbf{V}_a = \overline{\mathbf{V}}_a, \mathbf{V}_b = \overline{\mathbf{V}}_b, \tag{16}$$

$$\mathbf{A}_{22}(\mathcal{Q}) \triangleq \frac{\partial \mathbf{f}_b(\mathbf{V}_a, \mathbf{V}_b; \boldsymbol{\mu})}{\partial \mathbf{V}_b} | \mathbf{V}_a = \overline{\mathbf{V}}_a, \mathbf{V}_b = \overline{\mathbf{V}}_b. \tag{17}$$

In the linearized cell state equation, $\mathbf{A}_{11}(\mathcal{Q})$ is an $m \times m$ matrix, $\mathbf{A}_{12}(\mathcal{Q})$ is an $m \times (n–m)$ matrix, $\mathbf{A}_{21}(\mathcal{Q})$ is an $(n–m) \times m$ matrix, and $\mathbf{A}_{22}(\mathcal{Q})$ is an $(n–m) \times (n–m)$ matrix. They are constant real matrices whose elements depend on the constant input $\mathbf{I}_a \in \mathbb{R}^m$, the control parameter $\boldsymbol{\mu} \in \mathbb{R}^\rho$, and the cell equilibrium point $\mathcal{Q}(\mathbf{I}_a, \boldsymbol{\mu})$.

We are now able to define local activity at a cell equilibrium point $\mathcal{Q}$. Given any continuous input function of time $\mathbf{i}_a(t)$ for $t \geq 0$ and assuming zero initial conditions $\mathbf{v}_a(0) = 0, \mathbf{v}_b(0) = 0$, a solution of the linearized cell state equations about cell equilibrium point $\mathcal{Q}$ is an infinitesimal cell state in the neighborhood of the cell equilibrium point $\mathcal{Q}$, denoted by $\mathbf{v}_a(t)$ and $\mathbf{v}_b(t)$ for $t \geq 0$. Let us define the local power flow $p(t) \triangleq \mathbf{v}_a(t) \cdot \mathbf{i}_a(t)$ as rate of change of energy at time t at cell equilibrium point $\mathcal{Q}\big(\mathbf{V}_a = \overline{\mathbf{V}}_a,\ \mathbf{V}_b = \overline{\mathbf{V}}_b\big)$. Mathematically, the term p(t) denotes the scalar (dot) product between the two vectors $\mathbf{v}_a(t)$ and $\mathbf{i}_a(t)$.

The principle of local activity is based on the idea that when operating in an infinitesimal neighborhood of a cell equilibrium point $\mathcal{Q}$, a locally-active cell must behave like a unit (e.g., a transistor in technology) operating at an active operating point whereby a small (low-power) input can be converted into a large (high-power) output at the expense of an energy supply (e.g., a battery in the case of a transistor). In general, a cell is said to be locally active at an equilibrium point $\mathcal{Q}$ if it is possible to find a local (i.e., infinitesimal) input $\mathbf{i}_a(t)$ such that by applying the global input $\mathbf{I}_a(t) \triangleq \overline{\mathbf{I}}_a + \mathbf{i}_a(t)$, we can extract more infinitesimal energy at $\mathcal{Q}$ over some time interval $0 < T < \infty$ than what the cell has taken from its external environment which consists of the coupling spatial grid of all the other cells.

Let $w(t)$ be the total energy (i.e., the "infinitesimal sum" or integral of power flow $p(t) \triangleq \mathbf{v}_a(t) \cdot \mathbf{i}_a(t)$) accumulated since the initial time $t = 0$ until $t = T$. It is convenient though arbitrary to distinguish the reference direction of the total energy entering and leaving the cell at $t = T$. If $w(t) > 0$, then there is a net total energy accumulated since the initial time $t = 0$ entering the cell at

$t = T$. Conversely, if $w(t) < 0$, then at $t = T$, the cell is actually delivering energy to the external circuit. In this case, at $t = T$, the cell behaves like a local source of energy, rather than a sink.

Definition of Local Activity. A cell is said to be locally active at a cell equilibrium point Q if and only if there exists a continuous input time function $\mathbf{i}_a(t) \in \mathbb{R}^m, t \geq 0$, such that at some finite time T, $0 < T < \infty$, there is a net energy flowing out of the cell at $t = T$, assuming the cell has zero energy at $t = 0$, namely

$$w(\mathrm{t}) = \int_0^T \mathbf{v}_a(t) \cdot \mathbf{i}_a(t)\, dt < 0, \tag{18}$$

where $\mathbf{v}_a(t)$ is a solution of the linearized cell state equation about Q with zero initial state $\mathbf{v}_a(0) = \mathbf{0}$ and $\mathbf{v}_b(0) = \mathbf{0}$.

Definition of Local Passivity. A cell is said to be locally passive at a cell equilibrium point Q if and only if it is not locally active at Q, namely

$$w(t) = \int_0^T \mathbf{v}_a(t) \cdot \mathbf{i}_a(t)\, dt > 0, \tag{19}$$

for all continuous input time functions $\mathbf{i}_a(t)$ and for all $t \geq 0$, under zero initial states $\mathbf{v}_a(0) = \mathbf{0}$ and $\mathbf{v}_b(0) = \mathbf{0}$.

Definition of Locally Active Reaction-Diffusion Equations. Reaction-diffusion equations are called locally active if and only if its associated cells are locally active at some cell equilibrium point. Otherwise, they are said to be locally passive.

Understanding the Local Activity Principle. In the definition of local activity, we need the assumption of zero energy at $t = 0$, because otherwise the cell may have some stored energy $t = 0$ and it could be discharging it to the outside circuit even though it is locally passive. The cell's ability to act as a source of small-signal energy implies that it can amplify an initially small input signal into a larger-energy signal. The increase in energy must, of course, come from some external energy supply, such as a kind of external pump or battery if the cell is a transistor, or "glucose" if the cell is a neuron. According to the conservation principle of energy, there is nothing coming from nothing, or, in economic terms, there is "no free lunch".

Mathematically, the signal must be infinitesimal small in order that we can model the cell by only the linear terms in its Taylor series expansion. This in turn allows us to apply well-known linear mathematics and derive explicit analytical criteria for the cell to be locally active at the equilibrium point where the Taylor series expansion is computed. This also proves that complexity originates from infinitesimal small perturbations, notwithstanding the fact that the complete system is typically highly nonlinear.

Intuitively, a cell is locally-active if it is endowed with some excitable "innate" potential, such that under certain conditions, it can become "mathematically alive", capable of exhibiting oscillation and chaos. The deepest and counter-intuitive property of local activity is that a "mathematically dead" but locally active cell can become explosive, even if it is interfaced with a locally-passive load, or "sink". That can never happen with a locally-passive cell, whose entropy must increase continuously.

Complexity Function and Complexity Matrix. In order to prove that a cell is locally-active at an equilibrium point Q, the definition of local activity requires that an input time function $\mathbf{i}_a(t)$ must be found which initiates a positive energy flow $\int_0^T \mathbf{v}_a(t) \cdot \mathbf{i}_a(t)\, dt$ out of the cell at some finite time $0 < T < \infty$, assuming $\mathbf{v}_a(0) = \mathbf{0}$ and $\mathbf{v}_b(0) = \mathbf{0}$. The definition is intuitively clear and mathematically precise, but misses a constructive procedure whether such an input time function exists or not. Therefore, computationally practical necessary and sufficient conditions must be found to test the local activity of some cell at an equilibrium point. In natural and engineering sciences, Laplace transforms are used for analysis of linear time-invariant systems (e.g., electrical circuits, harmonic oscillators, optical devices, and mechanical systems). In this analysis, the Laplace transform is sometimes interpreted as a transformation from the time-domain, in which inputs and outputs are functions of time, to the frequency-domain, where the same inputs and outputs are functions of complex angular frequency (in radians per unit time). In any way, given a simple mathematical or functional description of an input or output to a system, the Laplace transform provides an alternative functional description that often simplifies the process of analyzing the behavior of the system, because it converts a system of linear differential equations to a system of linear algebraic equations. Therefore, we consider the Laplace transforms of each component of the vectors $\mathbf{v}_a(t)$, $\mathbf{v}_b(t)$, and $\mathbf{i}_a(t)$.

By this procedure, a complexity function $Y_Q(s)$ can be derived for complex number $s = a + ib$ with $\text{Re}[s] = a$ and $\text{Im}[s] = b$, in short, $s = \text{Re}[s] + i\ \text{Im}[s]$. The complexity function can be illustrated as mapping on a complex plane. It is said to be a positive-real function iff (1) $Y_Q(s)$ is a real number whenever s is a real number, (2) $\text{Re}[Y_Q(s)] \geq 0$ for all s with $\text{Re}[s] \geq 0$ where $Y_Q(s)$ is not singular. Since $Y_Q(s)$ is assumed to be a rational function, condition (1) is always satisfied. In this case, $Y_Q(s)$ is a positive-real function iff the closed right-half s-plane is mapped into the closed right-half Y_Q-plane.

In the simplest case with only one diffusion coefficient ($m = 1$), one port state variable $\hat{\mathbf{v}}_a(s)$ and one non port variable $\hat{\mathbf{v}}_b(s)(n = 2)$, the complexity function $Y_Q(s)$ reduces to a scalar rational function. For $m = 1$ and $n = 2$, we get a rational function of complex variable s. For $m > 1$, we get a $m \times m$ complexity matrix $Y_Q(s)$ whose elements are rational functions of complex variable s.

In the scalar case (one port state variable), the Local Passivity Principle can be proved ([7], 3442):

The Local Passivity Theorem. A uncoupled cell with one port state variable $m = 1$ is locally passive at a cell equilibrium point $Q(V_1 = \overline{V}_1)$ if, and only if, the complexity function $Y_Q(s)$ is a positive real function.

It follows from the Local Passivity Theorem that a cell is locally passive if, and only if, the closed right-half s-plane maps into the closed right-half Y_Q-plane. These insights deliver a practical test for local passivity ([7], 3443).

Test for Local Passivity. A cell with one port state variable is locally passive at equilibrium point $Q(V_1 = \overline{V}_1)$ if, and only if, all four conditions are satisfied:

(i) $Y_Q(s)$ has no poles in the open right plane $\mathrm{Re}[s] > 0$.
(ii) $Y_Q(s)$ has no multiple poles on the imaginary axis.
(iii) If $Y_Q(s)$ has a simple pole $s = i\omega_P$ on the imaginary axis, then $K_Q(i\omega_P) \triangleq \lim_{s \to i\omega_P} (s - i\omega_P) Y_Q(s)$ must be a positive real number.
(iv) $\mathrm{Re}[Y_Q(i\omega)] \geq 0$ for all $\omega \in (-\infty, \infty)$ where $s = i\omega$ is not a pole.

Since all four conditions must be satisfied for $Y_Q(s)$ to be locally passive, the negation of any one of these conditions gives us the desired

Test for Local Activity of Complexity Function $Y_Q(s)$. A cell with one port state variable is locally active at equilibrium point $Q(V_1 = \overline{V}_1)$ if, and only if, any one of the following conditions is true:

(i) $Y_Q(s)$ has a pole in the open right plane $\mathrm{Re}[s] > 0$.
(ii) $Y_Q(s)$ has a multiple pole on the imaginary axis.
(iii) $Y_Q(s)$ has a simple pole $s = i\omega_P$ on the imaginary axis and $K_Q(i\omega_P) \triangleq \lim_{s \to i\omega_P} (s - i\omega_P) Y_Q(s)$ is either a negative real number, or a complex number.
(iv) $\mathrm{Re}[Y_Q(i\omega)] \geq 0$ for some $\omega \in (-\infty, \infty)$.

By the same procedure, one can prove the general test for local activity of complexity matrix $\mathbf{Y}_Q(s)$ for any $m \geq 2$.

The complexity matrix $\mathbf{Y}_Q(s)$ depends not only on the cell equilibrium point $Q(\mathbf{V}_a = \overline{\mathbf{V}}_a)$, but also on the cell control parameters $\boldsymbol{\mu} \in \mathbb{R}^\rho$. For each cell state $(Q, \boldsymbol{\mu})$, we can test whether any one of the four conditions in our Local Activity Test is satisfied. This explicit procedure can be derived analytically in simple cases, or numerically by a computer. We can partition therefore the parameter space into a locally-passive domain $\mathcal{P}$ and a locally-active domain $\mathcal{A}$, over all possible cell equilibrium points corresponding to $\mathbf{V}_a \in \mathbb{R}^m$ with $\mathcal{P} \cup \mathcal{A} = \mathbb{R}^m$.

It can be proven that no reaction-diffusion equation can exhibit complexity if its cell parameters lie in the locally-passive domain $\mathcal{P}$ ([7] 3447). The larger the size of the locally-active domain $\mathcal{A}$, the more chances are for the reaction-diffusion equation to exhibit complexity. Since the number of non-state variables can be increased by setting more diffusion coefficients to zero, it follows that $\mathcal{A}_1 \subset \mathcal{A}_2 \subset \ldots \subset \mathcal{A}_m \ldots \subset \mathcal{A}_n$, where $\mathcal{A}_m$ denotes the local-activity domain of a cell with m port state variables, and m is equal to the number of positive diffusion coefficients. The local activity domain $\mathcal{A}_m$ is defined as the union of all locally-active parameter domains at cell equilibrium points corresponding to all possible port state variables $\mathbf{V}_a \in \mathbb{R}^m$.

By definition, a cell is locally passive iff it is not locally active. Therefore, to prove that local activity is the origin of complexity, it suffices to prove that the reaction-diffusion equation cannot exhibit any form of complexity if the cells are locally passive. But what does complexity mean here?

Definition of Complexity. A spatially continuous or discrete medium made of identical cells which interact with all cells located within a neighborhood (called sphere of influence) with identical interaction laws is said to manifest complexity iff the homogeneous medium can exhibit a non-homogeneous static or spatio-temporal pattern, under homogeneous initial and boundary conditions.

It follows that a reaction-diffusion medium is capable of exhibiting complexity if, and only if, the corresponding continuous reaction-diffusion partial differential equations, or their discretized version, have at least one non-homogeneous static or spatio-temporal solution for some homogeneous initial and boundary conditions. The initial condition is required to be homogeneous since otherwise, we can consider a system made of only cells which are not coupled to each other, such as a system of reactive-diffusion equations with zero diffusion coefficients. This system can exhibit a non-homogeneous static pattern by choosing the initial condition to correspond to any pattern of cell equilibrium states, assuming each cell has two or more equilibrium states.

The main result is that if the cells are strictly locally passive, then all solutions of the reaction-diffusion differential equations must converge to a unique steady state as $t \to \infty$. Since the homogeneous steady state consisting of all uncoupled cells at the same equilibrium state is one such solution, it must be the only solution due to the uniqueness property. Therefore, the corresponding medium cannot exhibit any form of complexity.

The local activity domain $\mathcal{A}_m$ is the union of four local activity parameter subsets $\boldsymbol{\mu}(\mathrm{I}), \boldsymbol{\mu}(\mathrm{II}), \boldsymbol{\mu}(\mathrm{III}), \text{and}\, \boldsymbol{\mu}(\mathrm{IV})$ of the parameter space $\boldsymbol{\mu} \in \mathbb{R}^{\rho}$, each one satisfying the local activity conditions (i) – (iv) in the test for local activity of complexity matrix $\mathbf{Y}_Q(s)$. Although the first three subsets are disjoint subsets of $\mathbb{R}^{\rho}$, the fourth subset $\boldsymbol{\mu}(\mathrm{IV})$ may intersect each of the other three subsets. The subset of $\boldsymbol{\mu}(\mathrm{IV})$ which does not intersect $\boldsymbol{\mu}(\mathrm{I}), \boldsymbol{\mu}(\mathrm{II}), \text{or}\ \boldsymbol{\mu}(\mathrm{III})$ has poles only in the open left-half plane and hence its associated cell equilibrium points are both locally active and asymptotically stable, for all port input vectors $\mathbf{I}_a \in \mathbb{R}^m$. Although reaction-diffusion equations with cell parameters chosen from this subset may exhibit complexity, the most interesting phenomena are observed from uncoupled cells with $\mathbf{I}_a = \mathbf{0}$.

Definition of the Edge of Chaos. An uncoupled cell (with $\mathbf{I}_a = \mathbf{0}$) of a reaction-diffusion equation is said to be on the edge of chaos iff all of its cell equilibrium points are locally active but asymptotically stable. The set ε of all locally active parameters $\boldsymbol{\mu} \in \mathbb{R}^{\rho}$ with this property is called the edge of chaos parameter set.

The edge of chaos parameter set ε can be expressed in terms of the restricted parameter subsets with $\mathbf{I}_a = 0$:

$$\varepsilon = \boldsymbol{\mu}(\mathrm{IV})\backslash[\boldsymbol{\mu}(\mathrm{IV}) \cap \boldsymbol{\mu}(\mathrm{I})] \cup [\boldsymbol{\mu}(\mathrm{IV}) \cap \boldsymbol{\mu}(\mathrm{II})] \cup [\boldsymbol{\mu}(\mathrm{IV}) \cap \boldsymbol{\mu}(\mathrm{III})] \tag{20}$$

where all subsets are restricted to the uncoupled condition $\mathbf{I}_a = 0$. The area ε which does not intersect $\boldsymbol{\mu}(\mathrm{I}), \boldsymbol{\mu}(\mathrm{II})$, and $\boldsymbol{\mu}(\mathrm{III})$ is the edge of chaos parameter set. The restricted local activity parameter subsets $\boldsymbol{\mu}(\mathrm{I}), \boldsymbol{\mu}(\mathrm{II}), \boldsymbol{\mu}(\mathrm{III})$, and $\boldsymbol{\mu}(\mathrm{IV})$ are calculated with $\mathbf{I}_a = 0$.

3 Local Activity and Edge of Chaos of the Brusselator Equations

The Brusselator model (named after its research group in Brussels) was one of the first systems of equations used to explain self-organizing chemical reactions of the reaction-diffusion type. Based on this model, a theory of dissipative structures operating far from thermodynamic equilibrium was developed by Ilya Prigogine [3, 19, 20]. Under stability theory techniques, Prigogine and his group derived a critical bifurcation boundary for the uncoupled cell. They studied stationary and dynamic patterns emerging in the neighborhood of this boundary. But, except for the stability boundaries, the far from thermodynamic-equilibrium theory is too coarse to predict sharper and more precise domain of emergent behavior. Especially, they ignored the relatively small subset of the edge of chaos where the emergence of complexity is most likely [21, 22].

Brusselator Equations. The mathematical model of the Brusselator is defined by two partial differential equations (PDE)

$$\frac{\partial V_1(x,y)}{\partial t} = a - (b+1)V_1(x,y) + V_2(x,y)(V_1(x,y))^2 + D_1\nabla^2 V_1(x,y) \tag{21}$$

$$\frac{\partial V_2(x,y)}{\partial t} = b\,V_1(x,y) - V_2(x,y)(V_1(x,y))^2 + D_2\nabla^2 V_2(x,y) \tag{22}$$

with two diffusion coefficients D_1 and D_2 and two state variables V_1 and V_2 characterizing the chemical dynamics. The cell parameters are denoted by a and b, the spatial coordinates are denoted by x and y. The coupling coefficients are assumed as diffusion with $D_1 \geq 0$ and $D_2 \geq 0$.

Local Activity Test of Brusselator. A local activity test can be applied in following steps of an algorithm:

1. In a first step, the Brusselator PDE is mapped into a discrete-space version with two diffusion coefficients. For the edge of chaos, it suffices to consider the zero-input current case $I_1 = I_2 = 0$.
2. In the second step, the equilibrium points Q_i with $I_1 = I_2 = 0$ are determined. In the case of a Brusselator, there is a unique equilibrium point Q_1.
3. In the third step, the Jacobian matrix of the discretized Brusselator equations at the equilibrium point Q_1 is determined.

4. In the fourth step, the input data of the Jacobian matrix are used to classify each cell parameter point (a, b) at the equilibrium point Q_1 with a test algorithm into one of the three disjoint categories:
(a) *Locally Active and Stable* $S\ (Q_1)A\ (Q_1))$: Because there is only one equilibrium point of a Brusselator, this region coincides with the edge of chaos domain. The edge of chaos domain is defined as the region in the cell parameter space where the isolated cell is locally active and stable at least at one equilibrium point.
(b) *Locally Active and Unstable* $A\ (Q_1)U\ (Q_1))$: This region corresponds to the oscillatory or unstable region of an isolated cell.
(c) *Locally Passive* $P\ (Q_1)$: This is the region in the cell parameter space where complex phenomena are unlikely to occur in reaction-diffusion systems.

Pattern formation. Most of the parameter points found by Prigogine as examples of self-organization are located nearby the bifurcation boundary separating the stable from the unstable region. The local activity test is a general check and also identifies the ignored domain of local passivity $b \leq 1$. The boundary $b = 1$ between the local passivity and the local activity domain cannot be determined via Prigogione's linear stability analysis, because cells with parameters in both $S(Q_1)A(Q_1)$ and $P(Q_1)$ are always stable. The various static and dynamic patterns generated by the Brusselator are totally predicted by the local activity principle. Static patterns are not reduced to Turing's stationary patterns with both diffusion coefficients non zero, but also mean the emergence of patterns with only one diffusion coefficient. Some of them are located in the edge of chaos domain.

4 Local Activity and Edge of Chaos of the Gierer-Meinhardt Equations

The local activity theory is applied, again, to the Gierer-Meinhardt-equations to illustrate the emergence of complexity. On the basis of autocatalysis and lateral inhibition, Gierer and Meinhardt proposed a mathematical model [12] to explain pattern formation (morphogenesis) in living systems. Using numerical integration, they were able to produce a number of patterns relevant to the formation of biological structures. But, it is still too coarse to predict the precise domain in the cell parameter space where emergent behavior may occur. The local activity theory offers a rigorous and effective tool for sharpening existing results in the sense that it can identify more precisely those regions in the cell parameter space which are capable of emergent behaviors, and also in fine tuning such regions into a relatively small subset called the edge of chaos where the emergence of complex phenomena is most likely [14].

Gierer-Meinhardt Equations. The Gierer-Meinhardt model is described by the following system of partial differential equations (PDE) with two diffusion coefficients D_1 and D_2

$$\frac{\partial V_1(x,y)}{\partial t} = a + \frac{(V_1(x,y))^2}{V_2(x,y)} - bV_1(x,y) + D_1 \nabla^2 V_1(x,y) \tag{23}$$

$$\frac{\partial V_2(x,y)}{\partial t} = (V_1(x,y))^2 - V_2(x,y) + D_2 \nabla^2 V_2(x,y) \tag{24}$$

where $V_1$and V_2 are the two state variables characterizing the cell dynamics. The cell parameters are denoted by a, and b, and the spatial coordinates are denoted by x and y. In keeping with the physical meaning of diffusion we will assume the coupling coefficients $D_1 \geq 0$ and $D_2 \geq 0$.

Local Activity Test of Gierer-Meinhardt Model. A local activity test can be applied, again, in following steps of an algorithm:

1. The Gierer-Meinhardt PDE is mapped into a discrete-space version with two diffusion coefficients. For the edge of chaos, it suffices to consider the zero-input current case $I_1 = I_2 = 0$.
2. In the second step, the equilibrium points Q_i with $I_1 = I_2 = 0$ are determined. In the case of the Gierer-Meinhardt model, it is sufficient to determine the equilibrium point Q_1 with $V_1(Q_1) = \frac{a+1}{b}$ and $V_2(Q_1) = \left(\frac{a+1}{b}\right)^2$. An equilibrium point Q_2 with $V_1(Q_2) = V_2(Q_2) = 0$ is also a solution of the system, when $I_1 = I_2 = 0$, $a = 0$ and for zero derivatives. It can easily be checked that this equilibrium is always unstable, independent of cell parameters.
3. In the third step, the Jacobian matrix of the discretized Gierer-Meinhardt equations at the equilibrium point Q_1 is determined.
4. In the fourth step, the input data of the Jacobian matrix are used to classify each cell parameter point (a, b) at the equilibrium point Q_1 with a test algorithm into one of the three disjoint categories:

(a) *Locally Active and Stable* S $(Q_1)A((Q_1)$: The edge of chaos domain is defined as the region in the cell parameter space where the isolated cell is locally active and stable at least at one equilibrium point.

(b) *Locally Active and Unstable* A $(Q_1)U$ (Q_1): This region corresponds to the oscillatory or unstable region of an isolated cell.

(c) *Locally Passive* P (Q_1): This is the region in the cell parameter space where complex phenomena are unlikely to occur in reaction-diffusion systems. The restricted local passivity region with $I_1 = I_2 = 0$ can be partially or totally included in the local activity region which is defined for all possible equilibrium points when $I_1, I_2\, \varepsilon(-\infty, +\infty)$.

Pattern Formation. The classification of a cell parameter point into one of these three categories depends on whether there is only one diffusion coefficient, or there are two nonzero diffusion coefficients. Gierer and Meinhardt only considered selected cell parameter points. All of them lie within the edge of chaos region.

Many other examples can be determined by the local activity principle. Again, the boundary between local passivity and local activity domain cannot be determined by a linear stability analysis (in the sense of Prigogine, Gierer, and Meinhardt). The dynamics of associated computer simulations is displayed with respect to points of the parameter space. For the same cell parameter point different sets of coupling coefficients and/or initial states may lead to completely different behaviors.

5 Local Activity and Edge of Chaos of the Hodgkin-Huxley Equation

The long-lasting action and pace-maker potentials of the Purkinje fiber of the heart were first described by the Hodgkin-Huxley equations of the cardiac Purkinje fiber model of morphogenesis in [11]. The bifurcation diagrams of the corresponding computer simulations supply a possible explanation for why a heart with a normal heart-rate may stop beating suddenly: The cell parameter of a normal heart is located in a locally active unstable domain and just nearby an edge of chaos. The membrane potential along a fiber is simulated in a Hodgkin-Huxley model by a computer [23]. Computer simulations show that oscillatory patterns, chaotic patterns, or divergent patterns may emerge if the selected cell parameters are located in locally active domains but nearby the edge of chaos. This research demonstrates once again the effectiveness of the local activity theory in choosing the parameters for the emergence of complex (static and dynamic) patterns in a homogeneous lattice.

In the previous examples, the analytical criteria for testing the local activity of models with one and two local state variables have been presented. The criteria have been used to describe the bifurcation diagrams of the corresponding computer simulations, in particular finding the edge of chaos domains. In this section, the analytic criteria for testing the local activity of computer simulations with four state variables, and one diffusion coefficient are set up. After mapping the cardiac Purkinje fiber (CPF) equations to each cell of a computer simulation, which is then called the Hodgkin-Huxley cell, we choose the sodium equilibrium potential (denoted by E_{N_a}) and the potassium equilibrium potential (denoted by E_K) as cell parameters for calculating the bifurcation diagrams since E_{N_a} and E_K depend on the corresponding ionic concentrations which are not constants in vivo.

It is interesting to find that the cell parameter of a normal heart is located in the locally active unstable domain but nearby an edge of chaos domain. Roughly speaking, a computer simulation shows that as the values of E_{N_a} and E_K are increased, the frequency of the heartbeat (corresponding to the periodic frequency of the membrane potential described via the CPF equations) also increases [11]. However the amplitude of the membrane potential decreases until the heart stops beating. Conversely as the values of E_{N_a} and E_K are decreased, the frequency of the heartbeat is also decreased until the heart stops beating. These phenomena can be explained well via the corresponding bifurcation diagrams. Extensive computer simulations show that if the chosen cell parameters are nearby the edge of chaos and are located in a locally active unstable region, the corresponding patterns may show chaotic, periodic, or unbounded characteristics.

Cardiac Purkinje Fiber (CPF) Equations. The cardiac Purkinje Fiber (CPF) equations introduced by [11] have been used to describe the action pacemaker potentials of the Purkinje fibers of the heart. The behavior of the equation corresponds quite well with the observed behavior of the Purkinje fibers. The original CPF equations have the form

$$\frac{dV}{dt} = -\frac{1}{C_m}((400\,m^3h + 0.14)(V - a) + 1.2\exp((-V - 90/50) + 0.015\exp((V + 90)/60) + 1.2n^4 + (V + b)) \tag{25}$$

$$\frac{dm}{dt} = \alpha_m(V)(1 - m) - \beta_m(V)m \tag{26}$$

$$\frac{dh}{dt} = \alpha_h(V)(1 - h) - \beta_h(V)h \tag{27}$$

$$\frac{dn}{dt} = \alpha_n(V)(1 - n) - \beta_n(V)n \tag{28}$$

where $a = E_{N_a} = 40$, $b = -E_k = 100$, $C_m = 12$ and E_{N_a}, E_k and C_m are sodium equilibrium potential, potassium equilibrium potential and membrane capacity, respectively. The other terms are defined as follows,

$$\alpha_m(V) = 0.1(-V - 48)/(\exp((-V - 48)/15) - 1) \tag{29}$$

$$\beta_m(V) = 0.2(V + 8)/(\exp((V + 8)/5) - 1) \tag{30}$$

$$\alpha_h(V) = 0.17(\exp((-V - 90)/20)) \tag{31}$$

$$\beta_h(V) = 1/(\exp((-V - 42)/10) + 1) \tag{32}$$

$$\alpha_n(V) = 0.0001(-V - 50)/(\exp((-V - 50)/10) - 1) \tag{33}$$

$$\beta_n(V) = 0.002\exp((-V - 90)/80) \tag{34}$$

where V is equal to the membrane potential E minus the resting potential E_r. V is called the membrane potential.

Hodgkin-Huxley Discrete Computer Model. In the next step [11], the original CPF equations are mapped into a four-dimensional 4×30 Hodgkin-Huxley computer simulation, which is a discrete version of the CPF partial differential equation with one diffusion coefficient D_1

$$\dot{V}_{i,j} = -\frac{1}{c_m}((400\,m_{i,j}^3 h_{i,j} + 0.14)(V_{i,j} - a) + 1.2\exp((-V_{i,j} - 90/50)$$
$$+0.015\exp((V_{i,j} + 90)/60) + 1.2n_{i,j}^4)(V_{i,j} + b)) + D_1(V_{i+1,j}$$
$$+V_{i-1,j} + V_{i,j+1} + V_{i,j-1} - 4V_{i,j})$$

$$\dot{m}_{i,j} = \alpha_m(V_{i,j})(1 - m_{i,j}) - \beta_m(V_{i,j})m_{i,j} \tag{36}$$

$$\dot{h}_{i,j} = \gamma\,\alpha_h(V_{i,j})(1 - h_{i,j}) - \beta_h(V_{i,j})h_{i,j} \tag{37}$$

$$\dot{n}_{i,j} = \alpha_n(V_{i,j})(1 - n_{i,j}) - \beta_n(V_{i,j})n_{i,j}\;(i = 1,2,3,4;\, j = 1,2,\ldots,30) \tag{38}$$

where parameters a and b are considered to be the relevant parameter space, and γ is an additional parameter to observe the effect of small disturbances on the bifurcation diagrams on the computer simulations.

In component form, these equations become

$$\dot{\mathbf{V}} = f_1(\mathbf{V},\mathbf{M},\mathbf{H},\mathbf{N}) + D_1\nabla^2\mathbf{V} \tag{39}$$

$$\dot{\mathbf{M}} = f_2(\mathbf{V},\mathbf{M},\mathbf{H},\mathbf{N}) \tag{40}$$

$$\dot{\mathbf{H}} = f_3(\mathbf{V},\mathbf{M},\mathbf{H},\mathbf{N}) \tag{41}$$

$$\dot{\mathbf{N}} = f_4(\mathbf{V},\mathbf{M},\mathbf{H},\mathbf{N}) \tag{42}$$

where

$$f_1(\mathbf{V},\mathbf{M},\mathbf{H},\mathbf{N}) = -\frac{1}{c_m}((400\,\mathbf{M}^3\mathbf{H} + 0.14)(\mathbf{V} - a) + 1.2\exp((-\mathbf{V} - 90/50)$$
$$+\,0.015\exp((\mathbf{V} + 90)/60) + 1.2\,\mathbf{N}^4)(\mathbf{V} + b))$$

$$f_2(\mathbf{V},\mathbf{M},\mathbf{H},\mathbf{N}) = \alpha_m(\mathbf{V})(1 - \mathbf{M}) - \beta_m(\mathbf{V})(\mathbf{M})$$

$$f_3(\mathbf{V},\mathbf{M},\mathbf{H},\mathbf{N}) = \alpha_h(\mathbf{V})(1 - \mathbf{H}) - \beta_h(\mathbf{V})(\mathbf{H})$$

$$f_4(\mathbf{V},\mathbf{M},\mathbf{H},\mathbf{N}) = \alpha_n(\mathbf{V})(1 - \mathbf{N}) - \beta_n(\mathbf{V})(\mathbf{N})$$

and ∇^2 corresponds to a 120×120 matrix.

Equilibrium Points of the Hodgkin-Huxley Equations. In the next step, the cell equilibrium points Q_i can only be solved numerically via the equations

$$f_1(\mathbf{V},\mathbf{M},\mathbf{H},\mathbf{N}) = 0 \tag{43}$$

$$f_2(\mathbf{V},\mathbf{M},\mathbf{H},\mathbf{N}) = 0 \tag{44}$$

$$f_3(\mathbf{V},\mathbf{M},\mathbf{H},\mathbf{N}) = 0 \tag{45}$$

$$f_4(\mathbf{V},\mathbf{M},\mathbf{H},\mathbf{N}) = 0 \tag{46}$$

The cell coefficients $\alpha_{m,n}(Q_i)$ are defined via the corresponding Jacobian matrix.

Local Activity and Passivity of the Hodgkin-Huxley Model. According to the local activity and passivity criteria, local activity and passivity of a cell are characterized by certain conditions of its complexity function $Y_Q(s)$ at equilibrium point Q [10]. Sometimes, it is convenient to use the cell impedance $Z_Q(s) \triangleq Y_Q^{-1}(s)$.

A cell with one port state variable is locally active at equilibrium point $Q(\overline{\mathbf{V}}_1, \overline{\mathbf{I}}_1)$ if, and only if, any one of the following conditions is true:

i. $Z_Q(s)$ has a pole in the open right plane $\text{Re}[s] > 0$.
ii. $Z_Q(s)$ has a multiple pole on the imaginary axis.
iii. $Z_Q(s)$ has a simple pole $s = i\omega_P$ on the imaginary axis and $K_Q(i\omega_p) \triangleq \lim_{s \to i\omega_P}(s - i\omega_P)Z_Q(s)$ is either a negative real number or a complex number.
iv. $Z_Q^*(i\omega) + Z_Q(i\omega) < 0$ for some $\omega = \omega_0$, where ω_0 is any real number.

The locally passive domains with respect to the equilibrium points for the Hodgkin-Huxley model can be numerically calculated via computer programs. These domains with respect to two cell equilibrium points (denoted by Q_1 and Q_2) with different cell parameters were calculated numerically and can be shown in computer simulations. Since the equilibrium points Q_1 and Q_2 cannot be expressed analytically, they are obtained via simulation tools, taking different initial iterate values.

Although the results are not analytically exact, it can be concluded from the bifurcation diagrams: The cell parameter value (40, 100) of a normal heart is located in the locally active (unstable) domain with respect to the cell equilibrium points Q_1 and Q_2, and nearby the edge of chaos with respect to the equilibrium point Q_2. For any fixed parameter value, the bifurcation diagrams with respect to the equilibrium points Q_1 and $Q_2$are quite different if the parameter b is larger than 100. Roughly speaking, our computer simulation shows that the cell equilibrium points which are located in the locally passive domains or in the edge of chaos domains with respect to the equilibrium point Q_1 seem to be globally attractive. This result may explain why a heart with an approximately normal rate can stop suddenly. A small perturbation of the parameter γ ($\pm$ 1:5 %) does not cause the bifurcation diagrams to be noticeably changed. The same conclusion also holds for the case where the parameter C_m is disturbed. These facts may be interpreted as an indication of the robustness of the heart.

Applications to Cardiology. Based on the four conditions for local activity of cells, analytical criteria for systems with four state variables and one diffusion coefficient can be presented. The criteria can be easily implemented by a computer program to produce bifurcation diagrams for the corresponding Hodgkin-Huxley model. Although no chaotic phenomenon is observed, the cell parameters which cause the heart to stop beating are always located nearby the edge of chaos domains. It can be shown that the changes in the sodium equilibrium potential E_{N_a} (corresponding to the parameter a) cause greater changes to the frequency of the heartbeat than those of the potassium potential E_k (corresponding to the parameter $-b$). The parameter value $(a, b) = (58, 100)$ seems to play an extraordinary role. At this value, the frequency of the heartbeat is about 80 beats/min. However, if the parameter is changed to $(a, b) = (58.5, 100)$, the heart will stop beating. This computer simulation seems to be able to provide some insight into why it is possible that a patient with non-normal electrocardiogram but approximate normal frequency of the heartbeat might suddenly die without warning.

So-called smoothed circuit equation and the corresponding computer simulations exhibit bifurcation diagrams which show that no locally passive domain exists. Oscillatory patterns, convergent (static) patterns, and divergent (unbounded) patterns can be obtained if the parameter sets are chosen on the edge of chaos domain. In particular, emergence of complex patterns may exist if the corresponding cell parameters are chosen in the locally active unstable domain but nearby the edge of chaos domains. In summary, this research confirms once again that the local activity theory provides a practical and explicit analytical tool for determining a subset of the cell parameter space where complexity may emerge [24].

6 The Local Activity Principle of Brains and Artificial Minds

Artificial Brain of Memristors. On the horizon of future chip technology is the vision of neuromorphic computers, mimicking the human brain with billions of neurons and synaptic connections. Brains are considered complex networks with local activities of cells such as CNNs and CAs. A technical unit modeling a living neuron with synapses needs features of memories, digital circuits, and a form of analog information processing. A strong candidate fulfilling all these requirements is the memristor, a new circuit element, which was suggested [25, 26] more than 40 years ago. Modern technology assumes that the memristor will bring a new wave of innovation in electronics, packing more bits into smaller volumes and equipped with a non-volatile memory preventing the loss of data.

The memristor device has generated immense interests among both device researchers and the memory-chip industry alike [27]. These interests were due to the high potential economic impact of the HP (Hewlett-Packard) breakthrough. Since the titanium-dioxide HP memristor could be scaled down to about 1 nm and is compatible with current IC technology, many industry experts are predicting that nano memristor devices would eventually replace both flash memories and DRAMS (Dynamic Random Access Memory). Indeed, a PC with no booting time, and which remembers all data prior to unplugging the power, could become standard features within a few years.

Memristor is an abbreviation for "memory resistor" und was predicted as the fourth missing circuit element with respect to the basic equations of electric circuits [27]. These equations are defined for the four quantities voltage (v), current

(i), charge (q), and magnetic flux (φ). Each equation determines a relation between two of these variables. The simplest relation between voltage and current is Ohm's law $v = Ri$, meaning that voltage is proportional to current. The constant of proportionality is given by the resistance R. If a current of I amperes flows through a resistance of R ohms, then the voltage with respect to the resistance is v volts. Geometrically, the graph of current versus voltage for a resistor is a straight line with slope R.

There are 6 possible pairings of 4 variables. The 2 pairs (v, φ) and (i, q) are already related by definition $v = \frac{d\varphi}{dt}$ and i $= \frac{dq}{dt}$. The 3 pairs (v, i), (φ, i), and (q, v) define a resistor, inductor, and capacitor, respectively. The missing equation relating charge q and magnetic flux φ is called memristor. Obviously, the memristor was found by arguments of symmetry and completeness. There are two classes of memristors, namely, locally-passive memristors and locally-active memristors. The conventional resistor, capacitor, inductor, and the locally-passive memristor are passive circuit elements, which must be distinguished from active devices, such as transistors, which can amplify signals and inject power into circuits. However, locally-active memristors are active devices and hence can also amplify signal, with a power supply, just like transistors. The memristor is a nonlinear device defined by the graph of a curve in the flux vs. charge plane, whose slope, called the memristance, varies from one point to another [28].

A transistor is a three-terminal device with three connections to a circuit. It acts as a switch or amplifier, with a voltage applied to one terminal controlling a current flowing between the other two terminals. Although a locally-active memristor has only two terminals, it can also realize these functions. On the other hand, locally-passive memristors can be used to build both memory and digital logic without the need for a power supply. This is why they are called non-volatile memories. Metaphorically speaking, the memristor has a built-in sense of history. A signal applied at one moment can affect another signal that travels the same path later. The first signal realizes this control by setting the internal state of the memristor to high or low resistance.

Therefore, in a neuromorphic computer, locally-passive memristors would not totally supplant a transistor, but supplement them in memory functions and locic circuits. Memristors could play the role of synapses. In biological neural networks, each nerve cell communicates with other cells through thousands of synapses. An important mechanism of learning is realized by adjusting the strength of the synaptic connections. In an artificial neural network, synapses must be small, but effective structures. Locally-passive memristors satisfy all the needed requirements. They change their resistance in response to the currents that flow through them. This operation suggests a direct way of modeling the adjustment of synaptic strength.

Recall that there are two qualitatively distinct kinds of memristors, namely locally passive memristors and locally active memristors. The HP memristor is locally passive because it does not require a power supply, and is said to be non-volatile. The potassium and sodium ion channels in the classic Hodgkin-Huxley nerve membrane circuit model can be considered locally-active memristors, powered by a sodium and a potassium pump whose energy derives from ATP molecules. In contrast, synapses are locally passive memristors capable of retaining their synaptic efficacies over long periods of time without consuming any power.

Since our brains process information using only synapses and axons, it follows that circuits made of both types of memristors should be able to emulate higher brain functions as well. The long-term potentiation (LTP) phenomenon associated with long-term memory can also be emulated by a memristor. Many associative memory phenomena, such as the Pavlovian dog behavior, can be emulated by a memristor circuit. If brains are made of memristors, then we can expect that electronic circuits made of both locally passive and locally active memristors may someday emulate human minds [29]. The key to this fundamental process is to uncover how local activity could lead to the emergence of complex patterns from a mass of homogeneous brain tissues. Formally, the local activity principle is realized in any cellular automaton. In neurons and memristors, the local activity principle is not only a formal model, but biological and technical reality.

Hodgkin-Huxley Electrical Circuit Model. The Hodgkin-Huxley electrical circuit model of a squid giant axon membrane and its associated Hodgkin-Huxley equations (henceforth referred to as HH equations) has stood the test of time and has served as a classic reference in neurophysiology and brain science research for 70 years. The squid was chosen by Hodgkin and Huxley because they are endowed with enormous axons, the largest of them in a large Atlantic squid (*Loligo pealii*) being as much as one millimeter in diameter ([30, 31, 32, 33, 34]).

In the Hodgkin-Huxley electrical circuit model, I, I_{Na}, I_K, and I_l denote the external axon membrane current, the sodium ion current, the potassium ion current, and the leakage current, respectively. Likewise, E, E_{Na}, E_K, and E_l denote the membrane capacitor voltage, the sodium ion battery voltage, the potassium ion battery voltage, and the leakage battery voltage, respectively. The basic assumption Hodgkin and Huxley made is that the squid giant axon can be modeled by a distributed circuit consisting of a line of identical 2-terminal electrical devices (henceforth referred to as HH cells) described by the HH Circuit Model, and coupled by identical passive resistors. By assuming the HH cells to be physically small and taking the limit as the length $\Delta x \to 0$, the dissipative couplings tend to a Laplacian, thereby modeling the standard diffusion mechanism. The equations describing the entire system then tends to a system of reaction diffusion equations in one spatial variable.

The HH Circuit Model [31, 35] contains 7 circuit elements that must be specified before it is possible to formulate the equations governing the circuit. These elements and their parameter values must be determined by meticulous experiments. For convenience in their measurement setups, Hodgkin and Huxley had opted to measure all voltages with respect to the resting potential E_r, whose value depends on the axon specimen and measurement temperature.

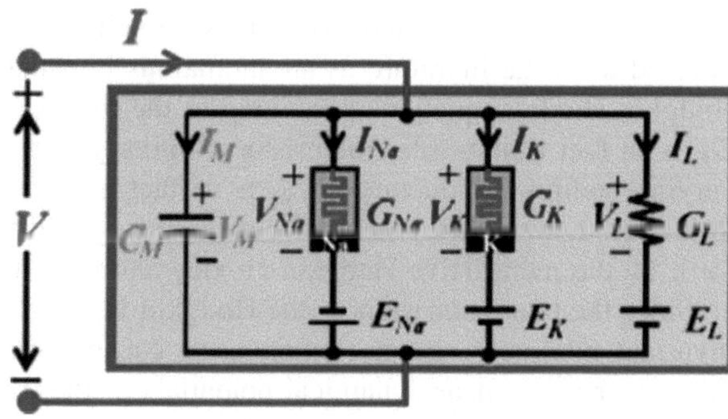

Fig. 1 Memristive Hodgkin-Huxley axon circuit model [10]

But, the time-varying resistances R_K and R_{Na} (resp., time-varying conductances g_K and g_{Na}) assumed by Hodgkin and Huxley are fundamentally wrong from a circuit-theoretic and scientific perspective. In particular, the two elements R_K and R_{Na} (resp., g_K and g_{Na}) in their HH axon Circuit model belong to the fundamentally different class of time-invariant circuit elements of memristors. After Hodgkin-Huxley's mis-identification error is corrected, all of the anomalies and confusions ([8, 23]) concerning anomalous inductances, rectification, frequency-dependent parameters, and the fundamental mechanisms responsible for the generation of action potentials can be clearly resolved in a simple and rigorous manner.

Memristive Hodgkin-Huxley Axon Circuit Model. Substituting the Hodgkin-Huxley time-varying potassium resistance R_K, and sodium resistance R_{Na} by the potassium ion-channel memristor and the sodium ion-channel memristor, respectively, we obtain the memristive Hodgkin-Huxley Axon Circuit Model. Observe that this circuit model contains only well-defined time-invariant circuit elements, as expected of any realistic physical model of the axon. With basic circuit theory, the Hodgkin-Huxley DC circuit model in Fig. 1 can be derived by simply deleting the axon membrane capacitor C_M, since $I_M = C_M \frac{dV_M}{dt} = 0$ at DC. For each $V = V_m = V_m = (Q)$ at an equilibrium point $Q(V_m, I_m)$, where $V = V_m$ and $I = I_m$ denote the axon membrane voltage and membrane current in Fig. 1, respectively, we can substitute the potassium ion-channel memristor by its small-signal circuit model, about the corresponding DC equilibrium point at $V_K = V_m(Q) - E_K = V_m(Q) - 12\,\mathrm{m}V \triangleq V_K(Q_K)$. Similarly, we can substitute the sodium ion-channel memristor by its small-signal circuit model, about the corresponding DC equilibrium point at $V_{Na} = V_m(Q) + E_K = V_m(Q) + 115\,\mathrm{mV} \triangleq V_{Na}(Q_{Na})$.

By recognizing that the two circuit elements R_K and R_{Na} in the Hodgkin-Huxley axon circuit model are not time-varying, but are rather time-invariant memristors, we were able to provide a firm circuit-theoretic foundation for analyzing, interpreting, and explaining various anomalous phenomena and paradoxes reported in the literature more than 70 years ago and which had remained unresolved [31]. For example, Hodgkin [31, 32] was quite shocked to find the small-signal impedance they measured from the axon membrane of squids had exhibited a positive reactance that suggested the presence of a gigantic inductance and an enormous magnetic field in the squid axon. This inexplicable phenomenon had since been referred to in the literature as an anomalous impedance [8]. But the terms in the Hodgkin-Huxley equations pertaining to the time varying potassium conductance G_K is in fact a first-order memristor. Similarly, the terms pertaining to the time-varying sodium conductance G_{Na} is in fact a second-order memristor. The historical Hodgkin-Huxley axon circuit model should therefore be replaced henceforth by the memristive Hodgkin-Huxley axon circuit model shown in Fig. 1.

Since the two memristors in the Hodgkin-Huxley Axon Circuit Model are time-invariant nonlinear circuit elements, we can exploit the local activity theory to uncover the nonlinear dynamical potentials of these two circuit elements. Classic circuit-theoretic concepts as small-signal admittance, small-signal impedance, pole-zero diagrams, etc. can be applied. All of these intrinsic linear circuit characterizations and their explicit analytical formulas can be derived from the memristive Hodgkin-Huxley axon circuit model. The circuit-theoretic properties represent definitive characterizations of the Hodgkin-Huxley Axon. They play a fundamental role in the research on the dynamics of ion channels.

The main theorem [16] asserts that the zeros of the scalar function $Y(\mathrm{s}, V_m) \triangleq \frac{1}{Z(\mathrm{S}, V_m)}$ called the small-signal admittance of the Hodgkin-Huxley memristor circuit model, are identical to the eigenvalues of the 4×4 Jacobian matrix of the HH equation, calculated at the equilibrium point $V_m = V(I_{ext})$ of the HH equations, for each constant DC excitation current I_{ext}. Here $s = \sigma + i\omega$ denotes the complex variable associated with the Laplace transform $\hat{v}(s)$ of a time function v(t) [16] and I_{ext} denotes an external current source applied to the Hodgkin-Huxley axon circuit model shown in Fig. 1. In chapter 2 it was explained that this theorem is valid not only for the 4-dimensional HH equation, but for any system of n differential equations. This theorem is a powerful tool because instead of calculating the eigenvalues of a high dimensional $n \times n$ matrix, one only has to calculate the roots (zeros) of a scalar polynomial equation of a single variable s, for any integer n.

Most deep insights concerning local activity and edge of chaos can be uncovered from an analysis of the linearized differential equations about the equilibrium points of its associated nonlinear dynamical system. The edge of chaos is typically a very small subset of the local activity domain. So small and yet so profound is the edge of chaos domain that we often dramatize its significance by dubbing it the pearl of local activity. Indeed, edge of chaos is the source of life, and we will show that neurons are poised near this pearl of local activity. It is rather enigmatic that while all complexity phenomena, including the generation of spikes, require strongly nonlinear dynamics, yet the mathematical genesis of such global phenomena is strictly local. The theory of local activity and edge of chaos is based entirely on linearized differential equations about an equilibrium point. Testing an equilibrium point Q for local activity in general, or edge of chaos in particular, involves examining the linearized Hodgkin-Huxley equations about Q.

Linearized Hodgkin-Huxley Equations. For each equilibrium point Q corresponding to a DC excitation current I_{ext}, let us superimpose an infinitesimally

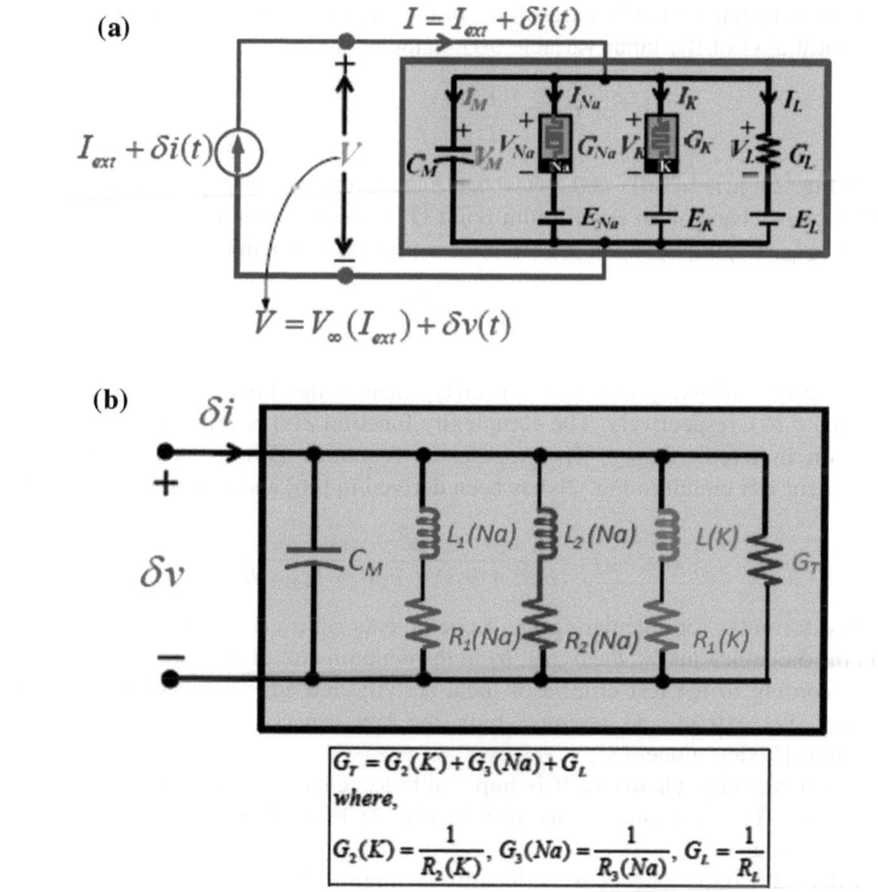

Fig. 2 Hodgkin-Huxley axon circuit model (**a**) and its linearized small-signal equivalent Hodgkin-Huxley circuit (**b**) [8]

small current signal $\delta i(t)$ and apply the composite signal $I_{ext} + \delta i(t)$ to the HH axon circuit model Fig. 2a. Whether a system (in this case the Hodgkin-Huxley axon) is locally active at Q or not is completely determined by the response $\delta v(t)$ to an infinitesimally small sinusoidal testing signal $\delta i(t) = A \sin \omega t$, where A denotes the amplitude, and $\omega = 2\pi f$ denotes the angular frequency. It follows from elementary circuit theory that the response $\delta v(t)$ to any small-signal current excitation $\delta i(t)$ can be predicted analytically from a small-signal equivalent circuit whose elements are calculated explicitly form the Jacobian matrix of the associated nonlinear differential equations, evaluated at the equilibrium point Q. Such an equivalent circuit is presented in Fig. 2b.

Complexity Function of Hodgkin-Huxley Equations. In general (chapter 2), the complexity function $C(s)$ for a single-input single-output system is defined by the ratio between the Laplace transform of the output variable $\hat{y}(t)$ and the Laplace transform $\hat{u}(s)$ of the input variable $u(t)$, namely,

$$\mathrm{C(s)} = \frac{L(y(t))}{L(u(t))} = \frac{\hat{y}(s)}{\hat{u}(s)}. \tag{47}$$

In Fig. 2a, $u(t) = \delta i(t)$ and $y(t) = \delta v(t)$, respectively. Testing for local activity and edge of chaos of an equilibrium point Q of the HH axon circuit in Fig. 2a at any $I = I_{ext}$ requires that we examine the complexity function defined by

$$\mathrm{Z(s)} = \frac{\hat{v}(s)}{\hat{i}(s)}, \tag{48}$$

where $\hat{v}(s) = L(\delta \mathrm{v(t)})$ and $\hat{\mathrm{i}}(\mathrm{s}) = L(\delta \mathrm{i(t)})$ denote the Laplace transform of δ $v(t)$ and δ $i(t)$, respectively. The complexity function $Z(s)$ is called the impedance function in circuit theory. The impedance functions $Z(s)$ for the small-signal equivalent HH circuit in Fig. 2b has been derived in [16] and is reproduced below:

$$\mathrm{Z}(\mathrm{s}, V_m) = \frac{a_3 s^3 + a_2 s^2 + a_1 s + a_0}{b_4 s^4 + b_3 s^3 + b_2 s^2 + b_1 s + b_0}. \tag{49}$$

The formulas for calculating the 4 coefficients a_0, a_1, a_2, a_3 in the numerator and the 5 coefficients b_0, b_1, b_2, b_3, b_4 in the denominator of Z(s) are listed in [16].

According to the test criteria of local activity and edge of chaos in previous chapter, we extract and rephrase only the key aspects that are essential for Hodgkin-Huxley model [8]:

Local Activity Theorem. It is impossible to generate a spike train unless the memristive Hodgkin-Huxley one-port in Fig. 2a is locally active at some equilibrium point.

Edge of Chaos Theorem. A locally asymptotically stable-equilibrium point Q of the Hodgkin-Huxley equation is poised on the edge of chaos if, and only if, $\mathrm{Re}\, Z(i\omega, V_m(Q)) < 0$ at some frequency ω. Re Z denotes the real part of the complex number $Z. Z(i\omega, V_m(Q))$ is the impedance function calculated at $s = 0 + i\omega$.

From action potentials to mental states. In general, it is assumed that all mental states are correlated to corresponding cell assemblies. But, they are not only sets of firing neurons, but hierarchical systems of neural subsystems of subsystems with different depth and degrees of complexity. Research hypothesis means that the corresponding cell assemblies must empirically be identified by observational and measuring instruments. In brain reading, for example, active cell assemblies correlated with words and corresponding objects can be identified. A single neuron is not decisive and may differ among different persons. There are typical distribution patters with fuzzy shapes which are represented in computer simulations. Brain research is still far from observing the activities of each neuron in a brain. Nevertheless, the formal hierarchical scheme of dynamics allows the explanation of complex mental states like, for instance, consciousness. Conscious states mean that persons are aware of their activities. Self-awareness is realized by additional brain areas monitoring the neural correlates of these human activities (e.g., perceptions, feeling, or thinking). Thus, even consciousness is no mysterious event, but observable, measurable, and explainable in this research framework. In the next step, the formal hierarchical model offers the opportunity to build corresponding circuits and technical equipments for technical brains and robots with these abilities.

Traditional terms "intelligence", "mind", "consciousness" etc. are historically overloaded with many meanings depending on different point of views, experience and historical positions. Therefore, their meaning depends on our definitions. Concerning intelligence, a simple working definition is suggested which does not depend on "human intelligence" (in the sense of Turing's AI-test). A system is called "intelligent" depending on its ability to solve problems. In that sense, a tick has a certain degree of intelligence, because it can solve the problem of finding blood. But, simple technical systems (e.g., a chip) can also have certain degrees of "intelligence", because there can solve certain problems. Thus, in philosophical terms, this position sympathizes with the pluralism of Leibniz who distinguished degrees of intelligence in nature instead of Descartes' dualism who believed in a "substance" called "intelligent mind" which was reserved to human beings. In that sense, there are already many intelligent functions of, e.g., robots [8].

Obviously, patterns of cell assemblies in the brain are not identical with our perceptions, feeling, and thinking. But, it is well confirmed in modern brain research that neural patterns of firing cells are correlated with mental states. "Mental states" can be defined and computationally modeled in state and parameter spaces with associated dynamical systems which allow us to test our models. With the technology of brain reading, an analysis of cell assemblies even allows us to recognize their represented meaning (e.g., pictures, words, phrases): Of course, there are only the first steps of research, but it seems to be possible at least in principle. Concerning computer science, semantics is technically realized in first steps and to certain degrees in a restricted and well-defined sense.

Motory, cognitive, and mental abilities are stored in synaptic connections of cell assemblies. A hard core of synaptic network is already wired, when a mammal brain is born. But many synaptic connections are generated during growth, experience and learning phase of mammals. Firing states of neurons with repeated action potentials enforce synaptic connections. Thus, during a learning phase, a cell assembly of simultaneously firing neurons creates a synaptic network storing the learnt information. Learning phases can be modeled computationally by learning algorithms [8]. As we all know, the learnt information can be forgotten, when learning is not repeated and the synaptic connections decay. Thus, on the microlevel, brain dynamics is determined by billions of firing and not firing neurons, and, on the macrolevel, by emerging and changing cell assemblies of neural networks coding different neural information.

The efficiency of neural networks depends on their number of hierarchical layers. They enable the brain to connect different neural states of, e.g., visual, haptic, and auditive information. But, there are also layers monitoring perceptual procedures and generating visual consciousness: A person is aware and knows that she perceives something. Even our emotions depend on specified neural networks which are connected with all kinds of brain activity. It is a challenge of brain research to identify the involved layers and networks of the brain during all kinds of mental and cognitive activities.

From action potentials to semantic understanding. Semantic understanding is made possible by hierarchical layers and learning procedures of brain dynamics. In formal semantics, a formal language gets its meaning by mapping its formal expressions onto expressions of another formal language. In computer science, several layers of symbolic languages are also used to generate semantic meaning and digital procedures. Natural languages of users refer to an advanced computer language. In a next step, a kind of behavioral language describes the functions that are intended to be executed in the system. They relate their arguments typically like mathematical functions. Then, an architectural description is needed that describes the available resources, the communication and the control. The relation between the behavioral description and the architecture is done by binding elementary functions to resources and scheduling them in time (a task that can be done by a compiler or an assembler). Finally, the architecture gets realized by a register transfer layer which is close to digital circuits that in term are realized by electrical components, voltages and currents. Each such layer has its own descriptive language, dynamics and syntactic rules. Semantics is provided by interaction between layers. The digital circuit provides the semantics for the electrical circuit (vice versa, the circuit layer provides timing information to the digital), the register transfer to the digital layer, the architecture to the register transfer layer and the behavioral description sits on top of that all, at least five layers of semantic abstractions, each expressed in a dedicated formal syntax. It is only because engineers can make such a strong taxonomy and abstraction of the semantic layers, that a digital system is understandable for the human user and the digital machine. The communication between man and machine is realized in an intermediate transfer process between layers.

The human brain has a much more complex architecture which, until nowadays, cannot be completely reconstructed in clearly distinguished layers. Obviously, the brain is not the design of an architect or engineer with specific technical purposes, but the result of a more or less random and blind evolution during millions of years. Thus, several functions of tissues and networks are still unknown, although topological layers can be identified and used for explanations of semantic processes. In the end, the brain should be totally scanned and modeled from its single neurons, synapses, and action potentials to cell assemblies, networks, and layers, in order to model the whole dynamics on the micro- and macrolevel. The "machine level" is already well known and described by the digital behavior of firing and non-firing neurons with emerging action potentials. The edge of chaos domains could clearly be identified in parameter spaces of the Hodgkin-Huxley equations. They are the origin of all kind of brain dynamics with attractors of neural states correlated with human cognitive and intelligent activities.

Compared with human brains, technical systems may be restricted, but they are sometimes much more effective with their specific solutions of cognitive and intelligent tasks. In computer science, semantic webs and i-phones can already understand questions to some extension and even answer in natural languages. The technology of applied (speech analysis) algorithms may be different from biological procedures which were developed during evolution. But, they solve the problem to some degree with their computer power, high speed, parallelism and storage which can be improved in the future. Human beings are hybrid complex systems with additional abilities (e.g., imaginations, feelings) which are always involved in our semantic acting. But, it cannot be excluded that these abilities could also be managed in a future technology.

7 Local Activity and Computational Universe

Can pattern formation in our universe be understood in the analytic framework of mathematical equations or is it too complex? Instead of the traditional analytical approach of mathematical physics, Steven Wolfram declared computer experiments with pattern formation of cellular automata as "new kind of science" (NKS). It is obviously a great merit of NKS to highlight the experimental approach in the computational sciences [6]. But we claim that even in the future quasi-empirical computer experiments are not sufficient [10]. Cellular automata must be considered complex dynamical systems in the strictly mathematical sense with corresponding equations and proofs. In short, we also need analytical models of cellular automata, in order to find precise answers and predictions in the universe of cellular automata. In this sense, the analytical approach goes beyond Wolfram's NKS.

In the analytical approach, cellular automata (CA) are defined as complex dynamical systems. Their (difference or differential) equations allow precise definitions of a complexity index and universal symmetries. It can be proved that the 256 one-dimensional cellular automata are classified by local and global symmetry classes of cellular automata. There is an exceptional symmetry group with universal computability which we called the "holy grail" in the universe of cellular automata [10]. Many analytical concepts of complexity research (e.g., attractors, basin of attractors, time series, power spectrum, fractality) are defined for cellular automata. The local activity of cells leads to pattern formation which can be analytically determined and completely classified in a rigorous mathematical manner.

Summing up all these insights, we are on the way to conceive the universe as a computational and dynamical system. The success of this research program depends on the digitization of physics. The question "*Is the Universe a computer*" leads to the question: How far is it possible to map the laws of physics onto computational digital physics? [36] Digitization is not only exciting for answering philosophical questions of the universe. Digitization is the key paradigm of modern research and technology. Nearly all kind of research and technical innovation depend on computational modeling. The emerging complexity of nature and society cannot be handled without computers with increasing computational power and storage.

In order to make this complex computational world more understandable, cellular automata are an educational tool. NKS and the analytical approach show that many basic principles of the expanding universe and the evolution of life and brain can be illustrated with cellular automata. The emergence of new structures and patterns depends on phase transitions of complex dynamical systems in the quantum, molecular, cellular, organic, ecological, and societal world [1, 37]. Cellular automata are recognized as an intuitive modeling paradigm for complex systems with many useful applications [38]. In cellular automata, extremely simple local interactions of cells lead to the emergence of complex global structures. The local principle of activity is also true in the world of complex systems with elementary particles, atoms, molecules, cells, organs, organisms, populations, and societies [8, 10]. Although local interactions generate a complex variety of being in the universe, they can be mathematically reduced to some fundamental laws of symmetry [39]. From a philosophical point of view, we cannot be sure in principle that computational and digitalized physics described the world completely, but its explanatory and predictive power is, until nowadays, overwhelming.

References

1. K. Mainzer, *Thinking in Complexity. The Computational Dynamics of Matter, Mind, and Mankind*, 5th edn. (Springer, Berlin, 2007)
2. E. Schrödinger, *What is Life? The Physical Aspect of the Living Cell & Mind and Matter* (Cambridge University Press, Cambridge, 1948)
3. I. Prigogine, *From Being to Becoming* (Freeman, San Francisco, 1980)
4. H. Haken, Synergetics, *An Introduction*, 3rd edn. (Springer, New York, 1983)
5. C.G. Langton, Computation at the edge of chaos. Phase transitions and emergent computation. Physica D **42**, 12–37 (1990)
6. S. Wolfram, *A New Kind of Science* (Wolfram Media, Champaign, 2002)
7. L.O. Chua, Local activity is the origin of complexity. Inter. J. Bifurcat. Chaos **15**(11), 3435–3456 (2005)
8. K. Mainzer, L.O. Chua, *Local Activity Principle. The Cause of Complexity and Symmetry Breaking* (Imperial College Press, London, 2012)
9. A.M. Turing, The chemical basis of morphogenesis. Philos. Trans. Roy. Soc. London, B **237**, 37–72 (1952)
10. K. Mainzer, L.O. Chua, *The Universe as Automaton. From Simplicity and Symmetry to Complexity* (Springer, Berlin, 2011)
11. D. Noble, A modification of the Hodgkin-Huxley equations applicable to Pukinje fibre action and peacemaker potentials. J. Physiol. **160**, 317–352 (1962)
12. A. Gierer, H. Meinhardt, A theory of biological pattern formation. Kybernetik **12**, 30–39 (1972)
13. H. Haken, H. Olbrich, Analytical treatment of pattern formation in the Gierer-Meinhardt model of morphogenesis. J. Math. Biol. **6**, 317–331 (1978)
14. R. Dogaru, L.O. Chua, Edge of chaos and local activity domain of the Gierer-Meinhardt CNN. Inter. J. Bifurcat. Chaos **8**(12), 2321–2340 (1998)

15. A. Mauro, Anomalous impedence, a phenomenological property of time-variant resistance. An analytic review. Biophys. J. **1**, 353–372 (1961)
16. L.O. Chua, V. Sbitney, and H. Kim, Neurons are poised near the edge of chaos. Inter. J. Bifurcat Chaos (forthcoming) (2012)
17. S. Smale, A mathematical model of two cells via Turing's equation. Lect. Appl. Math. **6** (American Mathematical Society) 15–26 (1974)
18. L.O. Chua, Passivity and complexity. IEEE Trans. Circ. Syst. **46**(1), 71–82 (1999)
19. I. Nicolis, *Prigogine. Exploring Complexity* (W.H. Freeman, New York, 1989)
20. R.J. Field, R.M. Noyes, Oscillations in chemical systems IV. Limit cycle behavior in a model of a real chemical reaction. J. Chem. Phys. **60**, 1877–1884 (1974)
21. R. Dogaru, L.O. Chua, Edge of chaos and local activity domain of the Brusselator CNN. Inter. J. Bifurcat. Chaos **8**(6), 1107–1130 (1998)
22. L. Min, K.R. Crounse, L.O. Chua, Analytical criteria for local activity and applications to the Oregonator CNN. Inter. J. Bifurcat. Chaos **10**(1), 25–71 (2000)
23. K.S. Cole, *Membranes, Ions and Impulses* (University of California Press, Berkeley, 1972)
24. L. Min, K.R. Crounse, L.O. Chua, Analytical criteria for local activity of reaction-diffusion CNN with four state variables and applications to the Hodgkin-Huxley equation. Inter. J. Bifurcat. Chaos **10**(6), 1295–1343 (2000)
25. D.B. Strukov, G.S. Snider, D.R. Duncan, R.S. Williams, The missing memristor found. Nature **453**, 80–83 (2008)
26. L.O. Chua, Memristor—the missing circuit element. IEEE Trans. Circ. Theor. **18**, 507–519 (1971)
27. B. Hayes, The memristor. Am. Sci. **9**(2), 106–110 (2011)
28. L.O. Chua, Resistance switching memories are memristors. Appl. Phys. A **102**(4), 765–783 (2011)
29. J. Mullins, Memristor minds: The future of artificial intelligence. *New Scientist* **7**(2009)
30. A.L. Hodgkin, A.F. Huxley, Currents carried by sodium and potassium ions through the membrane of the giant axon of Loligo. J. Physiol. **116**, 449–472 (1952)
31. A.L. Hodgkin, A.F. Huxley, B. Katz, Ionic currents underlying activity in giant axon of the squid. Arch. Sci. Physiol. **3**, 129–150 (1949)
32. A.L. Hodgkin, R.D. Keynes, Experiments on the injection of substances into squid giant axons by means of microsyringe. J. Physiol. London **131**, 592 (1956)
33. R. FitzHugh, Mathematical models of excitation and propagation nerve, in *Biological Engineering*, ed. by H. Schwan (McGraw-Hill, New York, 1969), pp. 1–85
34. J.Z. Young, Structure of nerve fibres and synapses in some invertebrates. Cold Spring Harb. Symp. Quant. Biol. **4**, 1–6 (1936)
35. R. Dogaru, L.O. Chua, Edge of chaos and local activity domain of the FitzHugh-Nagumo equation. Inter. J. Bifurcat. Chaos **8**(2), 211–257 (1998)
36. D. Deutsch, Quantum theory, the church-turing principle and the universal quantum computer. Proc. Roy. Soc. London A **400**, 97–117 (1985)
37. K. Kaneko, *Life: An Introduction to Complex Systems Biology* (Springer, Berlin, 2006)
38. A.G. Hoekstra, J. Kroc, P.M.A. Sloot (eds.), *Simulating Complex Systems by Cellular Automata* (Springer, Berlin, 2010)
39. K. Mainzer, *Symmetries in Nature*, (De Gruyter, Berlin, 1996) (German: Symmetrien der Natur, De Gruyter, Berlin, 1988)

[illegible]

Complexity Fits the Fittest

Joost J. Joosten

Abstract In this paper we shall relate computational complexity to the principle of natural selection. We shall do this by giving a philosophical account of complexity versus universality. It seems sustainable to equate universal systems to complex systems or at least to *potentially complex systems*. Post's problem on the existence of (natural) intermediate degrees (between decidable and universal Σ_1^0) then finds its analog in the Principle of Computational Equivalence (**PCE**). In this paper we address possible driving forces—if any—behind **PCE**. Both the natural aspects as well as the cognitive ones are investigated. We postulate a principle **GNS** that we call the *Generalized Natural Selection* principle that together with the Church-Turing thesis is seen to be in close correspondence to a weak version of **PCE**. Next, we view our cognitive toolkit in an evolutionary light and postulate a principle in analogy with Fodor's language principle. In the final part of the paper we reflect on ways to provide circumstantial evidence for **GNS** by means of theorems, experiments or, simulations.

Keywords Computational complexity · Intermediate degrees · Principle of computational equivalence · Natural selection · Dynamical systems

1 Complexity and Computation

It is a standard definition in the literature to call a computational process Π *universal* if it can simulate any other computational process Θ. In other words, Π is universal if (see for example [2] or any other basic text book on computability

J. J. Joosten (✉)
Department of Logic, History and Philosophy of Science,
University of Barcelona, Carrer Montalegre 6, 08001 Barcelona, Spain
e-mail: jjoosten@ub.edu
URL: http://www.phil.uu.nl/~jjoosten/

I. Zelinka et al. (eds.), *How Nature Works*, Emergence,
Complexity and Computation 5, DOI: 10.1007/978-3-319-00254-5_3,
© Springer International Publishing Switzerland 2014

theory) for any other computational process Θ, we can find an easy coding protocol $\mathcal{C}$ and decoding protocol $\mathcal{C}^{-1}$ so that we can encode any input x for Θ as an input $\mathcal{C}(x)$ for Π so that after Π has performed its computation we can decode the answer $\Pi(\mathcal{C}(x))$ to the answer that Θ would have given us. In symbols: $\mathcal{C}^{-1}(\Pi(\mathcal{C}(x))) = \Theta(x)$.

One can formalize what it means for a protocol to be easy but for the sake of this presentation that is not too relevant. Thus, if a process is universal, it can mimic all other processes if we just prepare the right input for it. It is certainly part of our intuition that complex systems can incorporate, mimic, or use, less complex systems. In this light it seems sustainable to *define* complex systems as those systems that are universal. Note that under this definition a complex system need not necessarily manifest itself in a complex appearance: a universal process can mimic *any* other process whence also the very easy ones.

2 Intermediate Degrees

In this section we study the complexity that falls in between decidable and universal in a sense to be specified below.

2.1 Turing Degrees

For sets of natural numbers, the notion of universality can also be defined. Contrary to real-world computations, for sets of natural numbers there are infinitely many ever-increasing notions of universality. The one that corresponds to the computational notion is that of Σ_1^0 universality. A set K is called Σ_1^0-universal if for any computably enumerable set X (that is a set whose values we can computably enumerate but not necessarily decide for each number if it is in the set or not) there is a computable function $f_X : \mathbb{N} \to \mathbb{N}$ so that

$$x \in X \iff f_X(x) \in K.$$

We call such a function f_X also a reduction. Post [10] raised the famous question of whether there is some natural computably enumerable set of natural numbers that is computationally more informative than a decidable set, but less informative than the universal set K.

Often, instead of speaking of sets directly one considers *degrees* also called *Turing degrees*. A Turing degree can be considered as the entity of all the sets that contain the same amount of information in the sense of the above considered reduction. Thus, two sets X and Y fall in the same degree—we write $X \sim Y$—whenever there is some computable $f : X \to Y$ such that $x \in X \Leftrightarrow f(x) \in Y$ and some computable $g : Y \to X$ such that $y \in Y \Leftrightarrow g(y) \in X$. For two Turing degrees X and Y we write $X<Y$ to indicate that there is some computable $f: X \to Y$ such that $x \in X \Leftrightarrow f(x) \in Y$ but no computable $g: Y \to X$ such that $y \in Y \Leftrightarrow g(y) \in X$.

It is common practice to denote the degree of decidable sets by $\emptyset$ and the degree of Σ_1^0-universal sets by $\emptyset'$. Post's question stated in terms of degrees now translates to whether there exists some degree X which falls strictly in between $\emptyset$ and $\emptyset'$ in terms of the above defined reduction, that is $\emptyset<X<\emptyset'$. It took the scientific community twelve years to find such an intermediate degree. However, it is generally held that this solution does not provide a *natural* intermediate degree.

Clearly the notion of being natural is rather vague and auto-determined by the scientific community itself. A clear indication for a mathematical notion to be natural is that it occurs in various other fields as well. Likewise, applicability to other kind of problems or admitting different proof methods are typically also considered an indication of naturalness. The canonical way of finding intermediate degrees is by what are called priority arguments with finite injury and it is generally held that they do not meet the above mentioned indications for being natural. We refer the reader to [13] for a more detailed account of priority arguments in the context of this paper.

2.2 *Church-Turing Thesis and* PCE

Post's question on intermediate degrees finds it real-world analog in the Principle of Computational Equivalence (**PCE**) which was postulated by Wolfram in his NKS book [14]:

> **PCE**: Almost all processes that are not obviously simple can be viewed as computations of equivalent and maximal sophistication.

The processes here referred to are processes that occur in nature, or at least, processes that could in principle be implemented in nature. Thus, processes that require some oracle or black box that give the correct answer to some hard questions are of course not allowed here.

As noted in the book, **PCE** implies the famous Church-Turing Thesis (again, see [2] for more details) (**CT**):

> **CT**: Everything that is algorithmically computable is computable by a Turing Machine.

Both theses—**PCE** and **CT**—have some inherent vagueness in that they try to capture/define an intuitive notion. While the **CT** thesis aims at defining the intuitive notion of algorithmic computability, **PCE** aims at defining what degrees of complexity occur in natural processes. But note, this is not a mere definition as, for example, the notion of what is algorithmically computable comes with a clear intuitive meaning. And thus, the thesis applies to all such systems that fall under our intuitive meaning.

As a consequence, the **CT** thesis would become false if some scientists were to point out an algorithmic computation that cannot be performed on a Turing Machine with unlimited time and space resources. With the development and progress of scientific discovery the thesis has to be questioned and tested time and again. And this is actually what we have seen over the past decades with the invention and systematic study of new computational paradigms like DNA computing [9], quantum computing [8], membrane computing [1], etc. Most scientists still adhere to the **CT** thesis.

But the **PCE** says more. It says that the space of possible degrees of computational sophistication between obviously simple and universal is practically void. In what follows we shall address the question what might cause this. We put forward two observations. First we formulate a natural candidate principle that can account for **PCE** and argue for its plausibility. Second, we shall briefly address how cognition can be important. In particular, the way we perceive, interpret and analyze our environment could be such that in a natural way it will not focus on intermediate degrees even if they were there.

3 Complexity and Evolution

In this section we shall dwell on the intimate relation between evolution and the emergence of complexity. We shall follow [6] in great lines citing certain passages but also adding new insights.

In various contexts but in particular in evolutionary processes one employs the principle of Natural Selection, often also referred to as Survival of the Fittest. These days basically everyone is familiar with this principle. It is often described as species being in constant fight with each other over a limited amount of resources. In this fight only those species that outperform others will have access to the limited amount of resources, whence will be able to pass on its reproductive code to next generations causing the selection.

We would like to generalize this principle to the setting of computations. This leads us to what we call the principle of Generalized Natural Selection:

> **GNS**: In nature, computational processes of high computational sophistication are more likely to maintain/abide than processes of lower computational sophistication provided that sufficiently many resources are around to sustain the processes.

If one sustains the view that all natural processes can be viewed as computational ones, this generalization is readily made. For a computation, to be executed, it needs access to the three main resources space, matter, and time. If now one computation outperforms the other, it will win the battle over access to the limited resources and abide. What does outperform mean in this context?

Say we have two neighboring processes Π_1 and Π_2 that both need resources to be executed. Thus, Π_1 and Π_2 will interfere with each other. Stability of a process is thus certainly a requirement for survival. Moreover, if Π_1 can incorporate, or short-cut Π_2 it can actually use Π_2 for its survival. As an analogy we mention a monkey that can predict and thereby use the behavior of an ant by inserting a stick into an ant colony waiting for ants to climb on the stick so that the monkey can eat the ants by pulling the stick out again.

A generalization of incorporating, or short-cutting is given by the notion of simulation that we have given above. Thus, if Π_1 can simulate Π_2, it is more likely to survive. In other words, processes that are of higher computational sophistication are likely to outperform and survive processes of lower computational sophistication. In particular, if the process Π_1 is universal, it can simulate any other process Π_2 and thus is likely to use or incorporate any such process Π_2.

Of course this is merely a heuristic argument or an analogy rather than a conclusive argument for the **GNS** principle. One can think of experimental evidence where universal automata in the spirit of the Game of Life are run next to and interacting with automata that generate regular or repetitive patterns to see if, indeed, the more complex automata are more stable than the repetitive ones. In setting up such experiments, much care needs to be taken to not run into hard philosophical problems of ontological nature like the question "what are the defining properties of a particular process". One can think of similar questions about a tree without leaves still being a tree etc. In particular, it seems more sensible to focus on some particular features, like for example entropy or other complexity measures. We will take up these considerations in more detail in Sect. 5.

Of course, one cannot expect that experiments and circumstantial evidence can substitute or prove the principle. A more detailed discussion of the principle can be found in [6].

Just like the theory of the selfish gene (see [4]) shifted the scale on which natural selection was to be considered, now **GNS** is an even more drastic proposal and natural selection can be perceived to occur already on the lowest possible level: individual small-scale computational processes.

In [6] it was noted that under some reasonable circumstances we may see **GNS** as a consequence of **PCE**. However, **GNS** only talks about computational processes in nature and not in full generality about computational processes either artificial or natural as was the case in **PCE**. Thus we cannot expect that **CT** + **GNS** is actually equivalent to **PCE**. However, if we restrict **PCE** to talk only about processes in nature, let us denote this by **PCE**′, then we do argue that we can expect a correspondence. That is:

$$\mathbf{PCE}' \approx \mathbf{CT} + \mathbf{GNS}.$$

But **PCE**′ tells us that almost all computational processes in nature are either simple or universal. If we have **GNS** we find that more sophisticated processes will outperform simpler ones and **CT** gives us an attainable maximum. Thus the combination of them would yield that in the limit all processes end up being complex. The question then arises, where do simple processes come from? (Normally, the question is where do complex processes come from, but in the formal setting of **CT** + **GNS** it is the simple processes that are in need of further explanation.)

Simple processes in nature often have various symmetries. As we have argued above these symmetries are readily broken when a simple system interacts with a more complex one resulting in the simple system being absorbed in the more complex one. We see two main forces that favor simple systems.

The first driving force is what we may call *cooling down*. For example, temperature/energy going down, or material resources growing scarce. If these resources are not available, the complex computations cannot continue their course, breaking down and resulting in less complex systems.

A second driving force may be referred to as *scaling* and invokes mechanisms like the Central Limit Theorem. The Central Limit Theorem is a phenomenon that creates symmetry by repeating a process with stochastic outcome a large number of times yielding the well-known Gaussian distribution. Thus the scale (number of repetitions) of the process determines the amount of symmetry that is built up by phenomena that invoke the Central Limit Theorem.

In analogy, we can mention that whilst various universal processes that are executed at cell level, a tree by itself can hardly be called a universal computational process.

In the above, we have identified a driving force that creates complexity (**GNS**) and two driving forces that creates simplicity: cooling down and scaling. In the light of these two opposite forces we can restate **PCE**′ as saying that simplicity and universality are the two main attractors of these interacting forces.

Note that we deliberately do not speak of an equivalence between **PCE**′ and **CT** + **GNS**. Rather we speak of a correspondence. It is like when modeling the movement of a weight on a spring on earth. The main driving forces in this movement are gravitation and the tension of the spring. However, this does not fully determine a final equilibrium if we do not enter in more details taking into account friction and the like. It is in the same spirit that we should interpret the above mentioned correspondence.

4 Complexity, Evolution and Our Cognitive Toolkit

Fodor has postulated a principle concerning our language. It says that (see [5]) the structure and vocabulary of our language is such that it is efficient in describing our world and dealing with the frame problem. The frame problem is an important problem in artificial intelligence which deals with the problem how to describe the world in an efficient way so that after a change in the state of affairs no entirely new description of the world is needed. See for example [11].

In particular, Fodor considers particles that can be either frigeons or nonfrigeons. A particle is a frigeon if Fodors refrigerator happens to stand open and otherwise it is a nonfrigeon. It is clear that we can perfectly well define such concepts and words. However, the mere availability of these concepts will not help us understand the world better. Nor are we likely to be able to act better in a competitive setting by having access to these concepts. And what is even worse, our description of the world becomes very cumbersome if we take these concepts into account. In particular of course at moments when Fodor's refrigerator is either opened or closed.

Based on this thought experiment Fodor posed the thesis that our language—an essential part of our cognitive toolkit—has evolved in such a way to efficiently describe the world and the important changes occurring therein.

On a similar page, we would like to suggest that our cognitive toolkit has evolved over the course of time so that it best deals with the processes it needs to deal with. Now, by **PCE** these processes are either universal or very simple. Thus, it seems to make sense in terms of evolution to have a cognitive toolkit that is well-suited to deal with just two kinds of processes: the very simple ones and the universal ones.

Taking these considerations into account, it can well be conceived that there actually are computational processes out there that violate **PCE** but firstly, by **GNS** these processes will be very scarce and secondly, even if they are out there, our cognitive toolkit is just not well-equipped enough to deal with them.

Actually, throughout mathematics and mathematical logic there are various indications present that seem to substantiate the claim that indeed many of our most commonly used intellectual and cognitive tools within these fields, although rather sophisticated, all fall in one of few classes of operational strength. In this paper we have already seen that it is very hard to get sets that are not computationally universal. In [6] we gave some more examples to this same phenomenon.

In this setting we would also like to mention the program of reverse mathematics (see [12]). Reverse mathematics tries to gauge the logical strength of important mathematical theorems. One starts out with some weak base theory T_0. Next, one considers some important mathematical theorem τ. These are typically mathematical theorems that are frequently used by the mathematical community.

We mention here some examples of such theorems without further reference, context or proof. They just serve to give the flavor of the kind of theorems considered:

- Every countable commutative ring has a prime ideal;
- A continuous real function on the closed unit interval is Riemann integrable;
- Uniqueness of algebraic closure (of a countable field);
- Gödel's completeness theorem: a formula φ in a countable language is provable from a set Γ of assumptions in that same language, if and only if φ is true in every model where all of Γ is true.

As said, we do not want to go into the details of these theorems. They merely serve the purpose of illustrating what kind of theorems are considered and how wildly divers the scope of these different theorems are. The next step in the recursive mathematics project is to consider the system $T_0 + \tau$, that is, the base system together with one of those particular mathematical theorems. We call two such systems $T_0 + \tau$ and $T_0 + \tau'$ equivalent and write $T_0 + \tau \equiv T_0 + \tau'$, if they prove exactly the same set of theorems, that is,

$$T_0 + \tau \vdash \varphi \iff T_0 + \tau' \vdash \varphi$$

for any formula φ. It turns out that almost all important mathematical theorems fall into one of five equivalent systems. That is to say, if you take six arbitrary important mathematical theorems $\{\tau_i \mid i \in \{1, \ldots, 6\}\}$, almost surely you will have that $T_0 + \tau_i \equiv T_0 + \tau_j$ for some $i \neq j$. We think that this is an important indication of the fact that our intellectual/cognitive toolkit is designed in such a way as to efficiently/naturally recognize, and deal with a limited set of problems that are most useful to us in our daily life and fight for survival. In particular we mention that all the above mentioned examples of important mathematical theorems are equivalent over some natural base theory T_0.

5 On Testing the Generalized Natural Selection Principle

In this final section we shall address the question on how to test the principle of Generalized Natural Selection **GNS** as put forward in [6] and discussed here in Sect. 3. As mentioned before, such tests can never substitute a full proof. Rather they can merely supply "circumstantial evidence" in favor of or against the principle. Let us start out by pointing out some subtleties underlying **GNS**.

5.1 General Observations

The principle **GNS** tells us that complex processes are more likely to be more successful than others and thus more likely to "survive". Let us recall the exact formulation of **GNS** and make some general observations that should always be taken into account when studying it.

> **GNS**: In nature, computational processes of high computational sophistication are more likely to maintain/abide than processes of lower computational sophistication provided that sufficiently many resources are around to sustain the processes.

In this formulation we see the following difficulties naturally emerge.

1. The first, most natural, and most fundamental question is "what is determining the identity of a process". For example, suppose some process Π undergoes some minimal change, should we still call it the same process Π after that minimal change? The same sort of question arises in all kinds of sciences: "is a human being without limbs still a human being?" or "when is a particular cloud a Nimbo Cumulus?". It turns out even to be difficult to classify life within very broad categories like "animal" versus "plant" etc. Naturally the question is

related to deep philosophical questions relating, amongst others, to issues like fuzziness (see e.g., [15]) in particular and the Sorites paradox more in general [7]. The Sorites paradox deals with questions like "how many trees should there be to form a forest, and what happens if I would cut one tree".
When trying to make quantified statements about **GNS** one should always first isolate a well defined entity that substitutes either "process" or some essential property representing the process. In the paradigm of *The selfish gene* this is easy but in the paradigm of **GNS** this is not.
2. A second fundamental question is concerning what is meant by complexity and in particular how to measure it. As mentioned before, there are various essentially different definitions throughout the literature and we have put forward our own proposal here in Sect. 1.
3. We think that the two problems mentioned so far are the more serious ones. Minor but not less fundamental problems arise in also defining the other mentioned concepts. Thus, how should we specify probability when saying that one process is more likely to maintain/abide than another. What is exactly understood by interaction, etc.

5.2 Mathematical Analysis

In principle one could try to formulate **GNS** in a fully formalized setting and then try to *prove* **GNS** as a theorem within that formal setting. In doing so all above mentioned points/problems should be taken into account. We shall shortly see how many choices such an analysis entails. That naturally raises the questions on how natural these choices are and in how much the final analysis says something about the physical reality at all.

For example, one could identify a process by a set, or better, by a Turing degree X. This would be a first choice. In a next choice one has to define some mathematical operation $\oplus$ between two degrees that models the notion of interaction between two processes aka degrees. Thus, the outcome of two processes X and Y that interacted would be denoted and computed by $X \oplus Y$.

Subsequently, we can answer the question whether $X \leq X \oplus Y$ and $Y \leq X \oplus Y$. However, if we wish to say something on how likely it is that either $X \leq X \oplus Y$ or $Y \leq X \oplus Y$ we need to make yet more choices like introducing some probabilistic tools on the space of degrees between $\emptyset$ and $\emptyset'$.

5.3 Testing in the Laboratory

Instead of mathematical modeling, one can also try to isolate some real-world processes that can be considered naturally as computational ones and have them interact and run in a laboratory setting. Also here, all the difficulties as discussed in Sect. 5.1 will manifest themselves in this setting.

In particular we will have to decide on the identity determining aspects of the processes involved. Moreover we have to decide on the measure of complexity that is to be applied to these processes. Probably that is the harder and more arbitrary task in this setting.

When working with living organisms some care has to be taken as to prevent that we are just testing the well-established principle of natural selection instead of **GNS**.

5.4 Computer Simulations

In this final section we shall discuss a possible approach to test **GNS** via computer simulations. Again we should settle upon choices for the modeling problems as posed in Sect. 5.1.

Just as with the mathematical modeling, we would like to stay as close as possible to the physical reality in our computer simulations. Due to the inherent parallel nature of physical reality (all goes on at the same time) and due to the locality of causality it seems a good idea to simulate parts of reality by cellular automata (CA).

For the sake of a simple presentation let us briefly recall the definition of one of the simplest CAs: a one-dimensional CA with radius 1 and two symbols. We shall depict the two different symbols by black and white respectively. Our CA acts on a one dimensional tape of discrete cells that extend infinitely both to the left and to the right. In CAs, time evolution is modeled by discrete time steps. An initial condition is given by telling what cell is of what color. The color of each cell will evolve over time. Basically our CA is just a look-up table with a rule how to compute the color of a particular cell at a next time step depending on its current color and the color of its two direct neighbors. An example of such a CA is depicted in Fig. 1.

The rule numbering is according to a numbering scheme as presented in [14] but not really relevant for the current presentation. Thus, for example, if a cell was white at time t and both its neighbors were black at time t, then at the next step $t+1$, the cell will turn black according to the defining look-up table of Rule 110.

As in [14] we depict the consecutive tape configurations from top to down. Thus, for example, if we start out with just a single black cell on an otherwise white tape, Rule 110 will give us the famous evolution as depicted in Fig. 2 below.

It is evident from Fig. 2 that complex behavior can already occur in these simple automata. As a matter of fact, it is know that Rule 110 is universal in that it

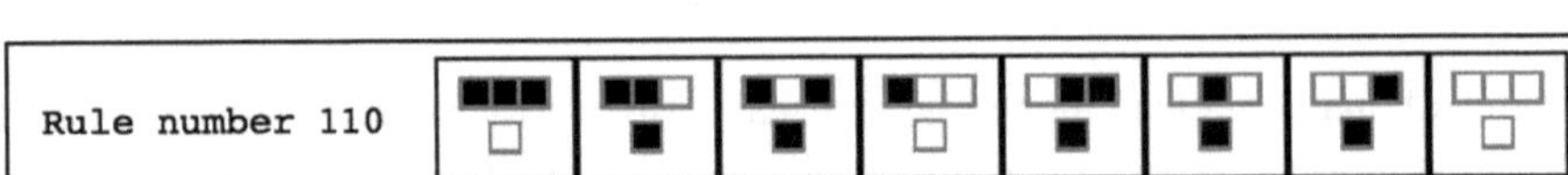

Fig. 1 Definition of rule 110

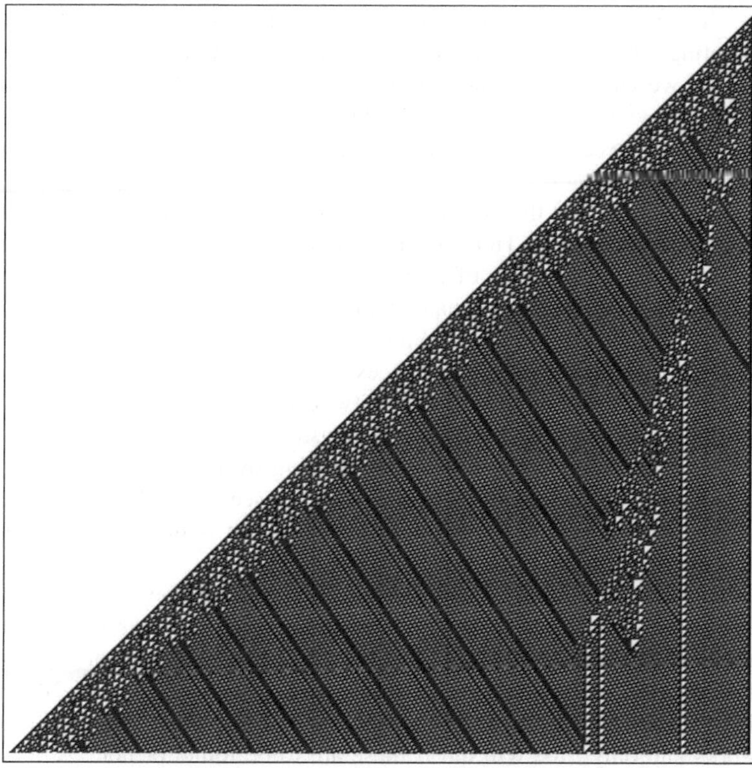

Fig. 2 Evolution of rule 110 starting with just one *black* cell and computed only for 700 steps. Figure generated with *Mathematica*

can -in some sense—emulate any other computational process. It is easy to see how CA can be generalized to more symbols, more dimensions and larger radii (taking more neighbors into account).

To come back to our test of **GNS** and in the setting of CAs, how should we model processes? Should a process be modeled by a particular CA? And if so, how should interaction be modeled? We think it is more natural to model a process by an initial condition. Let us briefly explain why.

We have reasoned before that there is a strong analogy between physical reality and CAs. Each cell in a CA with its respective symbol can be seen as a particular property of physical reality at some particular locus or region. The interaction between these properties at these regions are governed by the same laws of nature everywhere throughout the universe. At least it is generally believed to be the case that the laws of physics are the same throughout the universe.

One could not wish to adhere to this believe and keep the possibility open that somewhere far away in extreme circumstances -for example close to a black hole- the laws of physics do change. However, it still seems reasonable to expect the laws of nature to be at least locally stable. And as we are interested on interacting processes it is mainly local interaction that we are interested in. Thus, if we have a CA simulating physical reality, it should be the same CA at every part of the tape. Moreover, if we wish to simulate reality in which universality clearly occurs, we better start out with a universal CA.[1] Pushing the analogy further we are lead to accept that processes correspond to the configuration of our symbols evolving over time.

It is in this setting that the question about the defining properties of a particular process becomes very hard. Thus, in simulations using CAs it seems more fruitful to focus on particular features of a process rather than to find a set of defining properties that sharply tells us what a particular process is. For the moment we shall not address the issue of limited resources.

So now that we have identified a process with an initial condition the problem of finding a suitable definition of complexity becomes clearly defined. Remember that we propose to work first with one-dimensional CAs with just two symbols. Thus, a process is nothing but a string developing over time for which there are suitable and effective complexity measures defined (see [3] or [16]). By brute force simulations one can now try to quantify how likely it is that the more complex processes maintain/abide.

Even if these simulations could provide circumstantial evidence in favor of **GNS**, one still has to be very careful in how to interpret the repercussions of these simulations on the physical reality. However, positive outcomes of such simulations, experiments, or theorems will certainly help gain credibility of **GNS**. Further credibility could be obtained by applications of **GNS** in related theoretical frameworks and only time will tell if these are to be found or not.

Acknowledgments I would like to thank Barry Cooper and Hector Zenil for fruitful discussions on the subject.

References

1. C.S. Calude, G. Paun, *Computing with Cells and Atoms: An Introduction to Quantum, DNA and Membrane Computing* (CRC Press, Leiden, 2000)
2. S.B. Cooper, *Computability Theory* (Chapman & Hall/CRC, 2004) ISBN 1-58488-237-9
3. J. Delahaye, H. Zenil, Numerical evaluation of algorithmic complexity for short strings: a glance into the innermost structure of randomness. Appl. Math. Comput. **218**(24) (2012). Elsevier
4. R. Dawkins, *The Selfish Gene* (Oxford University Press, New York, 1976). ISBN 0-19-286092-5
5. J.A. Fodor, *Modules, Frames, Fridgeons, Sleeping Dogs, and the Music of the Spheres*, ed. by Z.W. Pylyshyn (1987)
6. J.J. Joosten, On the necessity of complexity. in *Irreducibility and Computational Equivalence: 10 Years After the Publication of Wolfram's A New Kind of Science*, ed. by H. Zenil (ed.), to be published in 2012/2013

[1] This requirement can be relaxed as a universal CA can mimic any other CA too.

7. R. Keefe, *Theories of Vagueness* (Cambridge University Press, Cambridge, 2000)
8. M.A. Nielsen, I.L. Chuang, *Quantum Computation and Quantum Information* (Cambridge University Press, Cambridge, 2000)
9. G. Paun, G. Rozenberg, A. Salomaa, *DNA Computing: New Computing Paradigms* (Springer, Berlin, 2010)
10. E. Post, Recursively enumerable sets of positive integers and their decision problems. Bull. Amer. Math. Soc. **50**, 284–316 (1944)
11. M. Shanahan, *Solving the Frame Problem; A Mathematical Investigation of the Common Sense Law of Intertia* (MIT Press, Cambridge, 1997)
12. S.G. Simpson, *Subsystems of Second Order Arithmetic*, 2nd edn. Perspectives in Logic (Cambridge University Press, New York, 2009)
13. K. Sutner, Computational processes, observers and Turing incompleteness. Theor. Comput. Sci. **412**(1–2), 183–190 (2011)
14. S. Wolfram, *A New Kind of Science* (Wolfram Media, 2002)
15. L.A. Zadeh, *Fuzzy Sets, Fuzzy Logic, Fuzzy Systems*, ed. by G.J. Klir, B. Yuan (World Scientific Press, 1996)
16. H. Zenil, Compression-based investigation of the dynamical properties of cellular automata and other systems. J. Complex Syst. **19**(1), 1–28 (2010)

7. [illegible] 2009
8. M.A. [illegible]
University Press, Cambridge, 2000
9. [illegible] Springer,
Berlin, 2010
10. [illegible]
11. [illegible]
12. [illegible]
13. [illegible]
14. [illegible]
15. [illegible]

Rugged Landscapes and Timescale Distributions in Complex Systems

D. L. Stein and C. M. Newman

Abstract We review a simple model of random walks on a rugged landscape, representing energy or (negative) fitness, and with many peaks and valleys. This paradigm for trapping on long timescales by metastable states in complex systems may be visualized as a terrain with lakes in the valleys whose water level depends on the observational timescale. The "broken ergodicity" structure in space and time of trapping in the valleys can be analyzed using invasion percolation, picturesquely in terms of "ponds and outlets" as water flows to a distant sea. Two main conclusions concern qualitative dependence on the spatial dimension d of the system landscape. The first is that the much-used example of a one-dimensional rugged landscape is entirely misleading for any larger d. The second is that once d is realistically large (above 6 seems to suffice), there are many nonmerging paths to the sea; this may be relevant for the issue of contingency vs. convergence in macro-evolution.

Keywords Broken ergodicity · Glasses · Spin glasses · Rugged landscape · Percolation · Invasion percolation · Random walk in random environment · Non-ergodicity · Aging · Biological evolution · Cambrian explosion

This paper is based on a talk at the Interdisciplinary Symposium on Complex Systems, Kos, Greece, September 19–25, 2012.

D. L. Stein (✉)
Department of Physics and Courant Institute of Mathematical Sciences,
New York University, New York, NY 10003, USA
e-mail: daniel.stein@nyu.edu

C. M. Newman
Courant Institute of Mathematical Sciences, New York University,
New York, NY 10012, USA
e-mail: newman@cims.nyu.edu

I. Zelinka et al. (eds.), *How Nature Works*, Emergence,
Complexity and Computation 5, DOI: 10.1007/978-3-319-00254-5_4,
© Springer International Publishing Switzerland 2014

1 Introduction

Complex systems, whether adaptive, like the immune system, or nonadaptive, like spin glasses, are commonly characterized by a high degree of metastability. That is, their dynamics possess multiple fixed point solutions which, while not globally optimal, can trap the system for long periods of time; how long depends on the depth of the metastable well and the magnitude of the noise (temperature in many physical systems, mutation rates in asexual biological speciation, etc.). Metastability can be desirable—after all, living organisms are all metastable—or undesirable, as when a folding protein or a combinatorial optimization algorithm finds itself trapped in a misfolded state or nonoptimal solution. In disordered systems theory, the trapping of a system on long timescales in a localized subset of its entire state space, absent a phase transition and broken symmetry, is known as "broken ergodicity" [1–5]. The presence of broken ergodicity can complicate the statistical mechanical analysis of such systems.

Mathematically, the noisy dynamics of a system with many metastable states can be recast as a random walk on a "rugged energy (or free energy) landscape" [6–9]. Doing so allows one to define and analyze a well-defined mathematical problem, make quantitative predictions, and compare with the observed phenomenology of a wide spectrum of complex systems. The drawback of this approach is that one starts not from a microscopic model of a given system, but instead from a coarse-grained, somewhat heuristic, *ansatz* about its state space.

Nevertheless, to understand a common, possibly universal, feature of complex systems, one that is independent of microscopic details, a serious analysis of diffusion on a rugged landscape is likely unavoidable. The standard picture invokes a high-dimensional surface with multiple "hills" and "valleys", the latter corresponding to the various metastable states and the former to the barriers that must be surmounted in order to escape those states and find another, perhaps deeper, valley. (Note that in the evolutionary biology setting, the fitness landscape is flipped relative to an energy landscape so that hills and valleys exchange roles.) As one adopts increasingly longer timescales, one expects to observe the system sampling an increasingly larger region of state space. The confinement here is structural rather than dynamic; there are no constants of the motion, aside from the fixed points, that confine the system to a particular dynamical orbit.

Of course, such a picture is meaningless unless one specifies a dynamics, such as a single-spin-flip dynamics for a spin glass or a single-site mutation mechanism for a genome [10]. We ignore those issues here, important though they are, and consider more simply a given rugged landscape with a specified diffusion rule.

An often-used picture is that of Fig. 1 in [3], shown below, where configurations in region A are surrounded by relatively low energy barriers, those in region B by higher barriers, and those in region C by even higher ones. This leads to the following picture: one can think of the energy surface as a mountainous terrain with isolated "lakes" in different valleys. The "water level" corresponds to the largest (free) energy scale that can be accessed at a given temperature T and observational timescale τ. As the water level rises, lakes merge into bigger lakes in a hierarchical fashion [2, 3, 5, 11]. At any T and τ, the system is confined to a given lake—or, to use the language of broken ergodicity, a given component [2]. Because the confining barrier increases as $T \ln \tau$, the portion of state space accessible to the system correspondingly increases. The system is therefore ergodic in ever larger components, and accordingly will continually revisit smaller components—now a subset of the accessible state space—to which it was once confined.

However, a 1D sketch as in Fig. 1 can be misleading, and the physical picture described above may be incorrect. The true configuration space is very high-dimensional, and in high dimensions new things can happen. In particular, a route around a mountain can often be found; one may not have to climb directly over the top [9]. This can have profound consequences for the system dynamics, as it can for hikers who encounter foul weather on the Tongariro Crossing [12].

2 Landscape Model

We consider a specific model with a continuum of possible free energy barriers, and therefore of activation timescales. Consider a non-sparse graph $\mathcal{G}$ in d dimensions; for specificity, we will take $\mathcal{G}$ to be the cubic lattice $\mathbf{Z}^d$, although this will not affect our results. We next assign random variables to all edges and all sites of $\mathcal{G}$. The exact nature of the distributions from which these random variables are chosen is not important, so long as four conditions are met: (1) the probability distributions are continuous; (2) edge values are always greater than the values assigned to their adjoining sites; (3) site values are bounded from below; (4) site and edge variables are all chosen independently.

The reasoning behind these conditions is as follows. We will consider the site values to correspond to the energies of metastable states, and edge values to

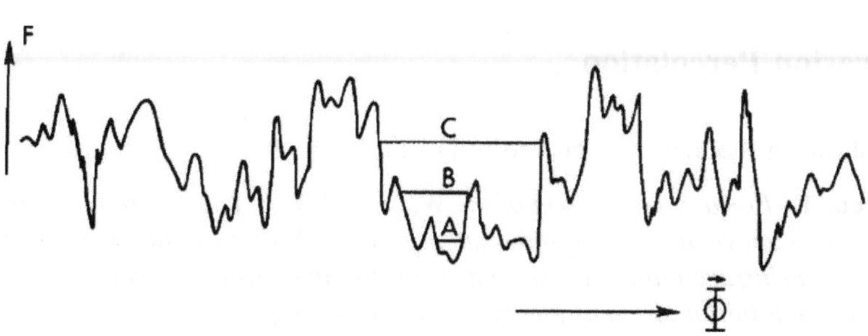

Fig. 1 Schematic picture of a rugged energy landscape, where is an abstract configuration coordinate. At zero temperature the system is stuck in one of many local minima; however, at positive temperature the system receives random "kicks" that can knock it out of its minimum. So at longer times or higher temperatures, more of the configuration space can be explored. From [3]

correspond to the free energy barriers separating them. These identifications lead naturally to conditions (2) and (3). An easy (but not necessary) way to implement (2) is to choose the edge variables from a distribution on the positive axis and the site variables from the negative axis. The last condition is included to simplify the analysis, but as we will see as we proceed, will not affect our main conclusions as applied to likely physical models. The first condition then eliminates ties. A further simplification, not explicitly mentioned but implicit in our using a simple graph like $\mathbf{Z}^d$, is choosing each distinct edge to connect a single pair of sites. Once again, we do not expect this to affect the conclusions for physically relevant models.

The model described above is Model B of [9].

We now treat this model as describing the state space of a physical system in equilibrium, held at inverse temperature β. This assumption provides us with a concrete model to analyze. In that case, if W_x is the variable assigned to site x, then the equilibrium probability density over sites $\pi_x(\beta)$ scales with β as

$$\pi_x(\beta) \sim \exp[-\beta W_x]. \tag{1}$$

Detailed balance then requires that

$$\pi_x(\beta) r_{xy}(\beta) = \pi_y(\beta) r_{yx}(\beta), \tag{2}$$

where r_{xy} is understood as the rate to make a direct transition from x to y. The rates, satisfying (2), are chosen so that

$$\begin{aligned} r_{xy}(\beta) &\sim \exp\left[-\beta(W_{xy} - W_x)\right] \\ r_{yx}(\beta) &\sim \exp\left[-\beta(W_{xy} - W_y)\right] . \end{aligned} \tag{3}$$

The assignment of random variables defines an ordering on the (undirected) edges of $\mathcal{G}$ in which $\{x, y\} < \{x', y'\}$ if $W_{xy} < W_{x'y'}$. That is, the barriers are ordered by increasing height, a central feature of broken ergodicity treatments. This will be central in what follows.

3 Invasion Percolation

The following theorem was proved in [13]:

Theorem 1 *For a.e. realization of the W_{xy}'s and W_x's, chosen from distributions as described above, as $\beta \to \infty$ the sequence in which sites are first visited, starting from some arbitrary initial site x_0, converges to the invasion percolation sequence with the same initial site and the same edge ordering.*

Before discussing the consequences of this theorem, a brief discussion of invasion percolation is in order. Invasion percolation [14–16] is a process that can be defined either on edges or on sites; for simplicity, we restrict this discussion to edges, but a similar construction holds for sites. Once again consider the lattice Z^d, and assign a random variable to each edge, chosen independently from a common continuous distribution, say the uniform distribution on $[0,1]$. Now order the edges by the values of their associated random variables; an edge $\{x,y\}$ is of lower order than an edge $\{x',y'\}$ if the random variable assigned to $\{x,y\}$ is less than that assigned to $\{x',y'\}$.

Beginning now from some arbitrary initial site x_0, choose the edge of lowest order connected to it. This results in a graph consisting of two sites and all $4d-1$ edges (including the previously chosen edge) connected to them. Now choose from among those edges not previously chosen (there are $4d-2$ of them) the edge of lowest order. One now has a cluster of three sites; one examines all (previously unchosen) edges connected to them, and again chooses the edge of lowest order. Repetition of this procedure generates in the limit an infinite cluster, called the *invasion region* of x_0. This cluster has several interesting properties; in particular, it exhibits the property of "self-organized criticality" [17] in that the invasion region of any site asymptotically behaves, as time increases, like the *incipient infinite cluster* of the associated *independent* bond percolation problem [18], which appears precisely at the critical percolation probability p_c. That is, the dimensionality of the invasion region far from x_0 approaches the fractal dimensionality of the incipient cluster at p_c in the independent bond percolation problem on the identical lattice. We will discuss other relevant properties of invasion percolation as we proceed.

Theorem 1 is therefore a rigorous statement about the *order in which sites are visited for the first time*, given an arbitrary starting site. This is a crucial result for a system in which broken ergodicity occurs, because it provides information about which sites are visited (or not visited) on a given timescale, and the nature of that process, which is what we're interested in.

Intuitively, Theorem 1 is quite reasonable. As the temperature is lowered, the timescales for transitions governed by different free energy barriers diverge from one another. Consequently, the system will, with probability approaching one as temperature approaches zero, make a transition over the lowest barrier available to it.

Although the idea is quite simple, it has important, and mostly unappreciated, consequences for systems whose dynamics are controlled by rugged landscapes, due to the geometry of invasion percolation. In this regard, an important result on the global connectivity structure of invasion percolation was derived in [19, 20] and says the following: for invasion percolation on the lattice Z^d in low dimensions, there is a *unique* invasion region. That is, given any two starting sites, their invasion regions will be the same except for finitely many sites (with probability one). However, in high dimensions, there are *infinitely* many disjoint invasion regions. That is, given two starting sites far from each other, their invasion regions will totally miss each other with high probability. The critical dimension d_c for the crossover was conjectured to be 8 in [19, 20], but more recent work suggests that $d_c = 6$ [21, 22]. For our purposes, the crossover dimension is unimportant, since the state space of all complex systems of interest is very high-dimensional.

What does this imply for diffusion on a rugged landscape? To simplify the discussion, consider the "degenerate" case where all site values are equal; this will not affect the main conclusion to be drawn below. Let p_c denote the critical value for independent bond percolation on Z^d and let w_c denote the energy level such that Prob $(W_{xy} \leq w_c) = p_c$. Then from any starting site, the invasion process will increasingly behave, as time increases, like the incipient infinite cluster at p_c in the corresponding independent bond percolation problem. Most importantly, once the process finds an infinite cluster of edges with energy levels $\leq w_0$, where $w_0 > w_c$, it *never again crosses an edge with* $W_{xy} > w_0$. One can then consider the invasion process from any point as following a "path" which eventually leads to "the sea" at infinity. In fewer than six dimensions, all invasion regions from different points eventually follow the same path to the sea. Along the way, all individual paths merge, some sooner, some later.

However, in greater than six dimensions, there are an infinite number of disjoint paths to infinity; that is, infinitely many distinct seas, each of which has many tributaries (invasion regions flowing into the sea, or equivalently, an infinite set of sites whose invasion regions eventually connect to the corresponding infinite cluster). A given process, starting from an arbitrarily chosen site, flows into one of these seas, and *it will never visit any sites which connect, via the invasion process, to any of the other seas.*

This seems to contradict the well-known result that the "random walk on a random environment" (hereafter referred to as the RWRE) asymptotically approaches ordinary diffusion at long times [23–25]. The resolution is that the older result and our result each corresponds to a different order in taking the limits time $\to \infty$ and $\beta \to \infty$, and the behavior of the RWRE is sensitive to this. When temperature is fixed and time goes to infinity, the RWRE will *eventually* exhibit normal diffusive behavior. How long this will take depends, of course, on the temperature and the specific model studied: there exists a temperature-dependent timescale—an ergodic time τ_{erg}, so to speak—beyond which our picture breaks down and normal diffusion takes over (or equivalently, ergodicity is restored). The ergodic time diverges as temperature goes to zero. A timescale of this type is a common feature in most systems which break ergodicity [2, 3].

We now return to the situation of a random walk in a strongly inhomogeneous environment, and apply these ideas and results to see how the process evolves.

4 Diffusion in a Random Environment

Theorem 1 and its consequences are interesting for RWRE's in dimension higher than one; in one dimension the conventional picture holds. That is, at fixed temperature $1/\beta$ and fixed observational timescale τ_{obs}, the particle is trapped between two barriers, neither of which are surmountable (with some prespecified probability) on a timescale of the order of τ_{obs}. If one is willing to wait considerably longer (on a logarithmic timescale), then the length of the line segment the particle explores is correspondingly larger, surrounded at each end by suitably large barriers. It is not hard to show that these grow in the manner specified in [2]: $\Delta F_{\rm esc} \sim (1/\beta) \log \tau_{\rm obs}$. Thus, as time proceeds, the particle explores an ever larger region in both directions, diffusing over previously travelled terrain infinitely often as $\tau_{\rm obs} \to \infty$.

In two and higher dimensions, the picture changes dramatically. In particular, as time increases, while components grow larger, *they do not contain previously visited portions of state space*. Perhaps more surprisingly, rather than the confining barriers increasing logarithmically with time as $\tau_{\rm obs} \to \infty$, a constant barrier value is approached (although not necessarily monotonically) [26]. To see why this is, we need to explore the properties of the incipient infinite cluster in ordinary percolation, which as noted above corresponds essentially to the asymptotic invasion region.

In particular, we need to examine the process whereby invasion percolation "finds a path" to infinity. This process is independent of whether there are one or many disjoint invasion regions. For any starting x_0 and realization of bond and site variables, the path to infinity is unique. To simplify the argument, suppose for the moment that all site variables have the same value and only the edge variables are chosen from a probability distribution (say, the uniform distribution in the interval [0, 1]).

Consider a starting site x_0, and all bonds with values smaller than some $p_1 > p_c$. In the associated independent bond percolation problem, these comprise a unique infinite cluster [27]. A little thought should then convince the reader that once the invasion process reaches *any* of the bonds within this infinite cluster, *it will never again cross any bond whose value is greater than* p_1. Consider next all bonds whose values are less than some p_2, where $p_c < p_2 < p_1$. These too form a unique infinite cluster which is a subset of the first, larger one. When the invasion process reaches any of the bonds within this newer infinite cluster, it will never again cross any bonds greater than p_2. It is easy to see that, as the process continues, the invasion region will focus itself down to infinite clusters of increasingly smaller maximum bond value, and will asymptotically behave like the incipient infinite cluster of the independent bond percolation problem at p_c. This, by the way, is an excellent natural example of "self-organized criticality" [17].

We now need to investigate the structure of the incipient infinite cluster. It is important to note that this is not an infinite cluster at all—there is no percolation at p_c—but rather consists of a sequence of increasingly large but disconnected clusters (see Fig. 2). A picturesque way of understanding its structure is the "ponds and outlets" picture of [13], which utilizes an earlier construction of Hammersley [28].

At the earliest stages of the process, a set of relatively smaller-valued bonds (depending on x_0) will be invaded before the process has to invade a relatively larger one to make its way toward infinity. Picturesquely, we think of the process as stranded on a pond; it needs to find a relatively high outlet before it can escape.

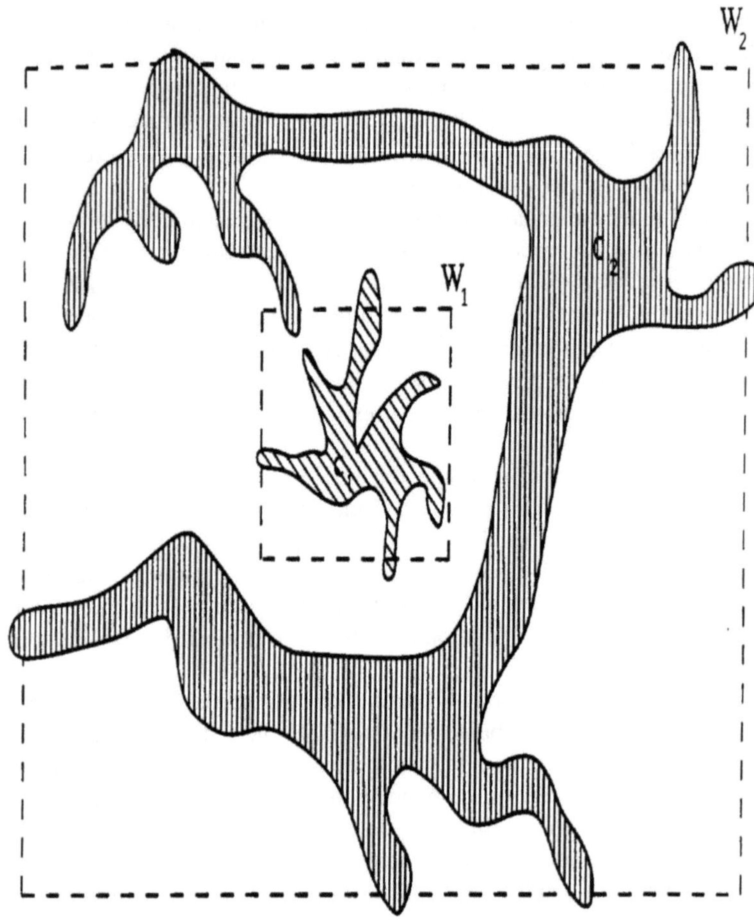

Fig. 2 A sketch of the so-called "incipient infinite cluster", which is not an infinite cluster at all but rather a sequence of infinitely many disjoint finite clusters. In the figure, the largest cluster seen in window W_i is (most of) C_i for $i = 1$ or 2. C_1 and C_2 are both finite. From [13].

The outlet corresponds to the bond whose value is larger than that of all others within the pond, but smaller than all others on the perimeter of the pond. The two important points following from this are first, this first outlet will be the *bond of largest value that the process will ever cross*, and second, once this outlet is crossed, *the process will not return to the first pond* [29]. That is, there is a sort of "diode effect" in the process.

After crossing the first outlet, the process will find itself on a second pond, and must invade an outlet of smaller value than the first one. In this way it invades a sequence of successively smaller outlets (with bond values larger than but tending toward p_c) on its way to the sea (see Fig. 3). The general trend is for the ponds to grow successively larger, but this need not be true monotonically.

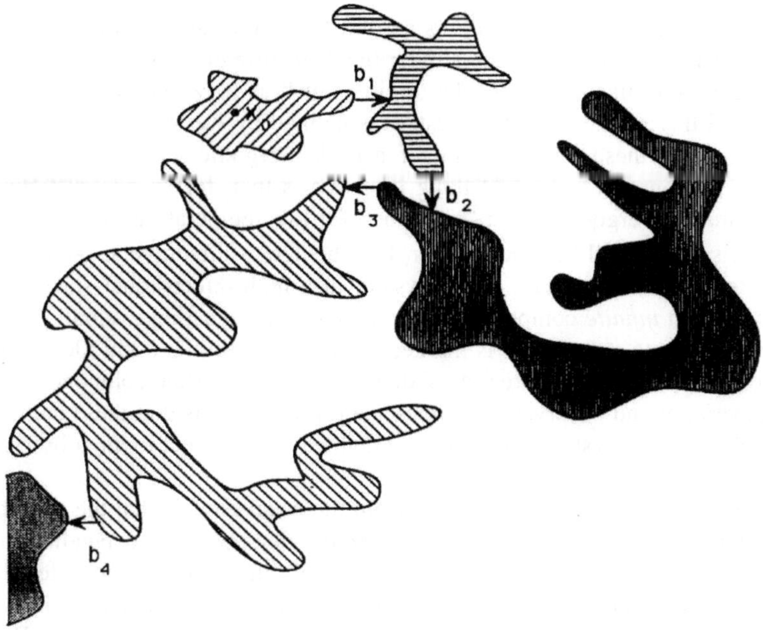

Fig. 3 A rough sketch of the "ponds and outlets" picture illustrating the diode effect. The first pond contains the starting site x_0. Arrows indicate the large-scale direction of motion; once the process leaves a given pond, it does not return. The values of the b_n decrease as n increases; b_n controls the height of the minimal barrier confining the system to pond n, as described in the text. From [13]

Ponds and outlets can be defined precisely. Consider all possible paths to infinity from the starting point x_0. Each such path $\mathcal{P}$ will contain some bond of maximum value; call it $b_{\mathcal{P}}$. The first "outlet" is then the bond b^* of *minimum* value from the set $\{b_{\mathcal{P}}\}$. The first "pond" is the finite cluster connected to x_0 consisting of all bonds whose values are strictly less than that of b^*. The second pond and outlet can be found using the same procedure from the starting point x_1, where x_1 is the site which touches b^* and is outside the first pond. This procedure can be repeated indefinitely to find ponds and outlets of any order.

Alternatively, one can define them in the following way. Starting from x_0, one considers the finite cluster connected to x_0 which consists of all bonds with values less than $p = p_c$. One then raises p in a continuous manner, causing the cluster connected to x_0 to grow. At some sharp value of p (depending on x_0) the cluster becomes infinite; it is not hard to see that there will be a single bond connecting the (previously finite) cluster containing x_0 with infinity. This bond is the first outlet, and all bonds in the interior finite cluster comprise the first pond.

A rigorous analysis of the distribution and other features of two-dimensional ponds and outlets was carried out by Damron and Sapozhnikov in a series of three papers [30, 31]. They obtained very precise estimates of the asymptotic behaviors of the radius and area of the kth pond and the bond value of the kth outlet as $k \to \infty$. They also analyzed the similarities and differences between the various ponds and the critical cluster of the origin as well as between the whole invasion region and the incipient infinite cluster (as defined by Kesten [32]).

So what do these results mean for a broken-ergodic view of diffusion in a random environment? The first thing to notice is that, for an observational time smaller than the ergodic time τ_{erg}—which for low temperatures and in glass-like complex systems will be much longer than the age of the universe—the diffusing "particle" (that is, the state of the system in a high-dimensional state space) is confined to an *infinite* component that comprises an exceedingly small proportion of the entire state space. This is markedly different from the usual idea of a finite component. But an even more radical departure from the usual component, defined by temperature and observational timescale (cf. Sect. 1), is that these components are *intrinsic* to the system itself, and not defined with respect to any observational timescale.

Perhaps the most serious departure from the usual view is that the evolution of the system (in terms of states visited) is largely deterministic, depending only on the starting point. Of course, the particle will still diffuse among the states allowed by the above confinement mechanism, but it does so in a manner which again differs considerably from all previous pictures of which we are aware. We now explore this in more detail.

While the notion of broken ergodicity components is realized in our approach through ponds and outlets on a local scale, with their union constituting a global component, there are some important differences as well. There is no recurrence on a large scale, though there is within a pond until an outlet is crossed. Once that outlet is crossed, however, *the system never returns to that pond.* On the scale of ponds, there is a "diode effect": the walker always moves forward, never backward (see Fig. 3). Some possible experimental manifestations of this will be listed in the Applications section below.

This diode effect is quite different from the nonreturn of ordinary random walks. For example, in $d = 2$, the diode effect remains valid even though ordinary random walks are recurrent. On the other hand, for large d, the invasion region dimension, which is four, is *twice* that of an ordinary random walk, which is two.

The other major difference is that in conventional BE, the system must surmount increasingly high barriers as time progresses. The opposite again holds here—the barriers which confine the system (that is, the outlets) diminish steadily as time increases. The landscape through which the diffusing system travels becomes increasingly flat. (The particle often sees many high peaks in its vicinity, but it avoids them.) This is true in all dimensions greater than one.

What role does entropy play? As it turns out, not much. Consider the system immediately after it has escaped the nth pond. We know already that almost all of the time expended up to that point was used in getting *to* the pond in the first place. More importantly, once inside the pond, the problem of finding its outlet is not a needle-in-the-haystack problem: the system is not wandering around aimlessly in state space, eventually finding the outlet by remote chance. Because the particle is confined to the pond on a certain timescale, the timescale for confinement within that pond is simply $\exp[\beta W_n^*]$, where W_n^* is the value assigned to the outlet for that pond. During this time the process thermalizes within the pond.

For entropy effects to counteract the decreasing values of W_n^* as n increases, it would seem to require exponentially increasing pond size, whereas pond size almost certainly increases much more moderately, probably as a power law. Consequently, the picture shown in Fig. 1 is a purely one-dimensional picture. In any higher dimension (irrespective of whether there are one or many global components), the "water level" does *not* rise as time increases at fixed temperature (or temperature increases at fixed time). Viewed from x_0, the water level initially rises, *but then stays forever fixed*, because it finds a path to the "sea" (that is, to infinity), into which it empties. Any additional water poured in simply escapes to infinity.

What about barrier heights as time increases? The height of the confining barrier, or outlet, is declining towards a limit w_c, corresponding to p_c in the independent percolation model. At the same time, the energy of the lowest "valley" within a pond generally becomes more negative as time progresses (and the system explores larger ponds), asymptotically (but slowly) approaching the minimum of the distribution; call it w_{min}. Therefore, the barriers which the system must surmount to escape successive ponds, rather than increasing without bound, asymptotically approach $w_c - w_{min}$.

5 Applications

Before considering specific applications, we need to consider how our picture is modified in the usual case of fixed temperature and time going to infinity. Our main result, that the dynamics of many RWRE models are governed by invasion percolation, is rigorous only when the limit $T \to 0$ is taken before $t \to \infty$.

What should we then expect to happen when temperature is fixed (at some value small compared to the majority of barrier heights) and time increases? As discussed earlier, there will be an ergodic timescale at which the system is likely to escape the global component (invasion region) in which it finds itself. Beyond this timescale we expect ordinary diffusive behavior, but of a specific sort: the system will mostly hop from component to component on the ergodic timescale, but after it finds itself in a new component, it stays there roughly for another ergodic time. So on shorter timescales, one will find the system within a global component; on longer timescales, it diffuses between components. Unless one is examining the system on timescales extremely large compared to $\tau_{\rm erg}$, often well beyond the reach of laboratory timescales, one will observe the basic picture described above.

Many experiments do not hold temperature fixed, and our picture can possibly shed light on these also. A well-known experiment on Ag:Mn spin glasses was performed some years ago by Chamberlin et al. [33] in which spin glass samples were cooled to low temperature in zero magnetic field, after which a DC magnetic field was turned on.

It was hypothesized in that paper that the application of this external field would "randomize" the energy surface. How might this occur? An initial guess might be that this change would take place in a continuous, gradual manner. If this were the case, then if the field were changed by a very small amount, a reversible change in magnetization would result. This was not observed, however, even for the smallest applied fields ($\sim$40 mOe); the magnetization always displayed an irreversible drift governed by a quasilogarithmic time dependence [33–36]. The explanation given was that any field, no matter how small, completely "scrambles" the energy surface, which would account for the irreversibility on all scales. This might be the case, but the explanation lacks a clear underlying model in which this occurs.

How might this same effect be understood in the picture presented in this paper? The simplest explanation is that that the surface is largely unchanged, aside from a "tilt" imposed by the field. The effect of this tilt is to lower the barriers in certain directions, making escape easier from the pond in which the system found itself before the field was turned on. Once the field is turned off, the system is in a new pond and one is simply observing the diode effect discussed earlier.

A similar explanation can be used to understand aging experiments in which the temperature is raised and then lowered, or vice-versa [37]. What is typically seen is that if temperature is first lowered and then raised, the system returns to its original state: the process is reversible. Raising the temperature first and then lowering it, however, is irreversible. In the ponds-and-outlets picture, this is easily understood: lowering the temperature simply confines the system to a smaller region of the pond it was in at the start of the process. Raising the temperature then simply restores the system to the entire pond. Raising the temperature first, however, will generally land the system in a new pond, if not an entirely new global component altogether, after which the diode effect insures irreversibility. (Of course, the one-dimensional picture of Fig. 1 makes essentially the same prediction, so these experiments do not distinguish the standard BE picture based on Fig. 1 from our picture. The experiments are consistent with both.)

A second interesting application of these ideas is to biological evolution, particularly in its early stages. A universal feature of systems governed by the scenario described in this paper is that the largest barrier the system must surmount is the outlet corresponding to the first pond the system finds itself in; once that barrier is surmounted, the system finds itself in ponds confined by increasingly smaller barriers. Qualitatively, this would correspond to the earliest stages of a process described by an RWRE as lasting the longest, and later stages having increasingly shorter lifetimes.

A well-known feature of life on earth is that it began roughly a half-billion years after the earth was formed; that is, relatively quickly on geological timescales. For the next three and a half billion years or so, life was confined mostly to simple, single-celled organisms, or colonies of such organisms. This epoch (which comprises most of the Earth's history) is known as the Precambrian Era. Then, roughly half a billion years ago, multicellular organisms rapidly evolved, an event known as the Cambrian explosion. Evolution thereafter proceeded rapidly.

On a qualitative level, at least, this corresponds well to the ponds-and-outlets scenario [38]. Our model would predict that this is a feature of evolution not confined only to earth, but that a similar distribution of timescales should occur anywhere life forms and evolves.

An issue of particular interest concerns contingency and convergence in evolution. That is, would slightly different starting situations lead to long term differences in the nature of evolving biomes (a contingency effect) or would it not (a convergence effect)? From the invasion percolation perspective, this corresponds to the question of whether different starting sites would or would not merge into a common path to the sea. The answer for invasion percolation, as discussed previously, depends on the dimension with convergence for low dimensions and contingency above a critical dimensions (now presumed to be six). Since the relevant space for biological evolution is presumably high dimensional, this picture supports the conclusion of contingency.

Acknowledgments This work was supported in part by NSF Grants DMS-0604869, DMS-1106316, and DMS-1207678.

References

1. J. Jäckle, On the glass transition and the residual entropy of glasses. Phil. Mag. B **44**, 533–545 (1981)
2. R.G. Palmer, Broken Ergodicity. Adv. Phys. **31**, 669–735 (1982)
3. R.G. Palmer, Broken ergodicity in spin glasses. in *Heidelberg Colloquium on Spin Glasses*, (Springer, Berlin, 1983), pp. 234–251
4. A.C.D. van Enter, J.L. van Hemmen, Statistical-mechanical formalism for spin-glasses. Phys. Rev. A **29**, 355–365 (1984)
5. R.G. Palmer, D.L. Stein, Broken Ergodicity in Glass. in *Relaxations in Complex Systems*, (U.S. GPO, Washington, 1985), pp. 253–259
6. S.A. Kauffman, *Origins of Order* (Oxford University Press, Oxford, 1993)
7. K. Binder, A.P. Young, Spin glasses: experimental facts, theoretical concepts, and open questions. Rev. Mod. Phys. **58**, 801–976 (1986)
8. P.F. Stadler, W. Schnabl, The landscape of the traveling salesman problem. Phys. Lett. A **161**, 337–344 (1992)
9. D.L. Stein, C.M. Newman, Broken ergodicity and the geometry of rugged landscapes. Phys. Rev. E **51**, 5228–5238 (1995)
10. D.L. Stein, C.M. Newman, *Spin Glasses and Complexity* (Princeton University Press, Princeton, 2013)
11. U. Krey, Amorphous magnetism: theoretical aspects. J. Magn. Magn. Mater. **6**, 27–37 (1977)
12. This happened to us
13. C.M. Newman, D.L. Stein, Random walk in a strongly inhomogeneous environment and invasion percolation. Ann. Inst. Henri Poincaré **31**, 249–261 (1995)
14. R. Lenormand, S. Bories, Description d'un mécanisme de connexion de liaison destiné à l'étude du drainage avec piègeage en milieu poreux. C.R. Acad. Sci. **291**, 279–282 (1980)

15. R. Chandler, J. Koplick, K. Lerman, J.F. Willemsen, Capillary displacement and percolation in porous media. J. Fluid Mech. **119**, 249–267 (1982)
16. D. Wilkinson, J.F. Willemsen, Invasion percolation: a new form of percolation theory. J. Phys. A **16**, 3365–3376 (1983)
17. P. Bak, C. Tang, K. Wiesenfeld, Self-organized criticality: an explanation of the 1/f noise. Phys. Rev. Lett. **59**, 381–384 (1987)
18. D. Stauffer, A. Aharony, *Introduction to Percolation Theory*, 2nd edn. (Taylor and Francis, London, 1992)
19. C.M. Newman, D.L. Stein, Spin glass model with dimension-dependent ground state multiplicity. Phys. Rev. Lett. **72**, 2286–2289 (1994)
20. C.M. Newman, D.L. Stein, Ground state structure in a highly disordered spin glass model. J. Stat. Phys. **82**, 1113–1132 (1996)
21. J.S. Jackson, N. Read, Theory of minimum spanning trees. I. Mean-field theory and strongly disordered spin-glass model. Phys. Rev. E **81**, 021130 (2010)
22. J.S. Jackson, N. Read, Theory of minimum spanning trees. II. Exact graphical methods and perturbation expansion at the percolation threshold. Phys. Rev. E **81**, 021131 (2010)
23. G. Papanicolaou, S.R.S. Varadhan, Diffusions with random coefficients. in *Statistics and Probability: Essays in Honor of C.R. Rao* (North-Holland, Amsterdam, 1982), pp. 547–552
24. C. Kipnis, S.R.S. Varadhan, Central limit theorem for additive functionals of reversible Markov processes and applications to simple exclusions. Comm. Math. Phys. **104**, 1–19 (1986)
25. A. De Masi, P.A. Ferrari, S. Goldstein, D.W. Wick, An invariance principle for reversible Markov processes. Applications to random motions in random environments. J. Stat. Phys. **55**, 787–855 (1989)
26. In ordinary BE there is assumed to exist some highest barrier, $\Delta F_{\max}$ that confines the system; after escape occurs over $\Delta F_{\max}$, dynamical processes are ergodic. Before this occurs (that is, when the system is still in the broken ergodic regime) barriers confining the system are assumed to grow logarithmically with time, which follows from the Arrhenius relation between escape time and barrier height. In contrast, there is no clear analogue in our models to $\Delta F_{\max}$; after initial transients, barriers are roughly constant, with expected small fluctuations owing to deviations in the diffusion process from the strict $T \to 0$ limit.
27. G. Grimmett, *Percolation* (Springer, New York, 1989)
28. J.M. Hammersley, A Monte Carlo solution of percolation in a cubic lattice. in *Methods in Computational Physics*, vol. I (Academic Press, New York 1963), pp. 281–298
29. A precise formulation which is valid for any $d \geq 2$, irrespective of whether there is one or many disjoint invasion regions, and which does not conflict with recurrence, is as follows: for any fixed N and with probability approaching one as β tends to ∞, the RWRE will visit all sites in the second through Nth ponds before it either returns to the first pond or crosses a bond of larger value than the first outlet.
30. M. Damron, A. Sapozhnikov, Outlets of 2*D* invasion percolation and multiple-armed incipient infinite clusters. Prob. Theory Rel. Fields **150**, 257–294 (2011)
31. M. Damron, A. Sapozhnikov, Relations between invasion percolation and critical percolation in two dimensions. Ann. Prob. **37**, 2297–2331 (2009); Limit theorems for 2D invasion percolation. arXiv:1005.5696v3 (2012)
32. H. Kesten, The incipient infinite cluster in two-dimensional percolation. Prob. Theory Rel. Fields **73**, 369–394 (1986)
33. R.V. Chamberlin, M. Hardiman, L.A. Turkevich, R. Orbach, $H - T$ phase diagram for spin-glasses: an experimental study of Ag:Mn. Phys. Rev. B **25**, 6720–6729 (1982)
34. J.-L. Tholence, R. Tournier, Susceptibility and remanent magnetization of a spin glass. J. Phys. (Paris) **35**, C4–229-C4-236 (1974)
35. C.N. Guy, Gold-iron spin glasses in low DC fields. I. Susceptibility and thermoremanence. J. Phys. F **7**, 1505–1519 (1977)

36. R.W. Knitter, J.S. Kouvel, Field-induced magnetic transition in a Cu-Mn spin-glass alloy. J. Magn. Magn. Mat. **21**, L316–L319 (1980)
37. P. Refrigier, E. Vincent, J. Hamman, M. Ocio, Ageing phenomena in a spin glass: effect of temperature changes below T_g. J. Phys. (Paris) **48**, 1533–1539 (1987)
38. S.J. Gould, C.M. Newman, D.L. Stein (unpublished)

Structural Complexity of Vortex Flows by Diagram Analysis and Knot Polynomials

Renzo L. Ricca

Abstract In this paper I present and discuss with examples new techniques based on the use of geometric and topological information to quantify dynamical information and determine new relationships between structural complexity and dynamical properties of vortex flows. New means to determine linear and angular momenta from standard diagram analysis of vortex tangles are provided, and the Jones polynomial, derived from the skein relations of knot theory is introduced as a new knot invariant of topological fluid mechanics. For illustration several explicit computations are carried out for elementary vortex configurations. These new techniques are discussed in the context of ideal fluid flows, but they can be equally applied in the case of dissipative systems, where vortex topology is no longer conserved. In this case, a direct implementation of adaptive methods in a real-time diagnostics of real vortex dynamics may offer a new, powerful tool to analyze energy-complexity relations and estimate energy transfers in highly turbulent flows. These methods have general validity, and they can be used in many systems that display a similar degree of self-organization and adaptivity.

Keywords Euler equations · Linear and angular momenta · Diagram analysis · Vortex knots and links · Topological fluid dynamics · Structural complexity

R. L. Ricca (✉)
Department of Mathematics and Applications, University of Milano-Bicocca,
Via Cozzi 53 20125 Milano, Italy
e-mail: renzo.ricca@unimib.it
URL: www.matapp.unimib.it/~ricca/

R. L. Ricca
Kavli Institute for Theoretical Physics, UCSB Santa Barbara, California, USA

I. Zelinka et al. (eds.), *How Nature Works*, Emergence, Complexity and Computation 5, DOI: 10.1007/978-3-319-00254-5_5,
© Springer International Publishing Switzerland 2014

1 Introduction

Networks of fluid structures, such as tangles of vortex filaments in turbulent flows or braided magnetic fields in magnetohydrodynamics, are examples of physical systems that by their own nature are fundamentally structurally complex [1]. This is the result of many contributing factors, among which the highly nonlinear character of the governing equations, the simultaneous presence of different space scales in the emerging phenomena, and the spontaneous self-organization of the constituent elements. In this respect ideal vortex dynamics offers a suitable theoretical framework to develop and apply methods of structural complexity to investigate and analyze dynamical properties and energy-complexity relations.

In this paper I present and discuss with examples new techniques based on the use of algebraic, geometric and topological information to quantify dynamical information and to determine new relationships between energy and complexity in coherent vortex flows. These flows, that arise naturally from spontaneous self-organization of the vorticity field into thin filaments, bundles and tangles of filaments in space, both in classical and quantum fluids, share common features, being the seeds and sinews of homogeneous, isotropic turbulence [2, 3]. Characterization and quantification of such flows is of fundamental importance from both a theoretical and applied viewpoint. From a theoretical point of view detailed understanding of how structural, dynamical and energetic properties emerge in the bulk of the fluid and change both in space and in time is at the basis of our analysis of how self-organization and non-linearities play their role in complex phenomena. For applications, understanding these aspects is also of fundamental importance to develop new real-time diagnostic tools to investigate and quantify dynamical properties of turbulent flows in classical fluid mechanics and magnetohydrodynamics.

The remarkable progress in the use of geometric and topological techniques introduced in the last decade [5–7], associated with continuous progress in computational power and visualization techniques [8, 9] in the light of the most recent developments in the field [10], is a testimony of the success of this novel approach. Here we shall confine ourselves to some new geometric and topological techniques introduced recently [11, 12] to estimate dynamical properties of complex vortex tangles of filaments in space. Most of the concepts presented here, being of geometric and topological origin, are independent of the actual physical model. We shall therefore momentarily drop the physics, and refer to the geometry and topology of the filament centerlines. These will be simply seen as smooth curves that may form knots, links and tangles in space, and it is to this system of curves that our analysis will be dedicated. Then, we shall apply this analysis to vortex dynamics, in order to get new physical information.

In Sect. 2 I start by introducing basic notions of standard and indented diagram projections to determine signed area and crossing numbers of knots and links in space. In Sect. 3 the definitions of linear and angular momenta of a vortex system in ideal conditions are introduced, providing a geometric interpretation of these quantities in terms of area. As illustration a number of examples are presented in

Sect. 4 to evaluate the impetus of some vortex knots and links; a general statement on the signed area interpretation of the momenta for vortex tangles is presented in Sect. 5. Then, in Sect. 6 knot polynomial invariants used to classify topologically closed space curves in knot theory are considered. I concentrate on the Jones polynomial, and, for illustration, the polynomial for several elementary knots and links (Sect. 7) are computed. By showing that it can be expressed in terms of kinetic helicity (Sect. 8), I show that the Jones polynomial can be re-interpreted as a new invariant of topological fluid mechanics (Sect. 9). In Sect. 10 I conclude with some considerations on possible future implementations of these concepts in advanced, adaptive, real-time diagnostics, to estimate energy and helicity transfers in real, turbulent flows.

2 Standard and Indented Diagrams: Signed Areas and Crossing Numbers of Knots and Links

To begin with, let us consider an isolated, oriented curve χ in $\mathbb{R}^3$; this can be thought of as the axis of a tubular neighborhood that constitute the support of the actual vortex filament in space; the orientation of the curve is then naturally induced by the orientation of vorticity. χ is taken sufficiently smooth (i.e. at least C^2) and simple (without self-intersections), given by the position vector $\mathbf{X} = \mathbf{X}(s)$, where $s \in [0, L]$ is arc-length and L the total length. A Frenet triad $\{\hat{\mathbf{t}}, \hat{\mathbf{n}}, \hat{\mathbf{b}}\}$, given by the unit tangent $\hat{\mathbf{t}} = \mathrm{d}\mathbf{X}/\mathrm{ds}$, normal and binormal vector, is defined on any point of χ and at each point of χ curvature $c = c(s)$ and torsion $\tau = \tau(s)$ are defined. From the fundamental theorem of space curves, any curve in space is prescribed uniquely, once curvature and torsion are given as known functions of s. For the purpose of this paper we confine ourselves to closed and possibly knotted curves. A closed curve is given by $\mathbf{X}(0) = \mathbf{X}(L)$ and smooth closure implies that this is also true for higher derivatives of $\mathbf{X}(s)$.

Under continuous deformations the geometric properties of χ change continuously, but the topological properties remain invariant. Any curve that can be continuously deformed to the standard circle (without going through self-intersections or cuts) is not knotted and it is called the *unknot*. The task of knot theory (and of topology in general) is precisely to classify curves according to the topological characteristics of their knot (or link) type, where a collection (disjoint union) of N such curves, knotted or unknotted, constitute a link. A link of two mathematical tubes, centered on the axes χ_1 and χ_2, is shown in Fig. 1a. A vortex tangle is thus an N-component link of vortex filaments, where vorticity is simply defined within the tubular neighborhood of each component.

Standard projection. Let us consider now the standard projection of an N-component link; for simplicity, let us take the case of the 2-component link of Fig. 1a, and consider the orthogonal projection p of this link onto the plane. The resulting graph $\Lambda = p(\chi_1 \cup \chi_2)$ is a nodal curve in $\mathbb{R}^2$ with 4 intersection points

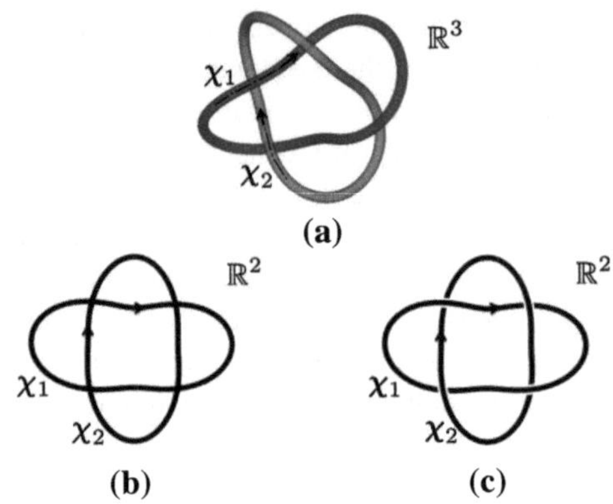

Fig. 1 **a** A 2-component link of tubes centered on the oriented curves χ_1 and χ_2 in space. **b** Standard projection of the link shown in (**a**) onto $\mathbb{R}^2$; the resulting graph $\Lambda = p(\chi_1 \cup \chi_2)$ is an oriented nodal curve in the plane, with 4 intersection points. **c** Indented projection of the same link onto $\mathbb{R}^2$: by small indentations in the plane, over-crossings and under-crossings are shown to preserve topological information of the original link in space

Fig. 2 A point P of multiplicity 3 can be reduced to 3 nodal points P_1, P_2, P_3 of multiplicity (degree) 2

(Fig. 1b). Each nodal point results from the intersection of 2 incident (oriented) arcs: these nodal points have therefore multiplicity (or degree) 2. In general, a nodal point of multiplicity μ is the intersection of μ incident arcs. By a small perturbation of the projection map p, a μ-degree point can be reduced to μ nodal points of degree 2 (see the example of Fig. 2). A *good*, standard projection is therefore a map that for any link of curves generates a planar graph Λ, with at most nodal points of degree 2. Let us restrict our attention to such good projections and consider the graph Λ, given by a collections of oriented arcs.

Indented projection. Let us focus now on the topological characteristics of a link, for instance the 2-component link of Fig. 1a. One way to analyze topological aspects of a link is to consider indented projections. One of these is given by projecting the link onto a plane by keeping track of the over/under-crossings by small indentations of the over/under-passes of the projected arcs (see the example of Fig. 1c). As above, we must ensure that at each apparent crossing only two arcs meet.

Signed areas from standard projections. The graph Λ, obtained from standard projections, determines a number of regions, say R_j ($j = 1, \ldots, Z$), in the plane.

Each region may be bounded by a number of oriented arcs, that may or may not have congruent orientation (see, for instance, the graph of Fig. 1b). In the latter case, the standard geometric area of a region R_j must be replaced by the *signed area*, to take account of the different orientation of the bounding arcs and area contributions from the graph Λ. With reference to Fig. 3, let us introduce a positive reference given by the pair of unit vectors $(\hat{\boldsymbol{\rho}}, \hat{\mathbf{t}})$. The radial vector $\hat{\boldsymbol{\rho}}$ is chosen arbitrarily to have a generic foot in the region R, pointing outwardly to the exterior of R, and $\hat{\mathbf{t}}$, the unit tangent to the boundary curve with direction induced by the projected orientation. Let us apply now this construction to the collection of regions R_j associated with a generic graph Λ, and consider the successive points of intersection given by the $\hat{\rho}$-line as it crosses the arcs of Λ in the $\hat{\rho}$-direction. To each intersection point let us assign the value $\epsilon = \pm 1$, according to the positive reference given by $(\hat{\boldsymbol{\rho}}, \hat{\mathbf{t}})$ determined by the orientation of each arc. For the simple case of Fig. 3, where there is only one region and one bounding curve, and therefore only one intersection point, we have $\epsilon = +1$, but in general we may have several intersection points (see examples of Sect. 4 below), each contributing ± 1 according to their signed crossing.

Definition. The *index* of a region R_j of $\Lambda \in \mathbb{R}^2$, is given by

$$\mathcal{I}_j = \mathcal{I}_j(R_j) = \sum_{r \in \{\hat{\rho} \cap \Lambda\}} \epsilon_r \ , \tag{1}$$

where $\{\hat{\boldsymbol{\rho}} \cap \Lambda\}$ denotes the set of intersection points given by the $\hat{\boldsymbol{\rho}}$ vector with all the arcs of Λ in that direction, and $\epsilon_r = \pm 1$, according to the sign of the reference $(\hat{\boldsymbol{\rho}}, \hat{\mathbf{t}})$ at each intersection point with Λ.

It can be easily proved that the index $\mathcal{I}_j$ is a topological property of the region R_j, independent of the choice of the position of the footpoints O_j and of the direction of the $\hat{\rho}_j$-lines. We can now define the signed area of a graph region according to the following definition.

Definition. The *signed area* of a region R_j is given by

$$\mathcal{A}_j(R_j) = \sum_j \mathcal{I}_j A_j(R_j) \ , \tag{2}$$

where $A_j(R_j)$ denotes the standard area of R_j.

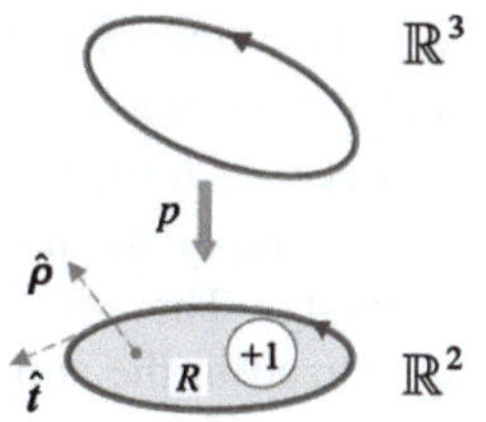

Fig. 3 Positive reference $(\hat{\boldsymbol{\rho}}, \hat{\mathbf{t}})$, given by the radial and tangent unit vectors

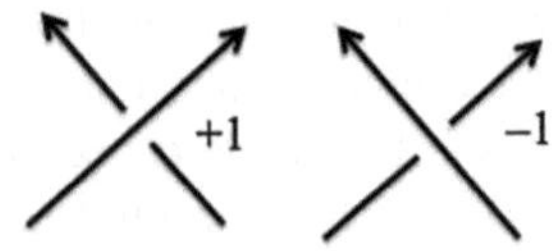

Fig. 4 Standard sign convention for a positive (over-) crossing and a negative (under-) crossing

The signed area extends naturally the concept of standard area for regions bounded by arcs of different orientations and it will be useful in the geometric interpretation of linear and angular momenta of vortex tangles.

Minimum number of crossings and linking number from indented projections. Two useful topological invariants can be extracted from indented projections. One is based on the minimum number of *apparent* crossings in an indented projection. Among the infinite number of possible indented projections, we consider the indented projection that gives the minimum number of crossings (*minimal diagram*) and, quite simply, we define this number as the *topological crossing number* of the knot or link type.

Another topological quantity can be computed from a generic indented projection. By standard convention (see Fig. 4) we assign $\epsilon_r = \pm 1$ to each apparent crossing site r. We can then introduce the following:

Definition. The *linking number* of a link type is given by

$$Lk = \frac{1}{2}\sum_r \epsilon_r \ , \tag{3}$$

where the summation is extended to all the apparent crossing sites, in any generic indented projection of the link.

In the example of Fig. 1c, there are 4 apparent crossings +1, hence the linking number $Lk(\boldsymbol{\chi}_1, \boldsymbol{\chi}_2) = +2$. Since the linking number is a topological invariant, its value is independent from the projection.

3 Linear and Angular Momentum of a Vortex Tangle from Geometric Information

First, let us consider a single vortex filament $\mathcal{K}$ in an unbounded, ideal fluid at rest at infinity, where vorticity is confined in the filament tube. Vortex filaments arise naturally in superfluid turbulence [3], where indeed vorticity remains localized for very long time on extremely thin filaments, with typical length of the order of 1 cm and vortex cross-section of the order of 10^{-8} cm.

Let $\mathcal{K} = \mathcal{K}(\chi)$ be centered on the filament axis $\boldsymbol{\chi}$. Let us assume that vorticity is given simply by $\boldsymbol{\omega} = \omega_0 \hat{\mathbf{t}}$, where ω_0 is a constant, and orientation is induced by vorticity. The vortex circulation (an invariant of the Euler's equations, and quantized in the superfluid case), is given by

$$\kappa = \int_A \omega \, \mathrm{d}^2\mathbf{X} = \text{constant}, \tag{4}$$

where A is the area of the vortex cross-section. Two fundamental invariants of ideal fluid mechanics are the linear and angular momenta. The linear momentum (per unit density) $\mathbf{P} = \mathbf{P}(\mathcal{K})$ corresponds to the hydrodynamic impulse, which is necessary to generate the motion of the vortex from rest; from its standard definition [13], it takes the form

$$\mathbf{P} = \frac{1}{2}\int_V \mathbf{X} \times \omega \, \mathrm{d}^3\mathbf{X} = \frac{1}{2}\kappa \oint_{L(\chi)} \mathbf{X} \times \hat{\mathbf{t}} \, \mathrm{d}s = \text{constant}, \tag{5}$$

where V is the filament volume. Similarly, for the angular momentum (per unit density) $\mathbf{M} = \mathbf{M}(\mathcal{K})$ (the moment of the impulsive forces acting on $\mathcal{K}$), given by

$$\mathbf{M} = \frac{1}{3}\int_V \mathbf{X} \times (\mathbf{X} \times \omega) \, \mathrm{d}^3\mathbf{X} = \frac{1}{3}\kappa \oint_{L(\chi)} \mathbf{X} \times (\mathbf{X} \times \hat{\mathbf{t}}) \, \mathrm{d}s = \text{constant}. \tag{6}$$

Now evidently, since $\hat{\mathbf{t}} \, \mathrm{d}s = \mathrm{d}\mathbf{X}$, the right-hand-side integrals in (5) and (6) admit an interpretation in terms of (twice) the geometric area. It is quite surprising that this geometric interpretation, recognized by Lord Kelvin in his early works on vortex motion, has remained almost unexploited to date, and it is this particular aspect that we want to exploit here. Since both $\mathbf{P}$ and $\mathbf{M}$ are vector quantities, each vector component can be related to the area of the graph resulting from the projection of χ along the direction of projection given by that component. By referring to the standard projection $\Lambda = p(\chi)$, we have

$$p(\chi) : \mathbb{R}^3 \to \mathbb{R}^2 , \qquad \begin{cases} p_x : \Lambda_{yz} , \; \mathcal{A}_{yz} = \mathcal{A}(\Lambda_{yz}) , \\ p_y : \Lambda_{zx} , \; \mathcal{A}_{zx} = \mathcal{A}(\Lambda_{zx}) , \\ p_z : \Lambda_{xy} , \; \mathcal{A}_{xy} = \mathcal{A}(\Lambda_{xy}) , \end{cases} \tag{7}$$

where, in the case of a simple, non-self-interesecting, planar curve Λ, $\mathcal{A}(\cdot)$ coincides with the standard geometric area bounded by Λ. Hence, we can write

$$\mathbf{P} = (P_x, P_y, P_z) = \kappa(\mathcal{A}_{yz}, \mathcal{A}_{zx}, \mathcal{A}_{xy}) , \tag{8}$$

$$\mathbf{M} = (M_x, M_y, M_z) = \frac{2}{3}\kappa(d_x\mathcal{A}_{yz}, d_y\mathcal{A}_{zx}, d_z\mathcal{A}_{xy}) , \tag{9}$$

where $d_x\mathcal{A}_{yz}$, $d_y\mathcal{A}_{zx}$, $d_z\mathcal{A}_{xy}$ are the areal moments given according to the following definition.

Definition. The *areal moment* around any axis is the product of the area $\mathcal{A}$ multiplied by the distance d between that axis and the axis $\mathbf{a}_G$, normal to $\mathcal{A}$ through the centroid G of $\mathcal{A}$.

Hence, d_x, d_y, d_z denote the Euclidean distances of the area centroid G of $\mathcal{A}$ from the axes x, y, z, respectively.

4 Impetus of Vortex Knots and Links: Some Examples

4.1 Single-Component Systems: Vortex Knots

Figure-of-eight knot. Let us consider the diagram of Fig. 5a and let us evaluate the indices of the graphs. Suppose that this diagram results from the projection of a figure-of-eight knot (in the diagram shown we kept track of over-crossings and under-crossings for visualization purposes only; in the standard, planar projection all the crossings become nodal points). For each region we arbitrarily choose a radial vector and for each vector we consider the intersections of the $\hat{\rho}$-line with the graph. At each intersection we assign a $+1$ or a -1, according to the positive reference defined in Sect. 2, and we sum up all the contributions according to (1), hence determining the index of that region. Their values are shown encircled in Fig. 5a. These values are topological in character, because they do not depend on the choice of the footpoint of $\hat{\rho}$, thus providing the necessary prefactor for the standard area. Using Eq. (8) we see that the central region of the figure-of-eight knot with index 0 does not contribute to the impetus in the direction normal to this plane projection, whereas the nearby regions, with relative indices $+1, -1$ and -2 will tend to contribute to the motion in opposite directions. The index -2 associated with the smallest region, then, may compensate for a modest area contribution.

Poloidal coil. Consider now the diagram of Fig. 5b, and suppose that this results from the projection of a poloidal coil in space. Since the central area has index $+1$ and the external lobes have all indices -1, by (8) we see that the resulting impetus component may amount to a negative value (depending on the

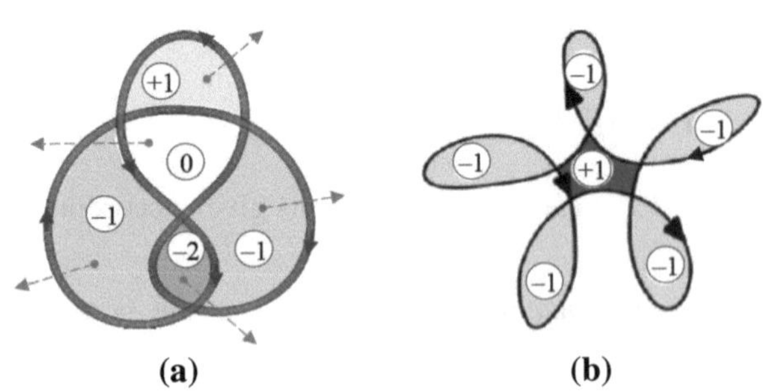

Fig. 5 **a** The figure-of-eight knot shown in an indented projection. In a standard plane projection all the apparent crossings collapse to nodal points. The encircled values denote the indices associated with their respective regions. **b** A poloidal coil in a standard plane projection

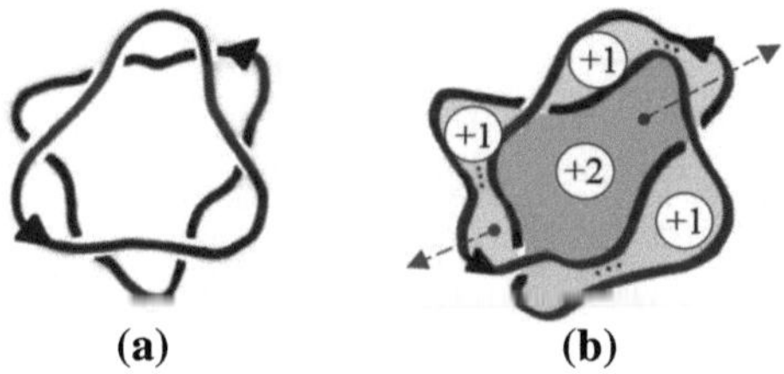

Fig. 6 **a** Two vortex ring interact and **b** reconnect to form a trefoil knot. The central region, having largest area and highest index, is likely to move more rapidly than the rest of the system in the normal direction to the projection plane

relative contributions from standard areas), giving rise to a backward motion in the opposite direction of the normal to the plane of projection. Such strange type of motion has been actually found by numerical simulation by Barenghi et al. in 2006 [14], and confirmed by more recent work by Maggioni et al. [15].

Trefoil knotting. A "thought experiment" to produce a trefoil vortex knot from the interaction and reconnection of vortex rings was conjectured by Ricca [16]. Upon collision (see Fig. 6a), two vortex rings propagate one after the other to reconnect, thus forming a trefoil knot (as in Fig. 6b). By assigning the indices to the different regions, it is possible to estimate the impulse associated with the different parts of the vortex in relation to the projected areas.

4.2 Multi-Component Systems: Vortex Links

Rings. Two vortex rings of equal but opposite circulation move towards each other to collide (see left diagram of Fig. 7). A finite number of reconnections take place on the colliding vortices, triggering the production of smaller vortex rings. Small rings are thus produced (right diagram of Fig. 7). This process has been actually realized by head-on collision of coloured vortex rings in water by Lim & Nickels [17]. Since at the initial state linear momentum $\mathbf{P} = 0$ (for symmetry reason), we expect that this remains so, till reconnections take place. The central diagram of Fig. 7 represents (schematically) the graph in the plane of collision, at the reconnection time. By applying the index computation, we can estimate the signed areas contributions. By using (8), we see that the central region does not contribute to the momentum of the system, whereas the outer regions, contribute with opposite sign to the momentum of the emerging small vortex rings. The alternating signs of the outer regions indicate the production of smaller rings of opposite polarity, thus ensuring $\mathbf{P} = 0$ throughout the process. The generation and shoot-off of smaller rings from the plane of collision in opposite directions seems in agreement with the experimental results of Lim and Nickels [17].

Hopf links. Finally, let us consider the projection of a Hopf link made by two vortex rings of circulation κ_1 and κ_2 (see Fig. 8). All indices have same sign,

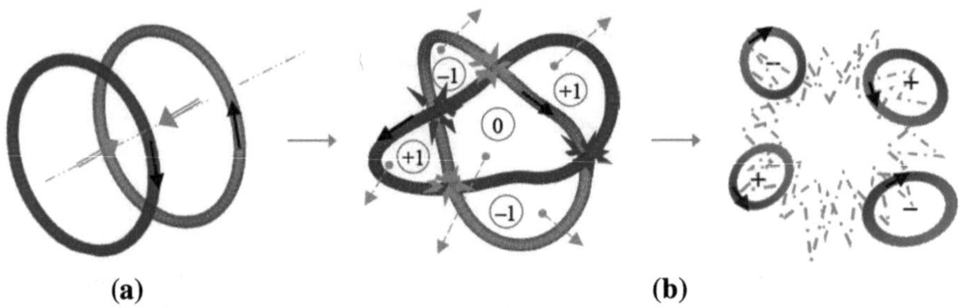

Fig. 7 **a** Two vortex ring interact and **b** reconnect to form a trefoil knot. The central region, having largest area and highest index, is likely to move more rapidly than the rest of the system in the normal direction of projection

indicating that all the graph regions in the projection contribute in the same direction to the impetus of the system. The central part, having the highest index, is likely to move faster than the rest of the system.

5 Momenta of a Tangle of Vortex Filaments

From the examples considered in the previous section it is clear that the geometric interpretation of the momenta based on Eqs. (8) and (9) can be easily extended to any complex graph resulting from the projection of a tangle $\mathcal{T} = \cup_i \mathcal{K}(\boldsymbol{\chi}_i)$ of filaments in space. We can now state the geometric criterium for the computation of the momenta from geometric diagram information.

Theorem (Momenta in terms of signed area interpretation) Let $\mathcal{T}$ be a vortex tangle under Euler equations. Then, the linear momentum $\mathbf{P} = \mathbf{P}(\mathcal{T}) = (P_x, P_y, P_z)$ has components

$$P_x = \kappa \sum_{j=1}^{Z} \mathcal{I}_j A_{yz}(R_j) \ , \quad P_y = \ldots \ , \quad P_z = \ldots \ , \tag{10}$$

and the angular momentum $\mathbf{M} = \mathbf{M}(\mathcal{T}) = (M_x, M_y, M_z)$ has components

Fig. 8 Projection of a Hopf link formed by two vortex rings of circulation κ_1 and κ_2

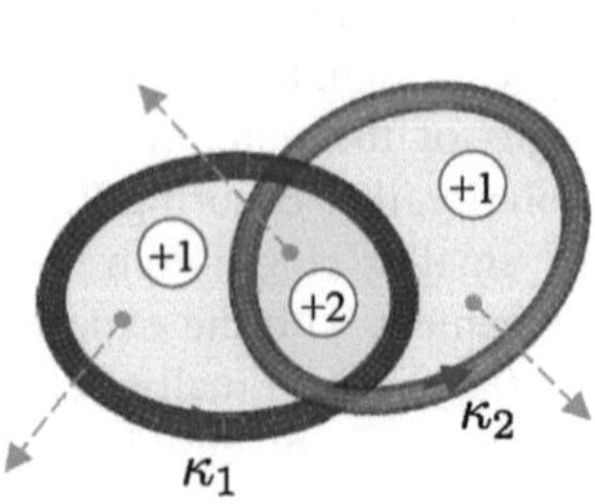

$$M_x = \frac{2}{3}\kappa d_x \sum_{j=1}^{Z} \mathcal{I}_j A_{yz}(R_j) \ , \quad M_y = \ldots \ , \quad M_z = \ldots \ , \tag{11}$$

where $A_{yz}(R_j)$, $A_{zx}(R_j)$, $A_{xy}(R_j)$ denote the standard areas of $R_j (j = 1, \ldots, Z)$, for any projection plane normal to the component of the momenta of $\mathcal{T}$,

Proof of the above Theorem is based on direct applications of (8), (9) and (2).

6 Knot Polynomial Invariants from Skein Relations: The Jones Polynomial

Indented diagrams of knot and link types are used to determine knot topology by extracting topological invariants known as *knot polynomials*. Several types of knot polynomials have been introduced as subsequent improvements, R-polynomials, Kauffman brackets [18], Jones polynomials [19] and HOMFLY-PT [20, 21] polynomials being such types of knot invariants. These polynomials are determined by skein relations derived from the analysis of crossing states in the indented diagrams of knots and links, given by un-oriented or oriented curves. With reference to Fig. 9, denoting by $\mathbf{L}_+$, $\mathbf{L}_-$, and $\mathbf{L}_0$ an over-crossing, an under-crossing and a non-crossing, respectively, we can derive skein relations for each polynomial.

6.1 Skein Relations of the Jones Polynomial

The Jones polynomial is a quite powerful knot invariant for oriented knots and links. It is therefore well-suited to tackle topological complexity of vortex tangles. The skein relations of the Jones polynomial are standardly derived by a technique called *local path-addition*, that consists of computing crossing states according to the analysis of the elementary states given by the over-crossing $L_+ =$ ⤧, the under-crossing $L_- =$ ⤨, and the disjoint union with a trivial circle. ↑ ⊔ ○. The skein relations of the Jones polynomial are given by [19, 22]:

$$V(\bigcirc) = 1 \tag{12}$$

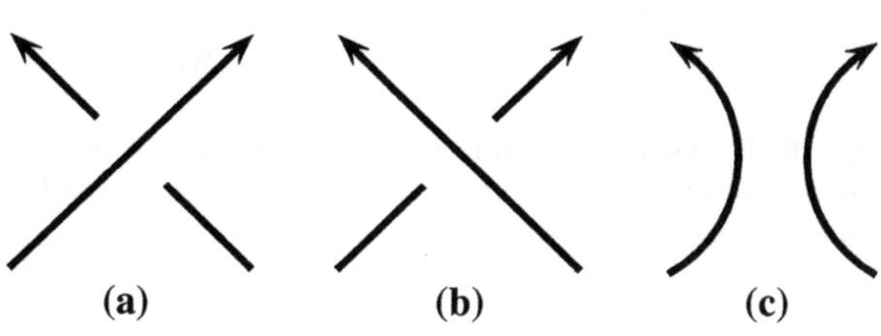

Fig. 9 **a** Over-crossing $\mathbf{L}_+(+1)$, **b** under-crossing $\mathbf{L}_-(-1)$, and **c** non-crossing $\mathbf{L}_0$ of oriented strands in an indented knot diagram

$$\tau^{-1}V\left(\times\right) - \tau V\left(\times\right) = \left(\tau^{\frac{1}{2}} - \tau^{-\frac{1}{2}}\right)V\left(\,)(\,\right) \tag{13}$$

Here we should stress that local path-additions are purely *mathematical* operations, performed *virtually* on the knot strands, the only purpose being simply the mathematical derivation of the polynomial terms, that give rise to the desired polynomial invariant. No actual physical process is therefore involved.

We can apply the skein relations (12)–(13) to determine the Jones polynomial of any given oriented knot/link. Calculations are based by applying reduction techniques performed recursively on (apparent) crossing sites, according to the diagrams shown in Fig. 10. These techniques resort to virtually split the over/under-crossing (Fig. 10a and b), by adding and subtracting local paths, so as to reduce each crossing site to a non-crossing plus a positive/negative writhe contribution, denoted respectively by γ_+ and γ_-.

7 Computation of the Jones Polynomial for Trefoil Knots and Whitehead Link

For the sake of illustration we compute the Jones polynomial by considering a number of elementary examples. First, let us consider the diagrams of Fig. 11. Evidently in (a) the writhe γ_+ and the writhe γ_- are both topologically equivalent to the unknot, i.e. the standard circle; hence by (12), we have $V(\bigcirc) = V(\gamma_+) = V(\gamma_-) = 1$. Now, by using (13), we have

$$\tau^{-1}V(\gamma_+) - \tau V(\gamma_-) = (\tau^{\frac{1}{2}} - \tau^{-\frac{1}{2}})V(\mathbf{l}_{cc}) \; , \tag{14}$$

hence,

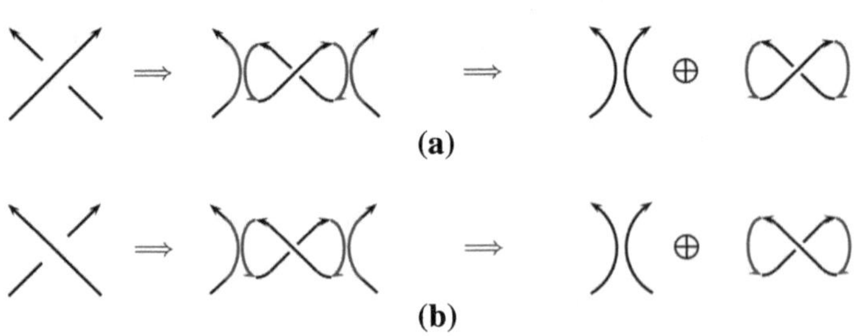

Fig. 10 By adding and subtracting local paths **a** an over-crossing $\mathbf{L}_+(+1)$ is reduced to a non-crossing plus a positive writhe γ_+, and **b** an under-crossing $\mathbf{L}_-(-1)$ is reduced to a non-crossing plus a negative writhe γ_-

$$V(\mathbf{l}_{cc}) = -\tau^{-\frac{1}{2}} - \tau^{\frac{1}{2}} \; . \tag{15}$$

Note that the orientation of any number of disjoint rings has no effect on the polynomial. As regards to the Hopf link $\mathbf{H}_+$ (Fig. 11b), we have

$$\tau^{-1}V(\mathbf{H}_+) - \tau V(\mathbf{l}_{cc}) = (\tau^{\frac{1}{2}} - \tau^{-\frac{1}{2}})V(\gamma_+) \; , \tag{16}$$

that gives

$$V(\mathbf{H}_+) = -\tau^{\frac{1}{2}} - \tau^{\frac{5}{2}} \; . \tag{17}$$

Similarly for the Hopf link $\mathbf{H}_-$ of Fig. 11c:

$$\tau^{-1}V(\mathbf{l}_{cc}) - \tau V(\mathbf{H}_-) = (\tau^{\frac{1}{2}} - \tau^{-\frac{1}{2}})V(\gamma_-) \; , \tag{18}$$

that gives

$$V(\mathbf{H}_-) = -\tau^{-\frac{1}{2}} - \tau^{-\frac{5}{2}} \; . \tag{19}$$

7.1 Left-Handed and Right-Handed Trefoil Knots

The left-handed trefoil knot $\mathbf{T}^L$ and right-handed trefoil knot $\mathbf{T}^R$ are shown by the top diagrams of Fig. 12a and b, respectively. By re-arranging (13), we can convert a crossing in terms of its opposite plus a contribution from parallel strands, that is

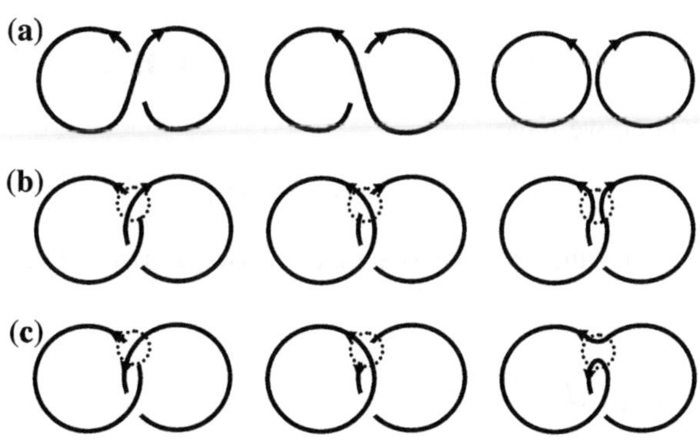

Fig. 11 **a** Writhes γ_+, γ_- and disjoint union of two trivial circles $\mathbf{l}_{cc}$. **b** Hopf link $\mathbf{H}_+$ with crossing $+1$, disjoint union of circles $\mathbf{l}_{cc}$ and writhe γ_+. **c** Hopf link $\mathbf{H}_-$ with crossing -1, disjoint union of circles $\mathbf{l}_{cc}$ and writhe γ_-. Note that the orientation of any number of disjoint circles does not influence the polynomial

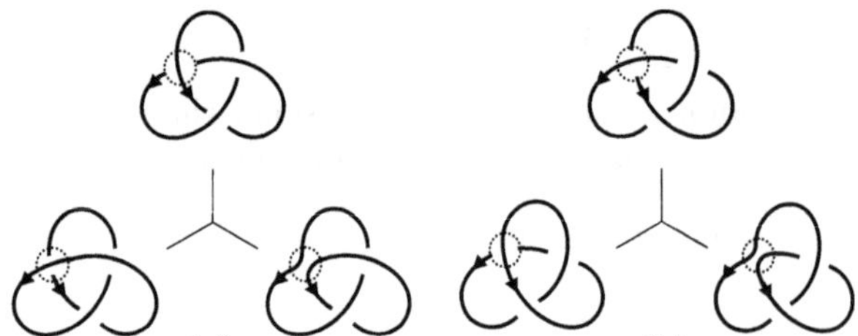

(a) (b)

Fig. 12 **a***Left-handed* and **b***Right-handed* trefoil knots (top diagrams) decomposed by applying standard reduction techniques (local path-addition) on crossing sites. Their Jones polynomials are obtained by analyzing the elementary states given by the diagrams of Fig. 11

$$V(L_+) = \tau^2 V(L_-) + (\tau^{\frac{3}{2}} - \tau^{\frac{1}{2}}) V(L_0) \tag{20}$$

By applying this relation to the encircled crossing of each trefoil knot we can transform the top diagrams of Fig. 12 into their relative decompositions given by a writhe and a Hopf link (bottom diagrams). With reference to the left-handed trefoil of Fig. 12a, we have a writhe γ_- and a Hopf link $\mathbf{H}_-$.

Hence, by using the elementary results above, we have: for the left-handed trefoil knot $\mathbf{T}^L$

$$V(\gamma_-) = \tau^2 V(\mathbf{T}^L) + (\tau^{\frac{3}{2}} - \tau^{\frac{1}{2}}) V(\mathbf{H}_-) , \tag{21}$$

that, by using (19), gives

$$V(\mathbf{T}^L) = \tau^{-1} + \tau^{-3} - \tau^{-4} . \tag{22}$$

For the right-handed trefoil knot $\mathbf{T}^R$, we have

$$\tau^{-1} V(\mathbf{T}^R) - \tau V(\gamma_+) = (\tau^{\frac{1}{2}} - \tau^{-\frac{1}{2}}) V(\mathbf{H}_+) . \tag{23}$$

Thus, by using (17), we have

$$V(\mathbf{T}^R) = \tau + \tau^3 - \tau^4 . \tag{24}$$

By comparing (22) with (24) we see that the two mirror knots have different polynomials.

7.2 Whitehead Link

A second example is provided by the Whitehead link $\mathbf{W}$ (see Fig. 13). With reference to the bottom diagrams of Fig. 13, by applying the skein relation (13) to the Whitehead link $\mathbf{W}_+$ (of crossing $+1$), we have the relation

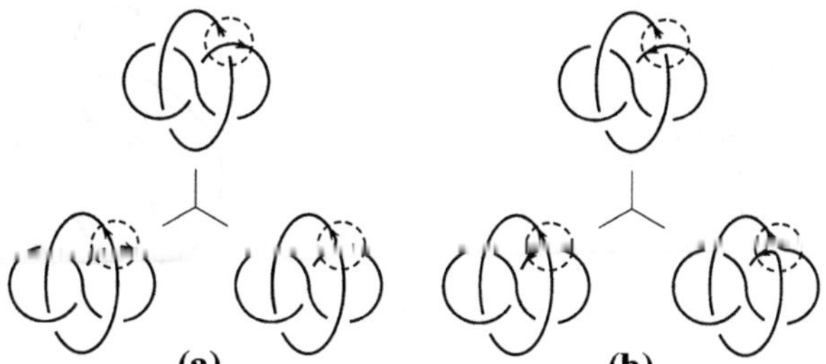

Fig. 13 Reduction schemes for Whitehead links $\mathbf{W}_+$ and $\mathbf{W}_-$. **a***Top*: Whitehead link $\mathbf{W}_+$ with crossing $+1$; *bottom*: Hopf link $\mathbf{H}_-$ and left-handed trefoil knot $\mathbf{T}^L$. **b***Top*: Whitehead link $\mathbf{W}_-$ with crossing -1; bottom: Hopf link $\mathbf{H}_+$, and figure-of-eight knot $\mathbf{F}^8$

$$\tau^{-1}V(\mathbf{W}_+) - \tau V(\mathbf{H}_-) = (\tau^{\frac{1}{2}} - \tau^{-\frac{1}{2}})V(\mathbf{T}^L) \ , \tag{25}$$

and application of (13) to the Whitehead link $\mathbf{W}_-$ gives

$$\tau^{-1}V(\mathbf{H}_+) - \tau V(\mathbf{W}_-) = (\tau^{\frac{1}{2}} - \tau^{-\frac{1}{2}})V(\mathbf{F}^8) \ , \tag{26}$$

where $\mathbf{F}^8$ denotes the figure-of-eight knot shown at the bottom of Fig. 13b. This latter can be further reduced according to the diagrams of Fig. 14. By applying (13) to the unknot with two writhes γ_-, denoted by $\gamma_=$, and to the Hopf link with writhe γ_+, denoted by $\mathbf{H}_-^+$, we have

$$\tau^{-1}V(\mathbf{F}^8) - \tau V(\gamma_=) = (\tau^{\frac{1}{2}} - \tau^{-\frac{1}{2}})V(\mathbf{H}_-^+) \ . \tag{27}$$

Now, since $V(\gamma_=) = 1$ and $V(\mathbf{H}_-^+) = V(\mathbf{H}_-) = -\tau^{-\frac{1}{2}} - \tau^{-\frac{5}{2}}$, we have

$$V(\mathbf{F}^8) = \tau^{-2} - \tau^{-1} + 1 - \tau + \tau^2 \ . \tag{28}$$

As can be easily verified, the mirror image of the figure-of-eight knot of Fig. 14 has the same Jones polynomial of Eq. (28).

Hence, by substituting (19) and (22) into (25), we have the Jones polynomial for $\mathbf{W}_+$. By similar, straightforward computation we obtain also the Jones polynomial for $\mathbf{W}_-$. The two polynomials coincide, that is $V(\mathbf{W}_+) = V(\mathbf{W}_-) = V(\mathbf{W})$, given by

$$V(\mathbf{W}) = \tau^{-\frac{7}{2}} - 2\tau^{-\frac{5}{2}} + \tau^{-\frac{3}{2}} - 2\tau^{-\frac{1}{2}} + \tau^{\frac{1}{2}} - \tau^{\frac{3}{2}} \ , \tag{29}$$

indicating that the two knots are actually the same knot type.

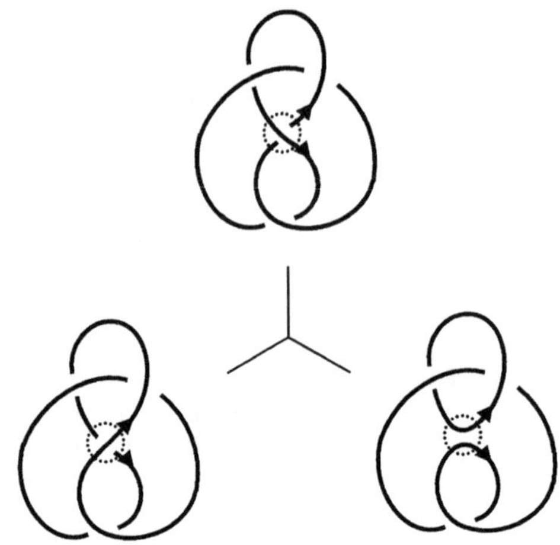

Fig. 14 Reduction scheme for (*top*) the figure-of-eight knot $\mathbf{F}^8$; *bottom* the unknot with two writhes γ_-, denoted by $\gamma_=$ (*left*), and a Hopf link with writhe γ_+, denoted by $\mathbf{H}_-^+$

8 The Jones Polynomial of Vortex Knots from Helicity

Fluid helicity is one of the most important conserved quantities of ideal fluid flows, being an invariant of the Euler equations, and a robust quantity of the dissipative Navier–Stokes equations [13]. In ideal conditions its topological interpretation in terms of Gauss linking number was provided by Moffatt [23] and extended by Moffatt & Ricca [24]. In the context of vortex dynamics (kinetic) helicity is defined by

$$H \equiv \int_\Omega \mathbf{u} \cdot \boldsymbol{\omega}\, d^3\mathbf{x}, \tag{30}$$

where $\mathbf{u}$ is the velocity field, $\boldsymbol{\omega} = \nabla \times \mathbf{u}$ is the vorticity, defined on Ω, and $\mathbf{x}$ the position vector. For simplicity we assume $\nabla \cdot \mathbf{u} = 0$ everywhere, and we request $\boldsymbol{\omega} \cdot \hat{\mathbf{n}} = 0$ on $\partial\Omega$, where $\hat{\mathbf{n}}$ is orthogonal to $\partial\Omega$, with $\nabla \cdot \boldsymbol{\omega} = 0$. For a thin vortex filament (30) reduces to a loop integral [25], given by

$$H = \kappa \oint_{\mathcal{K}} \mathbf{u} \cdot d\mathbf{l}\ , \tag{31}$$

where now $\mathbf{u}$ denotes the vortex velocity induced by the Biot-Savart law. On the other hand, for a single tangle component Eq. (31) can be written in terms of the well-known contributions due to the Călugăreanu-White formula [24], i.e. (by dropping the index)

$$H(\mathcal{K}) = \kappa^2 \mathbf{Lk} = \kappa^2 (\mathbf{Wr} + \mathbf{Tw})\ , \tag{32}$$

where helicity is decomposed in terms of writhe (*Wr*) and twist (*Tw*) contributions.

As a topological invariant of the knot $\mathcal{K}$, the Jones polynomial $V = V(\mathcal{K})$ is merely a function of a dummy variable (say τ), that in general has no physical meaning: thus $V(\mathcal{K}) = \mathbf{V}_\tau(\mathcal{K})$. Since in our case $\mathcal{K}$ is a vortex knot, during evolution it carries topological as well as dynamical information. Following [12], we can encapsulate this dual property by combining the two Eqs. (31) and (32) into the variable τ (by an appropriate transformation), showing that this new τ satisfies the skein relations of the Jones polynomial. Indeed, by using (31) and (32) and the transformation $e^H \to t^H \to \tau$, we have the following [12]:

Theorem ([12]): Let $\mathcal{K}$ denote a vortex knot (or an N-component link) of helicity $H = H(\mathcal{K})$. Then

$$t^{H(\mathcal{K})} = t^{\oint_{\mathcal{K}} \mathbf{u}\cdot d\mathbf{l}} , \tag{33}$$

appropriately re-scaled, satisfies (with a plausible statistical hypothesis) the skein relations of the Jones polynomial $V = V(\mathcal{K})$.

Full proof of the above Theorem is given in the reference above, where the skein relations are derived in terms of the variable

$$\tau = t^{-4\lambda H(\gamma_+)} , \qquad \lambda \in [0,1] , \tag{34}$$

where λ takes into account the uncertainty associated with the writhe value of γ_+ (see the first diagram of Fig. 11a) and $H(\gamma_+)$ denotes the helicity associated with γ_+ [12].

9 The Jones Polynomial as a New Fluid Dynamical Knot Invariant

For practical applications it is useful to go back to the original position, i.e. $\tau \to t^H \to e^H$, by referring to e^H rather than τ. By (34) we can write knot polynomials for a vortex tangle of filaments as function of topology and $H(\gamma_+)$, where the latter can be interpreted as a reference mean-field helicity of the physical system. Indeed, by (32), we can think of $H(\gamma_+)$ as a gauge for a mean writhe (or twist) helicity of the background flow. Since in any case this contributes in terms of an average circulation κ, we can re-interpret the Jones polynomial as a new invariant of topological fluid dynamics, i.e.

$$V_\tau(\mathcal{K}) \to V_t(\mathcal{K},\kappa) \to V(\mathcal{K},\kappa) . \tag{35}$$

In the case of a homogeneous, isotropic tangle of superfluid filaments, all vortices have same circulation κ; thus, by normalizing the circulation in dimensionless form, we can set

$$\bar{\lambda} = \langle\lambda\rangle = \frac{1}{2} , \qquad \langle H(\gamma_+)\rangle = \frac{\kappa^2}{2} , \tag{36}$$

where angular brackets denote average values. Hence,

$$\tau = t^{-\kappa^2} \to e^{-\kappa^2} . \tag{37}$$

In this case, we have

$$V(\bigcirc) = V(\gamma_+) = V(\gamma_-) = V(\gamma_=) = 1 \tag{38}$$

$$V(\mathbf{l}_{cc}) = -e^{\frac{\kappa^2}{2}}(1 + e^{-\kappa^2}) , \tag{39}$$

$$V(\mathbf{l}_{c,\dots,c}) = [-e^{\frac{\kappa^2}{2}}(1 + e^{-\kappa^2})]^{N-1} , \qquad \text{(N vortex rings)} , \tag{40}$$

$$V(\mathbf{H}_+) = -e^{\frac{-\kappa^2}{2}}(1 + e^{-2\kappa^2}) , \tag{41}$$

$$V(\mathbf{H}_-) = -e^{\frac{\kappa^2}{2}}(1 + e^{2\kappa^2}) , \tag{42}$$

$$V(\mathbf{T}^L) = e^{\kappa^2} + e^{3\kappa^2} - e^{4\kappa^2} , \tag{43}$$

$$V(\mathbf{T}^R) = e^{-\kappa^2} + e^{3\kappa^2} - e^{-4\kappa^2} . \tag{44}$$

$$V(\mathbf{F}^8) = e^{2\kappa^2} - e^{\kappa^2} + 1 - e^{-\kappa^2} + e^{-2\kappa^2} , \tag{45}$$

$$V(\mathbf{W}) = e^{-\frac{3}{2}\kappa^2}\left(-1 + e^{\kappa^2} - 2e^{2\kappa^2} + e^{3\kappa^2} - 2e^{4\kappa^2} + e^{5\kappa^2}\right) . \tag{46}$$

These results are obtained by straightforward application of the transformation (37) to the computations carried out in Sect. 7. For more complex physical systems knot polynomials can be straightforwardly computed by implementing skein relations and diagram analysis into a numerical code.

10 Concluding Remarks

Complex tangles of vortex filaments are ubiquitous in turbulent flows, and are key features of homogeneous isotropic turbulence in both classical and quantum systems. Detecting their structural complexity and attempting to relate complexity to dynamical and energetic properties are of fundamental importance for both theoretical and practical reasons. Moreover, vortex tangles represent a good paradigmatic example of complex systems displaying features of self-organization and adaptive behavior largely independent of the space scale of the phenomena; they therefore offer a perfect test case to study and tackle aspects of structural complexity in general.

In this paper I reviewed new results obtained following an approach based on the exploitation of geometric and topological information. Preliminary information on standard and indented diagrams have been given in Sect. 2. Then, a new method to compute linear and angular momenta of a tangle of vortex filaments in ideal fluids has been presented (Sects. 3 and 4). This method relies on the direct interpretation of these quantities in terms of geometric information (Sect. 5). Indeed, the technique proposed here is based on a rather straightforward analysis of planar graphs, and its direct implementation to analyze highly complex networks seems amenable to more sophisticated improvements. Similar considerations hold for the implementation of skein relations to quantify topological properties by knot polynomials, introduced in Sects. 6 and 7. Here, our attention has been restricted to the Jones polynomial, and its interpretation in terms of the helicity of fluid flows (Sect. 8). This, in turn, has led us to re-consider and present this polynomial as a new invariant of ideal fluid mechanics, and a number of elementary examples have been presented to demonstrate both the straightforward application of computational techniques associated with the implementation of the relative skein relations, and the possibility to extend this approach to more complex systems (Sect. 9).

These results can be extended to real fluid flows as well. For these systems viscosity play an important role, by producing continuous changes in the tangle topology, leading to the gradual dissipation of all conserved quantities, momenta, helicity and, of course, energy. This is certainly reflected in the continuous change of geometric, topological and dynamical properties. Therefore, a real-time implementation of an adaptive analysis that takes account of these changes can provide a useful tool for real-time diagnostics of the exchange and transfer of dynamical properties and energy between different regions in the fluid. This, together with an adaptive, real-time implementation of a whole new set of measures of structural complexity based on algebraic, geometric and topological information [25–27] will prove useful to investigate and tackle open problems in classical, quantum and magnetic fluid flows, as well as in many other systems that display similar features of self-organization.

Acknowledgments This author wishes to thank the Kavli Institute for Theoretical Physics at UC Santa Barbara for the kind hospitality. This research was supported in part by the National Science Foundation under Grant No. NSF PHY11-25915.

References

1. R.L. Ricca, Structural Complexity, in *Encyclopedia of Nonlinear Science*, ed. by A. Scott (Routledge, New York, 2005), pp. 885–887
2. M.A. Uddin, N. Oshima, M. Tanahashi, T. Miyauchi, A study of the coherent structures in homogeneous isotropic turbulence. Proc. Pakistan Acad. Sci. **46**, 145–158 (2009)
3. A.I. Golov, P.M. Walmsley, Homogeneous turbulence in superfluid 4He in the low-temperature limit: experimental progress. J. Low Temp. Phys. **156**, 51–70 (2009)

4. A.W. Baggaley, C.F. Barenghi, A. Shukurov, Y.A. Sergeev, Coherent vortex structures in quantum turbulence. EPL **98**, 26002 (2012)
5. V.I. Arnold, B.A. Khesin, *Topological Methods in Hydrodynamics*. Applied Mathematical Sciences, vol. 125, (Springer, Berlin, 1998)
6. R.L. Ricca, (ed.), *An Introduction to the Geometry and Topology of Fluid Flows*. NATO ASI Series II, vol. 47 (Kluwer, Dordrecht, 2001)
7. R.L. Ricca (ed.), *Lectures on Topological Fluid Mechanics*. Springer-CIME Lecture Notes in Mathematics 1973 (Springer, Heidelberg, 2009)
8. J. Weickert, H. Hagen (eds.), *Visualization and Processing of Tensor Fields* (Springer, Heidelberg, 2006)
9. H. Hauser, H. Hagen, H. Theisel (eds.), *Topology-based Methods in Visualization* (Springer, Heidelberg, 2007)
10. H.K. Moffatt, K. Bajer, Y. Kimura (eds.), *Topological Fluid Dynamics: Theory and Applications* (Elsevier, 2013)
11. R.L. Ricca, Momenta of a vortex tangle by structural complexity analysis. Physica D **237**, 2223–2227 (2008)
12. X. Liu, R.L. Ricca, The Jones polynomial for fluid knots from helicity. J. Phys. A: Math. & Theor. **45**, 205501 (2012)
13. P.G. Saffman, *Vortex Dynamics* (Cambridge University Press, Cambridge, 1991)
14. C.F. Barenghi, R. Hänninen, M. Tsubota, Anomalous translational velocity of vortex ring with finite-amplitude Kelvin waves. Phys Rev. E **74**, 046303 (2006)
15. F. Maggioni, S.Z. Alamri, C.F. Barenghi, R.L. Ricca, Velocity, energy, and helicity of vortex knots and unknots. Phys. Rev. E **82**, 026309 (2010)
16. R.L. Ricca, New developments in topological fluid mechanics. Nuovo Cimento C **32**, 185–192 (2009)
17. T.T. Lim, T.B. Nickels, Instability and reconnection in head-on collision of two vortex rings. Nature **357**, 225–227 (1992)
18. L.H. Kauffman, *On Knots* (Princeton University Press, Princeton, 1987)
19. V.F.R. Jones, Hecke algebra representations of braid groups and link polynomials. Ann. Math. **126**, 335–388 (1987)
20. P. Freyd, D. Yetter, J. Hoste, W.B.R. Lickorish, K. Millett, A. Ocneanu, A new polynomial invariant of knots and links. Bull. Am. Math. Soc. **12**, 239–246 (1985)
21. J.H. Przytycki, P. Traczyk, Conwav algebras and skein equivalence of links. Proc. Amer. Math. Soc. **100**, 744–748 (1987)
22. L.H. Kauffman, *Knots and Physics* (World Scientific Publishing Co., Singapore, 1991)
23. H.K. Moffatt, The degree of knottedness of tangled vortex lines. J. Fluid Mech. **35**, 117–129 (1969)
24. H.K. Moffatt, R.L. Ricca, Helicity and the Călugăreanu invariant. Proc. R. Soc. A **439**, 411–429 (1992)
25. C.F. Barenghi, R.L. Ricca, D.C. Samuels, How tangled is a tangle? Physica D **157**, 197–206 (2001)
26. R.L. Ricca, On simple energy-complexity relations for filament tangles and networks. Complex Syst. **20**, 195–204 (2012)
27. R. Ricca, New energy and helicity lower bounds for knotted and braided magnetic fields. Geophys. Astrophys. Fluid Dyn. Online. doi:10.1080/03091929.2012.681782

Two Conceptual Models for Aspects of Complex Systems Behavior

Burton Voorhees

Abstract This chapter presents two toy models dealing with trade-off issues arising in complex systems research. After comparison of modeling in classical physics and complex systems theorizing these models are discussed in detail. The first examines the trade-off between stability and flexibility in an environment subject to random fluctuations. The second compares possible response strategies in cases of potential risk and reward. The first model illustrates the general complex systems concept of virtual stability, defined as a condition in which a system maintains itself in an unstable state between attracting response states in order to gain flexibility in the face of random environmental fluctuations. The second model considers the trade-off between quickness and accuracy in cases of bounded decision time and information. Both models relate to decision processes in complex adaptive systems and some of their implications in this regard are discussed.

1 The Value of Toy Models

Traditionally, systems chosen for scientific analysis are simple. Simple systems have either a small or very large number of interacting agents (or particles) that undergo local interactions determined by fixed rules. If there are a large number of elements in the system they are treated thermodynamically or statistically.

The way such systems are studied goes back to the Platonic method of division: a system is decomposed into elementary components ("carved at the joints"), properties of each component are analyzed, and the system is reconstructed taking account of the possible interactions between the components and their now known properties. A clock, for example, can be decomposed into component parts which are directly understood in terms of the way that each part contributes to the overall

B. Voorhees (✉)
Center for Science, Athabasca University, 1 University Drive, Athabasca, AB T9S 3A3, Canada
e-mail: burt@athabascau.ca

I. Zelinka et al. (eds.), *How Nature Works*, Emergence, Complexity and Computation 5, DOI: 10.1007/978-3-319-00254-5_6,
© Springer International Publishing Switzerland 2014

time-keeping function of the clock, understood as the result of a mechanical implementation of the analogy between the position of hands on the face of the clock and the position of the sun in the sky.[1]

Saying that such systems are simple, however, does not mean that their study is easy. The insolvability of the three-body problem in classical mechanics, and the powerful mathematical techniques required in studies of statistical equations such as the Boltzmann equation show that simple systems can be difficult indeed. Sophisticated perturbation methods have been devised in order to deal with systems that are conceptually simple but mathematically intractable.

In contrast, the invention of the digital computer has made possible a science of complex systems having an intermediate number of variables, conceptualized in terms of interactions between agents with learning and even self-referential capacities. Jackson [1] refers to this as "the second metamorphosis of science." When interactions between system elements do not follow simple mechanical rules, when they may depend on the global output of the system itself, then the value of the method of decomposition is strongly curtailed. "By their very nature complex systems resist analysis by decomposition. ...The very essence of the system lies in the interaction among all its parts, with the overall behavior of the system emerging from these interactions. So by throwing away the interactions, one also throws away any hope of actually understanding the workings of the system" [2].

For such analytically recalcitrant systems, toy models are often illuminating. Some might question the value of toy models, models so simple that they seem to blur matters by overly wide categorization. And there is a warning that does have to be given before embarking on construction of these sorts of models, best captured in the story of the wealthy horse owner who hired a pure mathematician and a theoretical physicist to come up with a system to predict winners in horse races. After a year the mathematician turned in results: the exact winner could not be predicted, but there is a proof that the winner is unique—up to equivalences. The physicist, on the other hand, provided a method to predict the winner for the case of spherical horses.

Moral: toy models are only useful if they have some practical or heuristic value.

The contention here is that toy models can play a role in complex system theorizing equivalent to the simple solvable models that are used in classical physics education. Kuhn [3] pointed out that a good part of the education of students in physics involves learning to solve simple models that in one way or another capture aspects of physical theory applicable to more general empirical and theoretical research. These simple models provide paradigmatic exemplars that anchor intuition; that can be recalled as patterns of approach when dealing with more complicated problems encountered in real practice; and which provide

[1] This ignores the fact that a clock imposes a fixed unit of time. The actual length of a day, hence the position of the sun, depends on season. Abstraction from reality always introduces a lack of fit with reality that must be compensated in practice. This is the difference between theoretical science and engineering.

basic models for perturbation analysis. In the same way, toy models can serve as exemplary cases for complex systems theorizing.

With complex systems, there is danger of getting lost in detail. We are creatures with bounded rationality, we don't carry out exhaustive searches of huge possibility spaces, or analyze in details so refined that their complexity exceeds our cognitive capacity. We don't play chess like Deep Blue. We do not, indeed cannot, make decisions through complete analytic evaluation of all possibilities and their consequences. In complex decision-making, human rationality is bounded. "[P]eople use a sequence of pattern recognition, hypothesis formation, deduction using currently-held hypotheses, and replacement of hypotheses as needed. This type of behavior... enables us to deal with complication: we construct plausible, simpler models we can cope with. It enables us to deal with ill-definedness: where we have insufficient definition, our working models fill the gap. It is not antithetical to 'reason,' or to science for that matter. In fact, it is the way science itself operates and progresses" [4].

Under conditions of bounded rationality, toy models provide points of reference, anchoring points, and test cases for understanding. Some of the uses of toy models in complex systems research are:

1. To give simple examples of general principles that would be obscured by the details of more complicated models.
2. Hypothesis testing, to explore implications of an idea in simple cases where falsification will be easy.
3. Exploration of the consequences of simple assumptions (even "fishing expeditions").
4. "Intuition pumps"—means of developing intuitive understanding and generating new ideas.

The sort of understanding given by a toy model is: if the assumptions of the model are accurate as general reflections of reality, then the results of the model ought to provide fits to the corresponding real world cases. Since the models are toy models, the fit can be qualitative rather than quantitative—pointing to directions of future development rather than definite answers. Toy models can represent important aspects of the Platonic ideal that our "messy" real world material systems emulate.

In what follows two toy models, designed to illustrate basic principles and issues in complex system theorizing, are presented and discussed. Both of these models deal with trade-off situations—between stability and flexibility in the first case and between quickness and accuracy in the second. Both of these trade-offs are found empirically, and both are significant for system behavior and survival.

2 A Virtual Stability Model

The concept of virtual stability was introduced in [5, 6]. It refers to the capacity of a complex adaptive system to utilize self-monitoring and adaptive control to maintain itself in a state that would otherwise be unstable. This requires a small but ongoing expenditure of energy. What is purchased by this expenditure is the ability to respond quickly to random environmental fluctuations.

2.1 Stability in Complex Systems Theory

Some degree of stability is perhaps the single most important requirement for a complex system. Because of this, the general theoretical assumption is that most of the time a complex system will be found in a stable, or at least metastable state, with only brief periods of transition between such states. Heylighen [7], for example, takes stability as axiomatic, formulated as a Principle of Selective Retention: "Stable configurations are retained, unstable ones are eliminated." May's [8] argument, that characteristic ecosystem parameters can take on only very specific values because only small regions in a system parameter space provide long-term stability, applies to many complex systems. But what sort of stability is involved is another question.

In accord with the emphasis on stability, complex systems are often thought of in terms of a state space partitioned into basins of attraction determined by system dynamics, with each basin representing a coarse-grained state of relative stability. State space trajectories lie on a fitness landscape and the instantaneous system state remains in a neighborhood of a fitness peak [9]. Applied specifically to models of evolution, a set of phenotypic parameters characterizes each species, and parameter values for any given species are expected to cluster in those relatively small regions satisfying constraints imposed by the selective fitness barriers defining the fitness peak. From a discrete point of view, state space trajectory are represented as a series of transitions on a finite, although perhaps large set of states, on which a hierarchical structure is defined [e.g., 10, 11]. This gives a course-grain version of the continuous representation, each basin of attraction corresponding to a distinct state in the discrete model.

The conceptual view of basins of attraction in a state space becomes more complicated when the possibility of metastability is added. Metastability is particularly associated to chaotic systems and systems in which attractor basins can be topologically complicated with highly interconnected boundaries [12]. Much current research exists studying such cases, with results obtained on average time to the boundary, optimal escape paths, and methods of control [13]. The question of control, in particular, raises an important point [14, 15].

State space trajectories of complex adaptive systems are not random. They represent adaptive responses to environmental contingencies, or (in systems with a cognitive component) goal directed action sequences. In either case, control mechanisms (which may utilize, but are not determined by random noise) determine the system trajectory. A number of researchers have shown how small pertubative control functions can provide overall control of chaotic dynamical systems [e.g., 16, 17], and there is an analogy in opponent process theory [18], where the balance between two opposing forces (e.g., two large muscle groups) is managed by a much weaker control function (e.g., small muscle controls) through inhibitory mechanisms.

This introduces another form of stability, called virtual stability. As already indicated, virtual stability involves self-monitoring and adaptive control mechanisms that are used to maintain a system in an otherwise unstable state. Such processes can occur well below the level of conscious awareness. An example is standing [19]. We learn the standing posture in early childhood and it becomes automated as an unstable state, maintained by a process of proprioceptive feedback and small muscular adjustments. The resulting flexibility shows up in the ease of walking. If standing were stable, the stability would act as an attractive force maintaining the state. Every step would require effort to overcome the stability and would feel as if one were walking uphill. As it is, taking a step is a controlled fall.

From an overall systems point of view, it could be said that virtual stability just is stability, so why introduce a new term. Taking that view, however, obscures the fundamental importance of control mechanisms that manage instability, as well as the potential value of instability for complex systems.

2.2 The Importance of Instability

Given the importance of stability, it might seem strange to suggest that instability is also important. Ashby's Law of Requisite Variety [20] states that the variety of responses available to a system must be at least as great as the variety of possible perturbations that the system can be expected to meet in its environment. In terms of management and control, however, it is not only a matter of maintaining sufficient variety in a set of possible responses, equally important is being able to switch between responses in a timely manner. This implies the existence of a trade-off between stability and flexibility. It is easy to change an unstable state, difficult to change a stable one: if different possible behavioral response states are stable, change consumes time and energy; unstable states are easy to change, but energy is required to maintain them for any length of time.

There are many instances in which behavioral choices must be made quickly in situations where there are minimal environmental cues. This is particularly so in fight-or-flight situations but the general principle, of keeping options open until choice is a necessity, is well established. If behavioral responses at the physiological, neural, and habitual levels are in relatively stable attractors, avoidance of commitment to an immediate response can be likened to remaining on an unstable boundary between the attracting behaviors. Maintaining such a state requires effort and so exacts a cost. The energy expended purchases increased behavioral flexibility and speed of response in the face of environmental uncertainty.

Recent studies of paradoxical games also indicate the potential advantage of behavioral flexibility. In exemplary cases, the ability to alternate between one of two losing gambling games can result in a net winning game [21, 22], even if the alternation is random. The general principle is that in many cases it is important to maintain the ability to switch quickly between differing strategies rather than fixating on a single strategy.

It is important to be clear that virtual stability is not the same as stability or metastability. Formally, a system is stable if there is a single global attractor, or if it is completely deterministic and once the system trajectory enters an attractor basin, it remains there. More generally, a system subject to noise is stable against perturbations that are sufficiently small and which do not resonate with system dynamics.[2] Such cases can be analyzed in terms of branching ratios and the expected lifetimes of course-grained states, determined by the frequency of disruptive perturbations and the strength of the fitness, entropy, or other barriers between states [25].

While such cases are often termed metastable, metastability technically refers to systems with multiple attractors, having intricately interwoven basins of attraction with fractal basin boundaries containing chaotic saddles, and with boundary dimensions that are close to the dimension of the full state space [26]. Under such conditions, even a small amount of noise in the system can cause transitions between attractor basins, or equivalently, transitions between course-grained states. The characteristic time scales involved are now the average time that the system trajectory is in an attractor basin and the average time that it is on basin boundaries.

Virtual stability, on the other hand, describes cases in which, through processes of self-monitoring and adaptive control, a system maintains itself on the boundary between two or more attractor basins. In other words, a meta-level control directs the expenditure of energy to maintain the system on an unstable trajectory, or in an unstable state. This expenditure purchases an increase in behavioral flexibility.

At a minimum, virtual stability requires that a system have the capacity to monitor its momentary state and produce responses at a frequency high enough that only small (i.e., inexpensive) corrective actions are required. Otherwise, the energetic cost will be too large. The advantage of virtual stability is that a small but ongoing energy expenditure allows a complex adaptive system to avoid large (if infrequent) expenditures that would be required if the system had to escape from a stable attractor. The high frequency of self-monitoring and adaptive control provides cheap and rapid episodic responses to environmental change.[3] An important point is that virtual stability allows response to random environmental fluctuations. Periodic fluctuations such as seasonal changes or night and day are adequately dealt with by genetic adaptation or habituation.

---

[2] If the spectrum of external noise resonates with system dynamics then even small fluctuations can grow to macroscopic size in a process of fluctuation enhancement [e.g., 23, 24].

[3] The technical questions that arise relate to how high this frequency needs to be. This, in turn, involves the time scale for falling out of the unstable state, and the energy required for corrective actions of varying strengths.

2.3 A Population Model

A toy model has been constructed to explore the trade-off between stability and flexibility [5]. A population of agents is divided into three behavioral sub-populations labeled A, B, and C. The environment for these agents can be in one of three states, labeled A, B, and N. Populations A and B, respectively, are favored in environmental states A and B. In the environment N, populations A and B are equally favored. Population C is not favored in any environment. Stability is modeled by transition probabilities between A and B types, and between the C type and the A or B types: it is difficult for members of the stable A and B populations to change type. The "virtually stable" C population can change to A or B types with relative ease.

The model is discrete and operates on two time scales, related by a constant m that determines how many iterations of the fast scale correspond to one slow scale iteration. A slow scale iterations begins with the choice of an environmental state, based on a specified probability distribution. This sets the probabilities in a state dependent transition matrix that specifies transition probabilities for each population over the subsequent series of m fast scale iterations. The environmental transition matrix has adjustable parameters with the mortality rates for each population type and the transition probabilities between population types as independent parameters.

At each fast scale iteration, called a "short-run," every population member has the possibility of changing type, remaining the same type, or dying. The complete process, carried out m times for every member of the population, is a "long run." The transition matrices associated to the environmental states are:

$$
T_A = \begin{bmatrix}
\rho_{AB} & \alpha_{AB}(1-\rho_{AB}-\varepsilon_A) & 0 & 0 & 0 & 0 \\
1-\rho_{AB}-\varepsilon_A & \alpha_{AB}\rho_{AB} & 0 & 0 & 0 & 0 \\
0 & 0 & \delta_A\rho_C & \alpha_C\delta_A(1-\rho_C-\varepsilon_{AC}) & \delta_A(1-\mu-\varepsilon_C) & 0 \\
0 & 0 & \delta_B(1-\rho_C-\varepsilon_{AC}) & \alpha_C\delta_B\rho_C & \delta_B(1-\mu-\varepsilon_C) & 0 \\
0 & 0 & q_B & \alpha_C q_A & \mu & 0 \\
\varepsilon_A & \varepsilon_B & \varepsilon_{AC} & \varepsilon_{BC} & \varepsilon_C & 1
\end{bmatrix}
$$

$$
T_B = \begin{bmatrix}
\alpha_{BA}\rho_{AB} & 1-\rho_{AB}-\varepsilon_B & 0 & 0 & 0 & 0 \\
\alpha_{BA}(1-\rho_{AB}-\varepsilon_B) & \rho_{AB} & 0 & 0 & 0 & 0 \\
0 & 0 & \frac{1}{\alpha_C}\delta_A\rho_C & \delta_A(1-\rho_C-\varepsilon_{BC}) & \delta_A(1-\mu-\varepsilon_C) & 0 \\
0 & 0 & \frac{1}{\alpha_C}\delta_B(1-\rho_C-\varepsilon_{AC}) & \delta_B\rho_C & \delta_B(1-\mu-\varepsilon_C) & 0 \\
0 & 0 & \frac{1}{\alpha_C}q'_B & q'_A & \mu & 0 \\
\varepsilon_A & \varepsilon_B & \varepsilon_{AC} & \varepsilon_{BC} & \varepsilon_C & 1
\end{bmatrix}
$$

$$
T_N = \begin{bmatrix}
\rho_{AB} & 1-\rho_{AB}-\varepsilon_A & 0 & 0 & 0 & 0 \\
1-\rho_{AB}-\varepsilon_A & \rho_{AB} & 0 & 0 & 0 & 0 \\
0 & 0 & \frac{1}{2}\rho_C & \frac{1}{2}(1-\rho_C-\varepsilon_{AC}) & \frac{1}{2}(1-\mu-\varepsilon_C) & 0 \\
0 & 0 & \frac{1}{2}(1-\rho_C-\varepsilon_{AC}) & \frac{1}{2}\rho_C & \frac{1}{2}(1-\mu-\varepsilon_C) & 0 \\
0 & 0 & \frac{1}{2}(1-\varepsilon_{AC}) & \frac{1}{2}(1-\varepsilon_{AC}) & \mu & 0 \\
\varepsilon & \varepsilon & \varepsilon_{AC} & \varepsilon_{AC} & \varepsilon_C & 1
\end{bmatrix}
\tag{1}
$$

Rows are labeled from top to bottom, and columns from left to right in the order A, B, AC, BC, C, and D. The state D represents death. From these matrices there can be transitions between the A and B populations but no A or B population member can make a transition to the C population. Introduction of the AC and BC populations is necessary to prevent loss of identity when a C type is imitating an A or B type.

Fundamental parameters involved are the mortality rates ε_A, ε_B, ε_{AC}, ε_{BC}, and ε_C; the transition probabilities ρ_{AB}, ρ_C, and μ; the number m of short runs per long run; and the exponent n, called directionality power. Equation (2) provides definitions of the remaining parameters.

$$\begin{aligned}
&\alpha_{AB} = \frac{1-\varepsilon_B}{1-\varepsilon_A}, \quad \alpha_{BA} = \frac{1-\varepsilon_A}{1-\varepsilon_B}, \quad \alpha_C = \frac{1-\varepsilon_{BC}}{1-\varepsilon_{AC}} \\
&\delta_A = \frac{\varepsilon_{BC}^n}{\varepsilon_{AC}^n + \varepsilon_{BC}^n}, \quad \delta_B = \frac{\varepsilon_{AC}^n}{\varepsilon_{AC}^n + \varepsilon_{BC}^n} \\
&q_A = (1-\varepsilon_{AC})(1-\delta_A) + \rho_C(\delta_A - \delta_B), \quad q_B = (1-\varepsilon_{AC})(1-\delta_B) - \rho_C(\delta_A - \delta_B) \\
&q'_A = (1-\varepsilon_{BC})(1-\delta_A) + \rho_C(\delta_A - \delta_B), \quad q'_B = (1-\varepsilon_{BC})(1-\delta_B) - \rho_C(\delta_A - \delta_B)
\end{aligned} \tag{2}$$

The probability ρ_{AB} is close to 1, modeling stability of the A and B types. The C type is relatively unstable, modeled by setting the probability ρ_C high (in our trials, equal to ρ_{AB}) and the probability μ low. An initial population is exposed to a long sequence of environments (e.g., "ABBNABABNNABBBNBA…") and population numbers of each type are tracked (the C population includes AC and BC subtypes). Parameters are varied to explore conditions under which the C population dominates.

In a long run, every individual member of the population goes through m short runs. At each iteration in a short run, the transition matrix for the given environmental state is used to determine probabilistically whether an agent changes population type, remains the same type, or dies. The population type corresponding to the preferred environmental state has the lowest mortality rate. Since the subpopulations labeled AC and BC are interpreted as members of population type C that are imitating types A and B respectively, the mortality rates ε_{AC} and ε_{BC} are set to ε_A and ε_B. At the end of a long run the resulting population distribution becomes the preliminary new population for the next long run.

Since the state D in the transition matrix is an attracting state, however, it is necessary to introduce a redistribution of those population members that have died off in order to avoid the eventual death of the entire population. To avoid this, a redistribution process is introduced: the total population is kept fixed by redistributing individuals that have died during a long run to the population types A, B, AC, BC, and C in proportion to the frequencies of these types in the preliminary new population. This gives the new initial population for the next long run.

The flexibility advantage of population C shows up in its ability to easily enter sub-populations AC and BC and from these to return to C.

For example, if state A is specified as preferred then on the fast time scale members of population A remain as they are with only a small chance of making a transition to B, and little cost is accrued. Members of population B, on the other hand, not only have a greater mortality rate, they will also accrue a high cost since their probability of making a transition to population A is small. Similarly, if B is the preferred state then members of population B accrue little cost but members of population A will accrue a high cost. In either case, members of population C accrue a relatively low cost. The goal of the model is to explore parameter values that distinguish between situations where the stable A and B populations dominate and those in which the "virtually stable" C population dominates.

To deal with redistribution, a new representation of the transition matrices can be introduced, in which the state D is excluded from the matrices and incorporated instead as a scalar multiplying factor. The total population is

$$N = N_A(t) + N_B(t) + N_{AC}(t) + N_{BC}(t) + N_C(t). \tag{3}$$

Let the pre-reincarnation populations at iteration t + 1 be $N_i^*(t + 1)$ for i = A, B, AC, BC, C. Then, with redistribution included,

$$N_i(t+1) = N_i^*(t+1) + \frac{N_D^*(t+1)}{N - N_D^*(t+1)} \sum_j T_{ij} N_j(t) \tag{4}$$

where

$$N_D^*(t+1) = \varepsilon_A[N_A(t) + N_{AC}(t)] + \varepsilon_B[N_B(t) + N_{BC}(t)] + \varepsilon_C N_C(t). \tag{5}$$

The summation at the end of Eq. (4) is just $N_i^*(t + 1)$, hence equation can be written as

$$N_i(t+1) = N_i^*(t+1) \left[\frac{N}{N - N_D^*(t+1)} \right] \tag{6}$$

Making use of Eq. (6) allows description of the process by the conformal Markov equations

$$\begin{aligned} \vec{N}(t+1) &= \left[\frac{N}{N - N_D^*(t+1)} \right] T\vec{N}(t) \\ \vec{N}(0) &= (N_A(0), N_B(0), 0, 0, N_C(0)) \end{aligned} \tag{7}$$

where the matrix T is the reduced transition matrix for environmental states A, B, or N obtained by omitting the final row and column from the matrices of Eq. (1).

Dividing both sides of the top Eq. in (7) by the total population N gives a frequency representation. If V(i) = (x,y,u,v,w) is the steady state frequency vector obtained for a preferred environmental state i, then Eq. (7) becomes

$$\begin{gathered} TV(i) = \left[1 - \frac{N * (V(i))}{N}\right] V(i) \\ N * (V(i)) = (\varepsilon_A, \varepsilon_B, \varepsilon_A, \varepsilon_B, \varepsilon_C) \cdot V(i) \end{gathered} \tag{8}$$

Two forms of test runs are used to determine cases in which the C population might be favored. The first set of tests varies specific parameter values against a background of standard parameter values. Table 1 shows the background parameter values used in these tests. In addition to the values given in Table 1, the standard values m = 16 and n = 1 are used. Variations of these standard settings produce different results, but in all cases the overall behavior remains the same.

Figures 1, 2, 3, 4 show examples of parameter test runs, with percentages of each population type plotted over time in barycentric coordinates. The frequency of the A population is 1 at the top of the triangle, it is 1 for the B population at the left vertex, and 1 for the C population at the right vertex.

Figures 1, 2 show the effect of varying the number of iterations in a long run (Fig. 1) and the directionality power (Fig. 2). The parameter n is important since it has a possible interpretation as a control parameter. The significance of this will appear when the environmental tests are discussed—if n is large enough it turns out that the C population can dominate even in environments that always favor the A or B populations.

Figures 3, 4 show the effect of varying the stability parameters. In Fig. 3 changes in μ correspond to changes in the behavioral flexibility of the C population while changes in the stability of the A and B populations involve changing the probability ρ_{AB} as illustrated in Fig. 4. The threefold structure appearing in Fig. 4 arises because at the parameter value used almost all population members are able to make a transition to the favored state in each long run.

In all cases in the parameter tests, the general pattern remains the same, there is a threshold value at which the system changes between A/B domination and C domination. For the value of m (given the parameters of Table 1) this occurs at m = 11 where Fig. 1 indicates that the system allows coexistence of all populations. For directionality power, it is at n = 10.55 where, again, all populations can coexist.

Table 2 gives threshold values found for the parameter test runs, holding other parameters fixed at the values given in Table 1.

The second set of test results, called environmental tests, are also shown in barycentric coordinates. In this case the figures show percentages of the A, B, and N environments in the environmental sequence. The percentage of the N

Table 1 Standard parameter values

Env.	ε_A	ε_B	ε_{AC}	ε_{BC}	ε_C	μ	ρ_{AB}	ρ_C
A	0.01	0.0125	0.01	0.0125	0.015	0.25	0.965	0.965
B	0.0125	0.01	0.0125	0.01	0.015	0.25	0.965	0.965
N	0.012	0.012	0.012	0.015	0.25	0.25	0.965	0.965

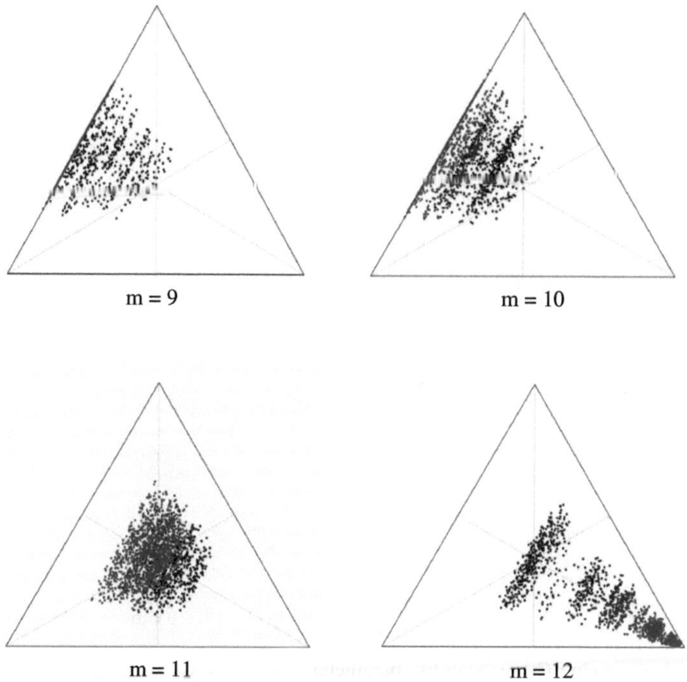

Fig. 1 Effect of varying number of short runs (2000 long runs: m = 9, 10 converges to A/B, m = 12 converges to C)

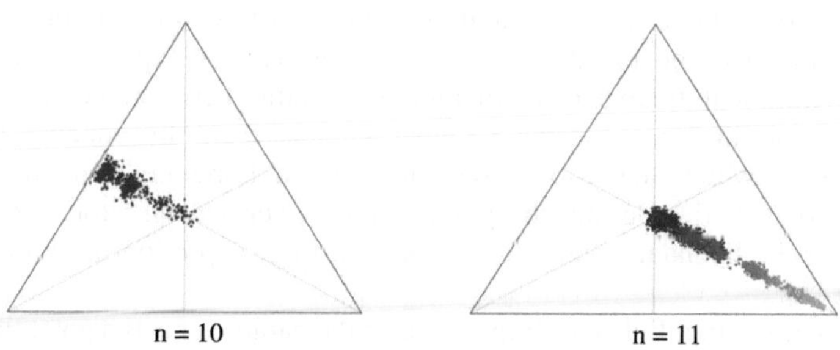

Fig. 2 Effect of varying directionality power n

environment is 1 at the top of the triangle, the percentage of the A environment is 1 at the left vertex, and the percentage of the B environment is 1 at the right vertex.

In the parameter tests, environmental frequencies were set at 40 % each for A and B and 20 % for N. The environmental test program runs the simulation for up to 5150 different sets of environmental percentages, determining for each case whether the simulation converges to A/B or to C, or does not consistently converge to either. Points that do not consistently converge in either direction are plotted,

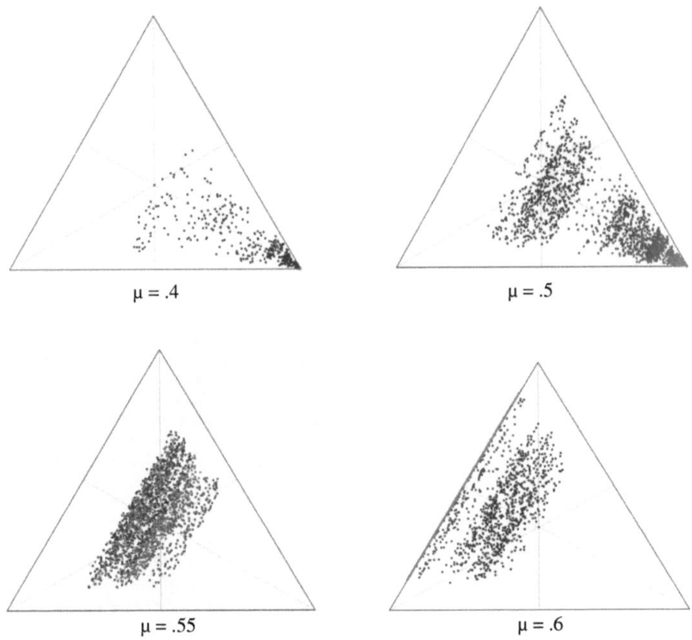

Fig. 3 Effect of varying C-type stability parameter

yielding figures as indicated in Fig. 5. In all environment test conducted, the A and B population types are treated symmetrically.

The most obvious result is that a high frequency of neutral environments leads to A/B dominance and the death of the C type. This is expected since, in the neutral environment, there is no distinction in mortality between the A and B types while the mortality rate for the C type remains high. Introducing Cartesian coordinates in which the x-axis coincides with a horizontal line through the barycenter of the triangle and the y-axis with its vertical bisector, the expected equation for the boundary region can be shown to be a portion of an ellipse or a hyperbola [5].

Test 2 in Fig. 5 highlights the importance of the parameter n as a potential control parameter. Here, so long as the frequency of the neutral environment is low the C type always wins, even if the environment always favors A or always favors B.

The explanation for this apparent paradox lies in the effect of the high directionality power ($n = 30$) assumed in this test. This leads to a situation in which, so long as the percentage of neutral states is not large, a C type agent mimicking the preferred type has a greater overall chance of remaining that type, or of returning to it should it make a transition away from it, than does an agent of the preferred type. For example, assuming that the environmental sequence is entirely A, a simple calculation using the transition matrix T_A with standard values other than n, indicates that the probability an agent beginning as an A type will again be an A

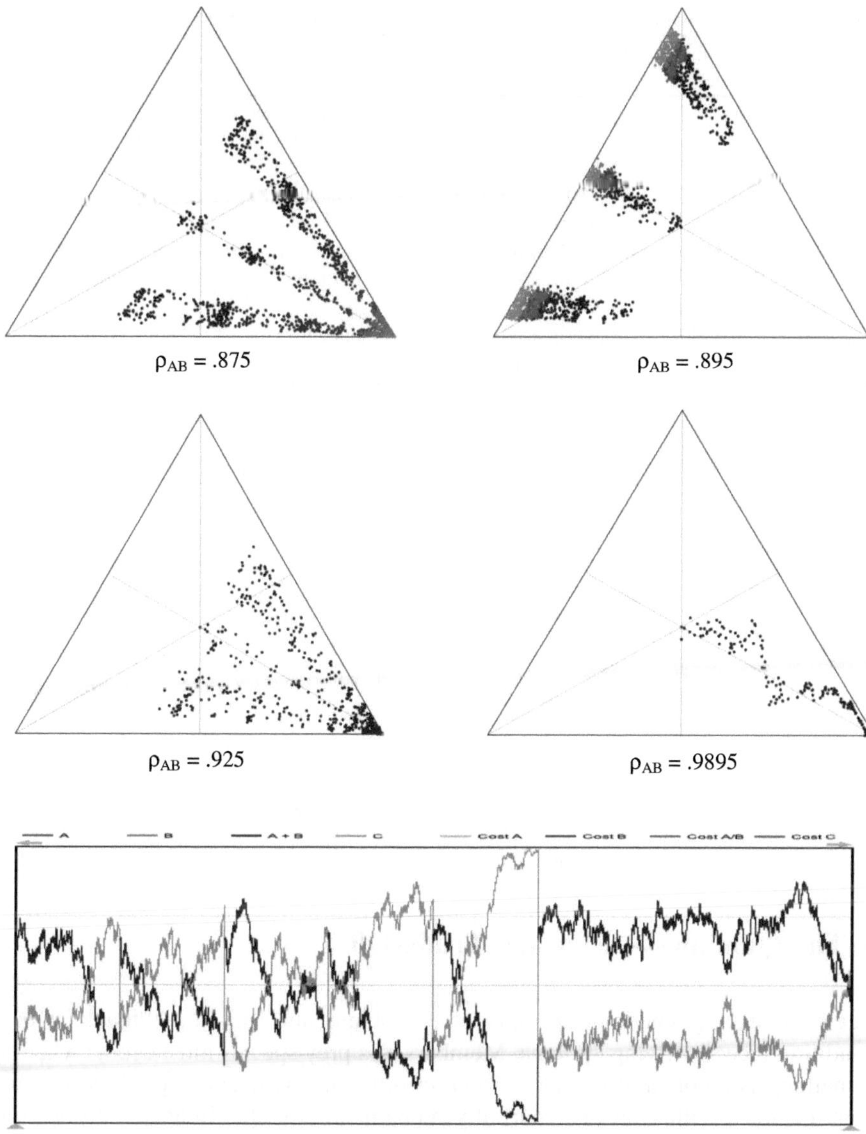

Eight Consecutive Trials (2000 long runs each) at $\rho_{AB} = .89362$

Fig. 4 Effect of varying the A/B stability parameter

type after three iterations is 0.9004, while the probability that an individual beginning as an AC subtype will again be an AC subtype after three iterations is 0.9371. In contrast, the same probability for the A type is still 0.9004 fo the first figure in Fig. 3 (n = 1.5), while the probability for the AC type is only 0.5190.

Table 2 Threshold values for parameter tests against standard background

Parameter varied	Threshold value	Notes
m	11	m ≤ 10 converges to A/B dominance, for m ≥ 12 convergence is to C dominance. m = 11 does not reliably converge to either A/B or C.
n	10.55	n ≤ 10 converges to A/B dominance, n ≥ 11 converges to C dominance.
μ	0.55	$\mu \leq .55$ converges to C dominance, $\mu \geq .55$ converges to A/B dominance. $\mu = .55$ does not reliably converge to either A/B or to C.
ρ_{AB}	0.89362	Convergence is to A/B dominance for smaller values, to C dominance for larger values.

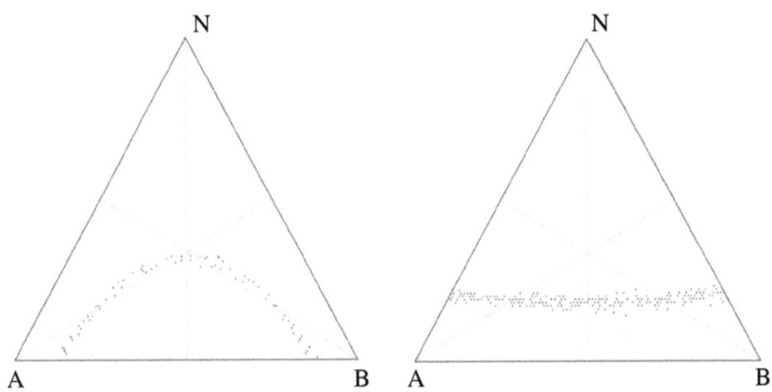

Fig. 5 Examples of environmental test cases

3 The Quickness–Accuracy Trade-Off

Decision making under uncertainty is a problem not only for humans, it is a condition faced by living systems. Members of a prey species must decide whether something is a potential source of food, or a threat. Members of predator species must decide whether an only partially recognized stimulus indicates a potential meal and, if so, whether it is worth the requisite expenditure of effort.

The model presented here considers agents acting in an environment in which threats and opportunities for resource acquisition arise at random. Further, these events may be disguised so that a threat appears as an opportunity and vice versa. In predator-prey terms, a predator may appear as such, or may be disguised as a prey and a prey may appear as a prey or be disguised as a predator. Thus, an agent faces four possible cases denoted DD (danger/danger), SD (safe/danger), DS (danger/safe), and SS (safe/safe). The first designation refers to appearance and the second to actuality.

Four possible behavioral strategies are given for response in such situations. The four strategies considered are:

1. HT (Hair Trigger): Immediately avoid apparent danger and immediately pursue apparent safety. In other words, this strategy reacts to immediate appearances.
2. SF (Safety First): Immediately avoid apparent danger and pause briefly to test apparent safety.
3. CS (Cautious Skeptic): Pause to test both apparent danger and apparent safety.
4. CO (Cautious Optimist): Pause to test apparent danger but immediately pursue apparent safety.

In the most basic terms, an agent who succumbs to danger has a high probability of becoming a meal while an agent who pursues actual safety has a high probability of obtaining a meal.

To quantify results, two functions are defined:

$$\begin{aligned} f_D(t) &= \frac{1}{1 - e^{a_D t} + e^{(b_D - c_D t^2)/t}} \\ f_S(t) &= 1 - \frac{1}{1 - e^{a_S t} + e^{(b_S - c_S t^2)/t}} \end{aligned} \tag{9}$$

Typical graphs of f_D and f_S are shown in Fig. 6.

$f_D(t)$ is the probability of being eaten in a situation of actual danger and $f_S(t)$ is the probability of obtaining a meal in a situation of actual safety. The form of these functions is not particularly important so long as this probability condition is met (normalized logistic functions could be used, for example). The probabilities of being eaten or of obtaining a meal are defined in terms of these functions according to Table 3.

Here t_0 is the immediate reaction time and $t_1 > t_0$ is the time required for testing the immediate appearance. From Eq. (9) this leads to the following inequalities:

$$\begin{aligned} f_D(t_0) &< f_D(t_1) \\ f_S(t_0) &> f_S(t_1) \end{aligned} \tag{10}$$

Taking extra time to test appearances increases the probability of being eaten in cases of actual danger and decreases the probability of obtaining a meal in cases of actual safety. This is counter-balanced by the loss of opportunity arising from avoidance of apparent danger when this appearance is deceptive, and the high probability of being eaten if apparent safety is pursued when this appearance is deceptive.

If (q_1,q_2,q_3,q_4) is a probability vector for the respective environmental situations DD, SD, DS, and SS while (p_1,p_2,p_3,p_4) is a strategy vector for choice of the strategies HT, SF, CS, and CO then the probabilities P_E of being eaten and P_M of having a meal are given by $P_E = \vec{p} \cdot E \cdot \vec{q}$, $P_E = \vec{p} \cdot M \cdot \vec{q}$ where

Fig. 6 Sample probability functions f_D and f_S

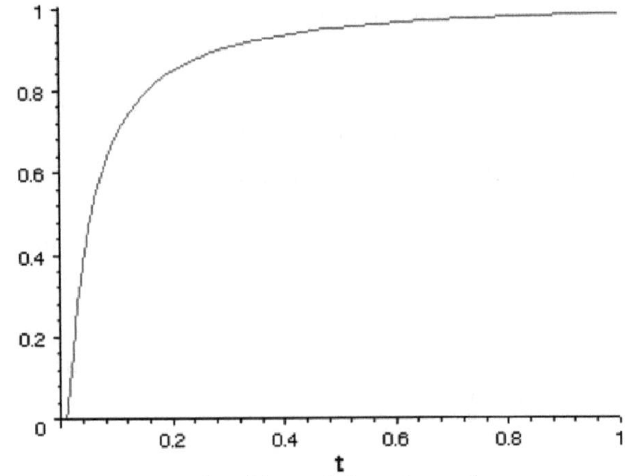

f_D with a = 1, b = .4, c = 1

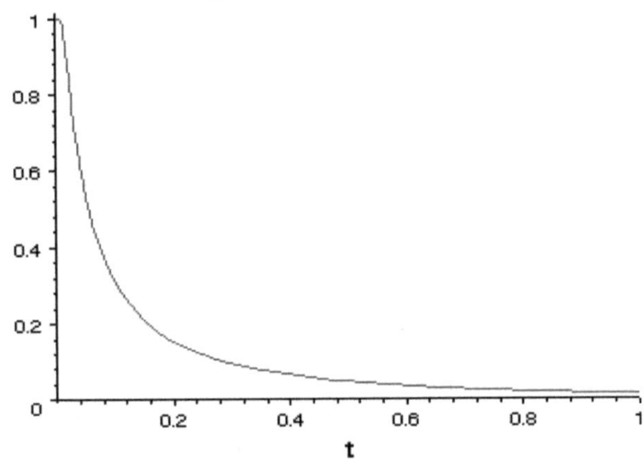

f_S with a = 1, b = .4, c = 1

$$E = \begin{pmatrix} f_D(t_0) & 1 - f_D(t_0) & 0 & 0 \\ f_D(t_0) & f_D(t_1) & 0 & 0 \\ f_D(t_1) & f_D(t_1) & 0 & 0 \\ f_D(t_1) & 1 - f_D(t_0) & 0 & 0 \end{pmatrix}, \quad M = \begin{pmatrix} 0 & 0 & 0 & f_S(t_0) \\ 0 & 0 & 0 & f_S(t_1) \\ 0 & 0 & f_S(t_1) & f_S(t_1) \\ 0 & 0 & f_S(t_1) & f_S(t_0) \end{pmatrix} \tag{11}$$

Hence

$$\begin{aligned} P_E = {} & p_1\{f_D(t_0)q_1 + [1 - f_D(t_0)]q_2\} + p_2[f_D(t_0)q_1 + f_D(t_0)q_2] + \dots \\ & + p_3 f_D(t_1)(q_1 + q_2) + q_4\{f_D(t_1)q_1 + [1 - f_D(t_0)]q_2\} \\ P_M = {} & [p_1 f_S(t_0) + p_2 f_S(t_1)]q_4 + p_3 f_S(t_1)(q_3 + q_4) + p_4[f_S(t_1)q_3 + f_S(t_0)q_4] \end{aligned} \tag{12}$$

Table 3 Probabilities of being eaten (f_D) or of obtaining a meal (f_S) for each possible strategy in each environmental situation

Strategy/situation	DD	SD	DS	SS
HT	$f_D(t_0)$	$1-f_D(t_0)$	0	$f_S(t_0)$
SF	$f_D(t_0)$	$f_D(t_1)$	0	$f_S(t_1)$
CS	$f_D(t_1)$	$f_D(t_1)$	$f_S(t_1)$	$f_S(t_1)$
CO	$f_D(t_1)$	$1-f_D(t_0)$	$f_S(t_1)$	$f_S(t_0)$

The function $R = P_M - P_E$ measures the advantage or disadvantage arising from choice of a particular mixed strategy given the environmental distribution of situations described by the vector q.

Since both p and q are probability vectors, the sum of their components is 1 and values of these components are represented in barycentric coordinates by points in the 3-simplices p-Δ_3 and q-Δ_3. Further, if q is fixed then P_E, P_M, and R are linear functions of p_1, p_2, p_3, and p_4 hence the maximum and minimum values will occur at faces, edges, or vertices of p-Δ_3.

The k-th vertex of this simplex is characterized by $p_k = 1$, $p_j = 0$, $j \neq k$. Vertex values of these functions are given in Table 4.

Comparisons of values from this table leads to the following conclusions:

1. An agent utilizing the p_4 (CO) strategy is the most likely to be eaten and also the most likely to obtain a meal.
2. An agent utilizing the p_2 (SF) strategy is the least likely to be eaten and the least likely to obtain a meal.
3. In most cases, ranking strategies from most to least likely to be eaten gives the order p_4, p_1, p_3, p_2 and ranking them from most to least likely to obtain a meal gives the order p_4, p_3, p_1, p_2.

To be more specific, the following strict inequalities hold:

$$\begin{aligned}
P_E(p_4 = 1) > P_E(p_1 = 1) > P_E(p_2 = 1) \\
P_E(p_4 = 1) > P_E(p_3 = 1) > P_E(p_2 = 1) \\
P_M(p_4 = 1) > P_M(p_1 = 1) > P_M(p_2 = 1) \\
P_M(p_4 = 1) > P_M(p_3 = 1) > P_M(p_2 = 1)
\end{aligned} \tag{13}$$

The relation between strategies p_1 and p_3 is more complex:

Table 4 Values of P_M, P_E, and R for pure strategies

=1	P_M	P_E	R
p_1	$f_S(t_0)q_4$	$f_D(t_0)q_1 + [1 - f_D(t_0)]q_2$	$f_S(t_0)q_4 - f_D(t_0)q_1 - [1 - f_D(t_0)]q_2$
p_2	$f_S(t_1)]q_4$	$f_D(t_0)q_1 + f_D(t_0)q_2$	$f_S(t_1)q_4 - f_D(t_0)q_1 - f_D(t_0)q_2$
p_3	$f_S(t_1)(q_3 + q_4)$	$f_D(t_1)(q_1 + q_2)$	$f_S(t_1)(q_3 + q_4) - f_D(t_1)(q_1 + q_2)$
p_4	$f_S(t_1)q_3 + f_S(t_0)q_4$	$f_D(t_1)q_1 + [1 - f_D(t_0)]q_2$	$f_S(t_1)q_3 + f_S(t_0)q_4 - f_D(t_1)q_1 - [1 - f_D(t_0)]q_2$

$$P_E(p_1=1)\ is \begin{cases} > P_E(p_3=1)\ if\ \dfrac{1-f_D(t_0)-f_D(t_1)}{f_D(t_1)-f_D(t_0)} > \dfrac{q_1}{q_2} \\ \le P_E(p_3=1)\ otherwise \end{cases}$$
$$P_M(p_1=1)\ is \begin{cases} < P_E(p_3=1)\ if\ \dfrac{f_S(t_0)-f_S(t_1)}{f_S(t_1)} < \dfrac{q_3}{q_4} \\ \ge P_M(p_3=1)\ otherwise \end{cases} \tag{14}$$

Since $1-f_D(t_0)-f_D(t_1)$will be of the order of $1-2\varepsilon$ while $f_D(t_1)-f_D(t_0)$ is of order ε and likewise $f_S(t_0)-f_S(t_1)$ will be of order ε and $f_S(t_1)$ will be of order $1-2\varepsilon$ ($\varepsilon \ll 1$), Eqs. (13) and (14) indicate that $P_E(p_1 = 1) < P_E(P_3 = 1)$ requires $q_1 \gg q_2$ while $P_M(p_1 = 1) > P_M(p_3 = 1)$ requires $q_4 \gg q_3$. That is, if there is little or no deception, so that appearances give a good reflection of actuality, then the HT strategy provides a better return than the CS strategy. On the other hand, if appearances are almost always deceptive then the p_3 strategy is favored over the other three strategies.

Table 5 gives conditions under which R will be a maximum at each vertex of the strategy representation simplex p-Δ_3:

This partitions points of q-Δ_3 into regions with boundaries determined by the functions $f_D(t)$ and $f_S(t)$. Conditions for maximum values of R to occur on p-simplex edges can also be determined from Table 3. These are shown in Table 6.

Note that $R(p_1 = 1) = R(p_3 = 1)$ if and only if $R(p_2 = 1) = R(p_4 = 1)$, which implies that if R is constant on any face of the p-simplex then it is constant in the interior as well, as it will also be if $R(p_1 = 1) = R(p_3 = 1)$ or $R(p_2 = 1) = R(p_4 = 1)$. The pairs (p_1,p_3) and (p_2,p_4) are composed of opposite strategies in the sense that while p_1 reacts immediately to danger and safety, p_3 pauses to test both; and while p_2 reacts immediately to danger but tests safety, p_4 tests danger but reacts immediately to safety. This general model can be treated from several different directions, described below:

3.1 A Discrete Version of the Model

For a discrete model a sequence of strategic decision are required in discrete time. Agents are presented with a string e(L),e(L−1),…,e(1), e(0) where e(t) represents the presented appearance at time t.

This string is presented to the population from right to left, one state at a time and within that presented state the entire population responds according to the mixture of strategies in the population. The numbers of individuals playing each strategy are counted as well as the number who have died (eaten or starved). Two assumptions are added: (a) If the presented state is DD or SD then individual agents who are not eaten survive without a meal; (b) If the presented state is DS or

Table 5 Conditions for R to be maximum for specified strategy

R max at	Conditions	Strong implications
$p_1 = 1$	$q_4 > \left[\frac{1-f_D(t_0)-f_D(t_1)}{f_S(t_0)-f_S(t_1)}\right] q_2, \quad q_3 < \left[\frac{f_D(t_0)-f_D(t_1)}{f_S(t_1)}\right] q_1$	$q_4 \gg q_2$ $q_1 \gg q_3$
$p_2 = 1$	$q_4 < \left[\frac{1-f_D(t_0)-f_D(t_1)}{f_S(t_0)-f_S(t_1)}\right] q_2, \quad q_3 < \left[\frac{f_D(t_0)-f_D(t_1)}{f_S(t_1)}\right] q_1$	$q_1 \gg q_3$
$p_3 = 1$	$q_4 < \left[\frac{1-f_D(t_0)-f_D(t_1)}{f_S(t_0)-f_S(t_1)}\right] q_2, \quad q_3 > \left[\frac{f_D(t_0)-f_D(t_1)}{f_S(t_1)}\right] q_1$	
$p_4 = 1$	$q_4 > \left[\frac{1-f_D(t_0)-f_D(t_1)}{f_S(t_0)-f_S(t_1)}\right] q_2, \quad q_3 > \left[\frac{f_D(t_0)-f_D(t_1)}{f_S(t_1)}\right] q_1$	$q_4 \gg q_2$

Table 6 Conditions for R to be maximum on p-simplex edges

Edge	Conditions for R maximum on edge
(p_1,p_2)	$q_4 = \left[\frac{1-f_D(t_0)-f_D(t_1)}{f_S(t_0)-f_S(t_1)}\right] q_2$
(p_1,p_3)	$q_4 = \left[\frac{1-f_D(t_0)-f_D(t_1)}{f_S(t_0)-f_S(t_1)}\right] q_2, \quad q_3 = \left[\frac{f_D(t_1)-f_D(t_0)}{f_S(t_1)}\right] q_1$
(p_1,p_4)	$q_3 = \left[\frac{f_D(t_1)-f_D(t_0)}{f_S(t_1)}\right] q_1$
(p_2,p_3)	$q_3 = \left[\frac{f_D(t_0)-f_D(t_1)}{f_S(t_1)}\right] q_1$
(p_2,p_4)	$q_4 = \left[\frac{1-f_D(t_0)-f_D(t_1)}{f_S(t_0)-f_S(t_1)}\right] q_2, \quad q_3 = \left[\frac{f_D(t_1)-f_D(t_0)}{f_S(t_1)}\right] q_1$
(p_3,p_4)	$q_4 = \left[\frac{1-f_D(t_0)-f_D(t_1)}{f_S(t_0)-f_S(t_1)}\right] q_2$

SS then no agents are eaten but agents who do not get a meal do not survive. With these assumptions the population evolution equation is:

$$n_i(t+1) = \left[\frac{\rho(i, e(t))N}{\sum\limits_j \rho(j, e(t))n_j(t)}\right] n_i(t) \tag{15}$$

where $N = n_1 + n_2 + n_3 + n_4$ with n_1 the number of agents playing strategy HT, n_2 the number of agents playing strategy SF, n_3 the number playing strategy CS, and n_4 the number playing strategy CO. N is to be kept constant as Eq. (15) is iterated by redistributing the number of dead in a round in proportional to the numbers of surviving agents in each of the four strategy subpopulations, before starting the next round. This equation can also be written in terms of frequencies: $x_i(t) = n_i(t)/N$ as:

$$x_i(t+1) = \left[\frac{\rho(i, e(t))}{\sum\limits_j \rho(j, e(t))x_j(t)}\right] x_i(t) \tag{16}$$

The coefficient ρ(i,e(t)) is defined by

$$\rho(i,e(t)) = \begin{cases} 1 - P_E(i,e(t)) & e(t) = DD \text{ or } SD \\ P_M(i,e(t) & e(t) = DS \text{ or } SS \end{cases} \tag{17}$$

Working with the frequency representation, iteration of Eq. (16) yields

$$x_i(t+1) = \left[\frac{\prod_{s=0}^{t} \rho(i,e(s))}{\sum_j \left[\prod_{s=0}^{t} \rho(j,e(t)) \right] x_j(t)} \right] x_i(0). \tag{18}$$

Writing this out in terms of the probabilities in Table 3 gives

$$\begin{aligned} x_1(t) &= \left[\frac{f_D^{k_1}(t_0)[1 - f_D(t_0)]^{k_2} f_S^{k_4}(t_0)}{W(t)} \right] x_1(0) \\ x_2(t) &= \left[\frac{f_D^{k_1}(t_0) f_D^{k_2}(t_1) f_S^{k_4}(t_1) x_2(0)}{W(t)} \right] x_2(0) \\ x_3(t) &= \left[\frac{f_D^{k_1+k_2}(t_1) f_S^{k_3+k_4}(t_1)}{W(t)} \right] x_3(0) \\ x_4(t) &= \left[\frac{f_D^{k_1}(t_1)[1 - f_D(t_1)]^{k_2} f_S^{k_3}(t_1) f_S^{k_4}(t_0)}{W(t)} \right] x_4(0) \end{aligned} \tag{19}$$

where k_1, k_2, k_3, and k_4 are the respective numbers of the states DD, SD, DS, and SS in the choice sequence and the denominator W(t) is given by

$$\begin{aligned} \sum_j \left[\prod_{s=0}^{t} \rho(j,e(t)) \right] x_j(t) \equiv\ W(t) &= f_D^{k_1}(t_0)[1 - f_D(t_0)]^{k_2} f_S^{k_4}(t_0) x_1(0) \\ &+ f_D^{k_1}(t_0) f_D^{k_2}(t_1) f_S^{k_4}(t_1) x_2(0) + f_D^{k_1+k_2}(t_1) f_S^{k_3+k_4}(t_1) x_3(0) \\ &+ f_D^{k_1}(t_0)[1 - f_D(t_0)]^{k_2} f_S^{k_4}(t_0) x_1(0). \end{aligned} \tag{20}$$

3.2 A Population Dynamics Version of the Model

Let n_s, s = 1,…,4 be the number of individual agents using strategy s out of a total population N = $n_1 + n_2 + n_3 + n_4$. Further, assume that only those agents who have had a meal can reproduce, and that members of population s reproduce at a rate a_s. Then the evolutionary dynamics of the total population is modeled by the equation

$$\frac{dn_s}{dt} = a_s P_M(s)[1 - P_E(s)]n_s \tag{21}$$

where $P_M(s)$ and $P_E(s)$ are the probabilities of obtaining a meal, or of being eaten for strategy p_s. What is of interest are the relative proportions of each strategy in the total population: $x_s = n_s/N$.

$$\frac{dx_s}{dt} = \frac{1}{N}\frac{dn_s}{dt} - \frac{n_s}{N^2}\frac{dN}{dt} \tag{22}$$

Substituting from Eq. (22) and the definition of x_s yields

$$\frac{dx_s}{dt} = a_s P_M(s)[1 - P_E(s)]x_s - x_s \sum_{k=1}^{4} a_k P_M(k)[1 - P_E(k)]x_k \tag{23}$$

which has the form of a simple replicator equation [27]. Writing a rescaled growth rate for each strategy group as $r_s = a_s P_M(s)[1 - P_E(s)]$ puts Eq. (23) into the form

$$\frac{dx_s}{dt} = r_s x_s - x_s \phi, \quad \phi = \sum_{k=1}^{4} r_k x_k \tag{24}$$

The relatively simple form of this equation allows direct solution. Consider ϕ as a function of time and define

$$\psi(t) = \exp\left[\int_0^t \phi ds\right] \tag{25}$$

With this definition, Eq. (24) becomes

$$\frac{dx_s}{dt} = x_s\left(r_s - \frac{d\psi}{dt}\right) \tag{26}$$

with solution

$$x_s(t) = x_s(0)e^{r_s t - \psi(t)} \tag{27}$$

Use of Eqs. (24) and (25) together with Eq. (27) yields

$$\frac{d\psi}{dt} = \sum_{k=1}^{4} r_k x_k(0)e^{r_k t - \psi(t)} = e^{-\psi(t)} \sum_{k=1}^{4} r_k x_k(0)e^{r_k t} \tag{28}$$

The solution of this equation is

$$\psi(t) = \ln\left[\sum_{k=1}^{4} x_k(0)e^{r_k t}\right] \tag{29}$$

Substituting from Eq. (29) into Eq. (27) gives the solution

$$x_s(t) = \frac{x_s(0)}{\sum_{k=1}^{4} x_k(0)e^{(r_k - r_s)t}} \tag{30}$$

By Eq. (30), if for any $k \neq s$ it is true that $r_k > r_s$ then $x_s(t)$ goes exponentially to zero while if $r_s > r_k$ for all $k \neq s$ then $x_s(t)$ goes to 1. If any mixed strategy is to persist the corresponding rescaled growth rates must be equal, and this places conditions on the growth rates a_s or on the functions f_D and f_S. In particular, if strategies s and k are to coexist then the condition $r_k = r_s$ yields the constraint on growth rates

$$a_k = \frac{P_M(s)[1 - P_E(s)]}{P_M(k)[1 - P_E(k)]} a_s \tag{31}$$

If all four strategies are to persist, this requires that

$$\begin{aligned} a_1 &= \left[1 + \frac{f_S(t_1)q_3}{f_S(t_0)q_4}\right] a_4 \\ a_2 &= \left[\frac{q_3}{q_4} + \frac{f_S(t_0)}{f_S(t_1)}\right] \left\{\frac{1 - f_D(t_1)q_1 - [1 - f_D(t_0)q_2]}{1 - f_D(t_0)q_1 - f_D(t_1)q_2}\right\} a_4 \\ a_3 &= \frac{[f_S(t_1)q_3 + f_S(t_0)q_4]\{1 - f_D(t_1)q_1 - [1 - f_D(t_0)q_2]\}}{f_D(t_1)f_S(t_1)(q_1 + q_2)(q_3 + q_4)} a_4 \end{aligned} \tag{32}$$

Results such as this might arise from an evolutionary process but Eq. (31) can also be satisfied if the probabilities P_M and P_E are allowed to depend on the choice of strategy. If this is the case, the form of the functions f_D and f_S becomes significant beyond their use to model qualitative probabilities.

3.3 Simulation of the Model

A simulation program was developed to model a population of agents partitioned into sub-populations playing fixed strategies. Initial results show that, as anticipated from the population dynamics model, a single strategy wins out. The sequences in this simulation were generated at random with only probabilities of each environmental state specified. Evolutionary paths are represented graphically, or traced out in the p-simplex, as shown in Figs. 7, 8.

The initial point for each trajectory shown in Fig. 7 is chosen near the center of the p-simplex and the environmental sequences are chosen at random. This figure

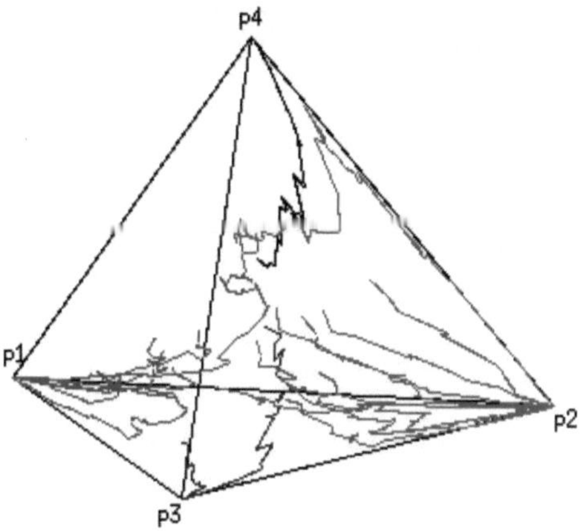

Fig. 7 Evolutionary trajectories: *Red* = HT, *Green* = SF, *Blue* = CS, *Purple* = CO

indicates that convergence to fixation at each of the possible strategies can occur. Examination of Fig. 7 indicates a general pattern of the approach to fixation with a trajectory often first going to a face, then an edge of the p-simplex before finally fixating at a vertex. This corresponds to the dying off of species in the simulation, with the loss of a species corresponding to the loss of a dimension in the simplex. Figure 8a–d shows cases in which each strategy goes to fixation as well as the sequential dying off of losing species.

Because the choice of starting point is the same in all cases, the behavior shown of convergence to a single fixed strategy would seem to depend on contingent aspects of the randomly generated sequence of environmental situations. The general implication arising from this is that each of the posited response strategies can come to dominate, depending on the particular environmental situations that are faced. Given that all of the listed environmental situations can arise, the evolutionary of different species can be expected to follow a path that relates behavioral strategies to other, biological characteristics of the organisms that adapt them to a particular strategy.

Preliminary studies seem to indicate that the q-simplex is partitioned into non-overlapping regions such that for any situation sequence with probabilities drawn from a given region the same p-strategy will win out.

There also appears to be a relation between foraging strategies and species growth rates. The population version of the model, for example, indicates that coexistence of strategies is possible only when very specific relations hold among reproductive rates. While this is likely an artifact of the simplicity of the model, it does raise the question of possible relationships between a species birth rate and it's foraging behavior. Do species where members produce few offspring exhibit different quickness-accuracy behavior than those in which members produce large numbers of offspring? Also, are there other biological characteristics of a species that are directly adapted to its behavioral characteristics?

Figure 9 shows two typical cases of the evolution of P_M, P_E. and R. The patterns in Fig. 9 both show an approach to a constant value but in the first graph P_E remains

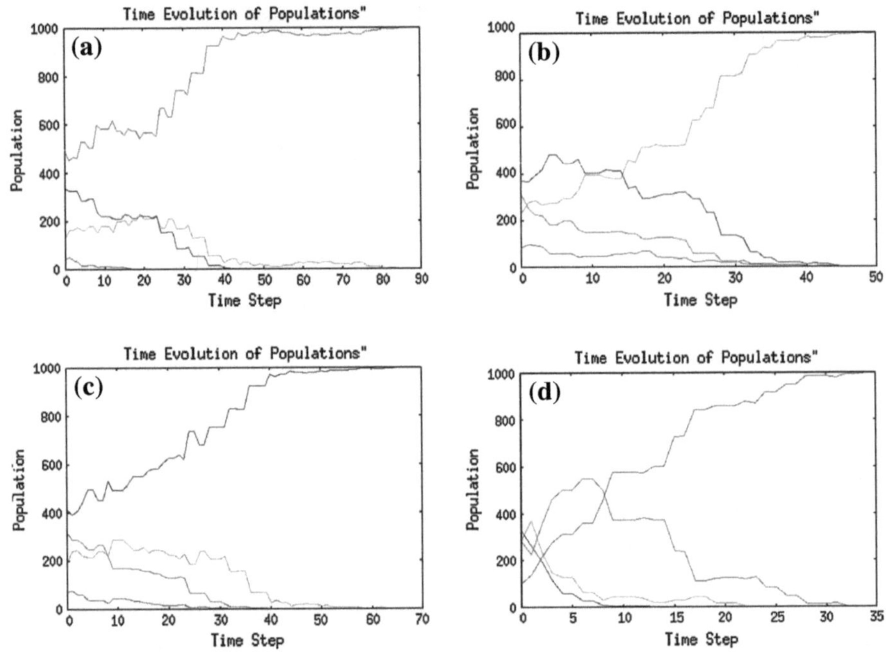

Fig. 8 CO Wins (**a**), SF Wins (**b**), CS Wins (**c**), HT Wins (**d**)

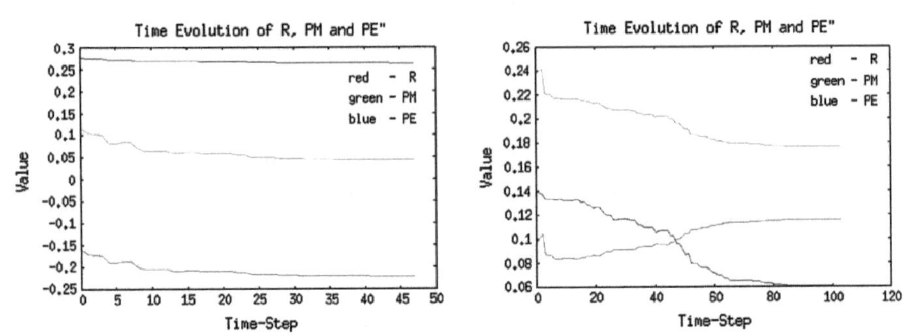

Fig. 9 Typical evolutions of P_M, P_E, and R

high while it decreases toward 0 in the second. The first graph is typical of cases where the HT strategy wins and the second is typical of cases in which SF wins.

One implication that can be drawn from examination of these results arises from an observation that while the initial populations are always started at the barycenter of the p-Δ_3 simplex, the results in any given run are undetermined: any strategy can win. This suggests that the actual winning strategy in each case is determined by random fluctuations in the probabilistically generated q-sequence. There is a connection to virtual stability in this—the initial conditions in the p-simplex are set at the barycenter as well, which appears to be an unstable equilibrium. This suggesting that in terms of control, maintain a virtually stable state at this equilibrium allows rapid adjustment to such environmental contingencies.

4 Discussion

As emphasized by MacIver, the class of potential behaviors available to a complex system within its environment is determined by a perceptual response horizon. This is the "perceptual" distance available to the system as compared to its capacity for motion. For organisms where these two distances are of the same order of magnitude, behavior is limited to responses based on a local spatial template. If the perceptual horizon is much greater than movement capacity, long term planning and the ability to optimize over many possible projected future actions becomes essential. In MacIver's view this is the evolutionary advantage provided by vision as life moved from the sea to the land [28, 29].

In the actual process of "optimizing over possible future actions," however, a variety of issues arise. When a leopard is chasing a baboon, each is acting to optimize over the same set of pursuit paths, but each is using different optimality criteria. The result of any individual pursuit is governed by how well each participant is able to maneuver in the rapidly changing space of future possibilities. If an animal sees a potential source of food, but also sees a predator, then there is a decision problem: is the time required to obtain the food and escape greater than the time it will take for the predator recognize that a prey is near and move into a capture position?

Two issues directly associated with success in such life or death situations are maneuverability and accuracy of perception and response. Evolution has provided a variety of answers that address these issues, in ways that are satisfactorily from the point of view of species survival. From the species point of view survival is a matter of statistics—in each generation enough members of the species must survive and reproduce. But selection acts on individual organisms; it is the cumulative effect of selection, acting as a function mediating between individual phenotypic expression and environmental contingency, that produces change in gene frequency distributions.

Thus, if a biological species is to survive its members must satisfy certain constraints imposed by the ecological niche occupied by that species. Species members need to be able to behave in ways that are adequate to insure that a sufficient proportion survive to reproductive age. Species that are observed in nature can, in general, be assumed to be well adapted—the species phenotype remains relatively stable while at the same time there is sufficient genetic variability to provide for accommodative response to environmental fluctuations.

A well-known ecological example is the British Peppered Moth (Biston belularia) [30]. Until the middle of the nineteenth century the overwhelming phenotype for this moth was called typica, characterized by white wings with dark speckles. As a result of smoke emitted from coal burning factories during the industrial revolution, trees in industrial areas were covered with dark soot. Against this dark background the white winged moth was easily spotted by predatory birds, introducing a strong selective pressure that led, by the end of the nineteenth century, to a population of moths almost completely dominated by the carbonaria phenotype—black wings with white speckles. As industrial emissions were cleaned up in the later twentieth century the typica phenotype reemerged.

While this is often pointed to as an example of evolution, it is important to recognize that survival of the species was contingent on there being sufficient variation available in the species gene pool, together with a generation time sufficiently short that phenotypic change takes place on a time scale matching that of environmental change. That is, the insight arising from the principle of virtual stability is that in such cases it is not only a matter of having a requisite variety of responses available, it is also necessary that these responses occur within a situation dependent temporal window, with a not too costly expenditure of energy or resources. This can motivate construction of formal models of such cases harmonizing mutational frequencies with generational and external time scales.

Species survive because a sufficient fraction of their members survive and reproduce. If an individual organism is to survive, on the other hand, it must be able to both acquire adequate resources and avoiding predation as it seeks these resources. The capacity to optimize over possible futures plays an important role in balancing the potentially conflicting goals of resource acquisition and predator avoidance. Similarly, if a social group is to survive it must be able to replenish its membership while maintaining the stability of those things that identify it as a distinct social collective. These may involve nationalistic, economic, religious, or other interests. The group must be able to maintain its core identity while adapting to change.

The toy models presented in this chapter address some of the trade-offs involved in such optimization. The model of the quickness-accuracy trade-off deals directly with the functional relation between the need for resources in order to survive and the cost (or danger) involved in obtaining those resources. Empirically, animal studies show the existence of a speed-accuracy trade-off that is consistent with the neural integration of sensory evidence until a decision threshold is reached [31, 32]. These studies have focused on cognitive and neural aspects involved, however, rather than the actual expression of trade-off behavior in a natural environment where the potential consequences of a mistake charge the decision with survival related tension.

With social groups the situation becomes more complicated. Ideological and religious groups typically favor assimilation over accommodation and either reject new ideas outright, or take an extended time to investigate them before either declaring them heretical, or allowing them into the group belief structure.

In a more detailed extension, the quickness-accuracy model would have a hierarchy of cues rather than just immediate appearance and actual reality, with corresponding temporal scales and graded levels of response.[4] Nevertheless, the model as presented provides insight into possible animal and social behaviors under conditions of uncertainty and points to potentially interesting lines of research. For example, it points to the question of whether there is a relationship between species fecundity and foraging behavior—do species that produce many offspring engage in riskier behavior than species that produce few offspring?

The virtual stability model illustrates the general concept of a stability-flexibility trade-off. Commitment to a fixed trajectory of future action provides stability but also exposes an organism to potential danger arising from the predictability of this trajectory to potential predators, as well as vulnerability to random, hence unpredictable, environmental fluctuations. It is easy to change an unstable state, difficult to change a stable one so the trade-off is between an ongoing energy expenditure required to maintain a state of virtual stability and the occasional need for a much larger episodic energy expenditure to escape from a stable state that is no longer preferred. A general mathematical formulation of this idea requires specification of the degree of instability, the relation between the frequency of adaptive corrections necessary to maintain the unstable state, and the average energy expenditure involved in each correction. The energy expenditure, averaged over a sufficiently long time scale must be less than the sum of expected energy expenditures involved in episodically moving from one stable state to another in response to environmental fluctuations. While it does not directly address these requirements, the model shows the existence of circumstances under which it is important to avoid commitment to any fixed course of action and indicates the value of being able to maintain a position on an unstable trajectory that, for example, moves on the boundary between competing basins of attraction.

A well-studied example of a virtually stable state is standing. Without ongoing self-monitoring and feedback the standing position is unstable. If standing were stable, walking would be difficult. As it is, taking a step is a controlled fall. Winter [32] notes this, following on an extensive description of the complexities involved in walking: "In order to accelerate our COG [center of gravity] in a forward direction we must voluntarily initiate the start of a forward fall to accelerate the COG ahead of the base of support." (p.205)

Extensive research has been done to model human and animal standing, walking and postural control [32, 33]. Mathematical analysis of human standing and walking uses an inverted pendulum model. Asai, et al. [17] use the equation

$$I\ddot{\Theta} = mgh\Theta - \left(K\Theta + B\dot{\Theta} + f_P\Theta_\Delta + f_D\dot{\Theta}_\Delta\right) + \sigma\xi \qquad (33)$$

where I is the moment of inertia of the body about the ankle, Θ is the tilt angle, g is gravitational acceleration, m the body mass, h the distance from the ankle joint to the body center of mass, K the passive ankle stiffness, B a viscosity measure, Δ the time delay for neural feedback ($\Theta_\Delta = \Theta(t - \Delta)$, $\dot{\Theta}_\Delta = \dot{\Theta}(t - \Delta)$) and ξ is a

[4] Such a model could use filtering and template matching of perceptions, for example, in order to find the best behavioral response fit.

Gaussian white noise of intensity σ. The terms f_P and f_D are control parameters. The equilibrium manifold is an unstable saddle and the control problem is to maintain the system trajectory within the stable manifold. Their analysis shows that this is best done with well-timed intermittent feedback. They conclude: "Overall the origin of such intermittency remains obscure and has, up to now been viewed mainly as a consequence of neurophysiological internal constraints that limit the computing power of the neural controller. However, there is the alternative possibility that intermittency has a functional role in the control strategy of human subjects: that of maintaining the stability of feedback control despite uncertainties about dynamic properties of the body or manipulated objects and the large neural delays in the transmission of the feedback signals."

Because of neural transmission delays on the order of 200 ms, the brain receives input of a situation as it was a short time in the past and, for effective control, it must extrapolate from this past state and project control signals to an anticipated future state. Since the standing position is unstable, the frequency of control feedback must be high enough that the system never deviates from the upright position by more than an easily correctable amount. While the mechanisms underlying control of standing and walking are still being worked out, evidence from animal experiments indicates that postural control appears to rely on maintaining a movable balance between antagonistic neural control systems with a small intermittent control function [17, 33]. In systems that are in a virtually stable state, the inherent instability of the state means that the system will tend to be "falling" out of the state into a more stable configuration and this implies that a threshold is required such that when it is crossed a control impulse must be introduced to return the system to the unstable configuration [34]. This fits well with theories of intermittent control and may point to the "functional role of intermittency" alluded to in [17].

Another broad area where the virtual stability model is potentially applicable is in studies of social groups. On one hand, it can point to the trade-off between intransigence and flexibility in business or political negotiations, where keeping options open can be advantageous. On the other, a person maintaining a state of non-commitment within a group may face sanctions because this smacks of opportunism. There is evidence of an innate human tendency not only to punish defectors, but also to punish those who are unwilling to punish defectors [35, 36], a tendency that seems to be directly linked to the evolution of cooperation [37].

Finally, the advantage in behavioral flexibility provided by the capacity to maintain states of virtual stability suggests that in any sufficiently energy rich environment with a moderate spectrum of random fluctuations, evolution will produce nervous systems have the necessary self-monitoring and adaptive control capacities to maintain virtually stable states. In other words, in a suitable environment, the evolution of sentient life may be the rule rather than the exception.

References

1. E.A. Jackson, A first look at the second "metamorphosis" of science. SFI Technical Report 95-01-001(1995a)
2. J. Casti, The Computer as Laboratory. Complexity **4**(5) 12 (1999)
3. T.S. Kuhn, in *The Essential Tension*. ed. by T.S. Kuhn. The Essential Tension, (University of Chicago Press, Chicago, 1977), pp. 225–239
4. W. Brian Arthur , Inductive reasoning and bounded rationality (The El Farol problem). American Economical Review (Papers and Proceedings), vol. 84 (1994), p.406
5. B. Voorhees, J. Senez, T. Keeler, M. Connors, A population model of the stability-flexibility tradeoff. Adv. Complex Syst. **11**(3), 443–470 (2008)
6. B. Voorhees, Virtual stability: constructing a simulation model. Complexity **15**(2), 31–44 (2009)
7. F. Heylighen, in *Principles of systems and cybernetics*. ed. by R. Trapel Cybernetics and Systems 92 (World Scientific, Singapore, 1992)
8. R.M. May, *Stability and Complexity of Model Ecosystems* (Princeton University Press, Princeton, 1973)
9. S. Gavrilets, *Fitness Landscapes and the Origin of Species* (Princeton University Press, Princeton, 2004)
10. B. Voorhees, Axiomatic theory of hierarchical systems. Behav. Sci. **28**, 24–34 (1982)
11. W. Stefan, Wu Yihren, Automata with hierarchical control and evolutionary learning. BioSystems **21**, 115–124 (1988)
12. S. Kraut, U. Feudel, C. Grebogi, Preference of attractors in noisy multistable systems. Phys. Rev. E **59**(5), 5253–5260 (1999)
13. U. Feudel, C. Grebogi, Multistability and the control of complexity. Chaos **7**(4), 597–604 (1997)
14. C. Foster Glenn, A. W. Hubler, Robuse and efficient interactions with complex systems. Proceedings of IEEE International Conference on Systems, Man & Cybernetics 2029–2034, (2003)
15. U. Feudel, C. Grebogi, Multistability and the control of complexity. Chaos **7**(4), 597–604 (1997)
16. H. Touchette, S. Lloyd, Information-theoretic limits of control. Phys. Rev. Lett. **84**, 1156–1159 (2000)
17. Y. Asai, Y. Tasaka, K. Nomura, T. Nomura, M. Casadio, P. Morasso, A model of postural control in quiet standing: robust compensation of delay-induced instability using intermittent activation of feedback control. PLoS ONE **4**(7), e169 (2009). doi:10.1371/journal.pone.ooo6169
18. E.R. Kandel, J.H. Schwartz, T.M. Jessell, *The Principles of Neural Science*, 4th edn. (McGraw-Hill, NY, 2000)
19. Laura N. Borodinsky, How fast can you go? Nature **440**, 158–159 (2006)
20. W.R. Ashby, *Introduction to Cybernetics* (Wiley, London, 1956)
21. Juan M.R. Parrondo, Gregory P. Harmer, Derek Abbott, New paradoxical games based on Brownian ratchets. Phys. Rev. Lett. **85**(24), 5226–5229 (2000)
22. J.S. Nicolis, T. Bountis, K. Togias, The dynamics of self-referential paradoxical games. Dyn. Syst.: Int. J. **16**(4), 319–332 (2001)
23. Yu. Itoh, Kei-ichi Tainaka, Spatial enhancement of population uncertainty in model ecosystems. J. Phys. Soc. Jpn. **73**(1), 53–59 (2004)
24. B. Voorhees, C. Luxford, R. Arthur, Emergence of cellular automata rules through fluctuation enhancement. Int. J. Unconventional Comput. **1**(1), 69–100 (2005)
25. Alfred W. Hubler, Predicting complex systems with a holistic approach. Complexity **10**, 11–16 (2005)
26. Suso Kraut, Ulrike Feudel, Celso Grebogi, Preference of attractors in noisy multistable systems. Phys. Rev. E **59**(5), 5253–5260 (1999)

27. R. Cressman, *Evolutionary Dynamics and Extensive Form Games* (MIT Press, Cambridge, 2003)
28. M.A. MacIver, N.A. Patankar, A.A. Shirgaonkar, Energy-information trade-offs between movement and sensing. PLoS Comput. Biol. **6**(5), e1000769 (2010)
29. M.A. MacIver, Neuroethology: from morphological computation to planning, in *The Cambridge Handbook of Situated Cognition*, ed. by P. Robbins, M. Aydede (Cambridge University Press, NY, 2009), pp. 480–504
30. B.S. Grant, Fine tuning the peppered moth paradigm. Evolution **53**(3), 980–984 (1999)
31. R. Bogacz, E.J. Wagenmaker, B.U. Forstmann, S. Nieuwenhuis, The neural basis of the speed-accuracy tradeoff. Trends Neurosci. **33**, 10–16 (2009)
32. D.A. Winter, Human balance and posture control during standing and walking. Gait & Posture **3**, 193–214 (1995)
33. A. Pastor-Bernier, E. Tremblay, P. Cisek, Dorsal premotor cortex is involved in switching motor plans. Frontiers Neuroeng. **5**, 1–15 (2012)
34. P. Gawthrop, I. Loram, M. Lakie, H. Golle, Intermittent control: a computational theory of human control. Biol. Cybern. **104**, 31–51 (2011)
35. H. Gintis, S. Bowles, R. Boyd, E. Fehr, Explaining altruistic behavior in humans. Evol. Hum. Behav. **24**, 153–172 (2003)
36. P. Richardson, R. Boyd, *Not by Genes Alone: How Culture Transformed Human Evolution* (University of Chicago Press, Chicago, 2004)
37. M. Nowak, R. Highfield, *Supercooperators* (Free Press, NY, 2011)
38. D. Rinberg, A. Koulakov, A. Gelperin, Speed-accuracy tradeoff in olfaction. Neuron **51**, 351–358 (2006)

Toward a Computational Model of Complex Human Systems Dynamics

Glenda H. Eoyang

Abstract A useful computational model of complex human systems dynamics could support advancements in theory and practice for social systems from intrapersonal experience to global politics and economics. Models of human interactions have evolved from Newtonian assumptions, which served a variety of needs in the past. Another class of models has been informed by nonlinear dynamics. None of the existing models, however, sufficiently represents the open, high dimension, and nonlinear self-organizing dynamics of social systems. A conceptual model, CDE Conditions for Self-organizing in Human Systems, is explored as an alternative. While the CDE overcomes the limitations of previous models, it also provides an explanatory base for prospective analysis to inform meaning making and action taking in response to complex conditions. An invitation is extended to engage in developing a computational model that incorporates the assumptions, meta-variables, and relationships of this conceptual model of the complex dynamics of human systems.

Keywords Human systems dynamics · CAS · CDE · Self-organizing · Complex adaptive systems · Complexity · Nonlinear dynamics

G. H. Eoyang (✉)
Human Systems Dynamics Institute, Circle Pines, MN, USA
e-mail: geoyang@hsdinstitute.org

I. Zelinka et al. (eds.), *How Nature Works*, Emergence, Complexity and Computation 5, DOI: 10.1007/978-3-319-00254-5_7,
© Springer International Publishing Switzerland 2014

1 Modeling Human Behavior

A computational model[1] that captures the nonlinear nature of the dynamics of human systems with fidelity would yield great benefits for scholars and practitioners who face emergent personal, professional, and political challenges. Scholars would use such a model to develop and test hypotheses about human behavior, institutional development, and evolution of industrial and political ecosystems. Practitioners would use a robust nonlinear model to inform decision making in real time and instructional programs to develop knowledge and skills required in complex environments. Individuals and groups would build adaptive capacity to see, understand, and influence complex and unpredictable patterns as they emerge.

1.1 Current Models of Human Systems

Many different quantitative and qualitative, rigorous and imaginative models are currently used for all of these functions. Rational choice theory, statistical analysis, systems dynamics modeling, adaptive leadership, Myers Briggs Type Indicator, Strength Finder, Technology of Participation, and so on are just a few examples. All of these models support useful methods of research and practice in a variety of contexts. Each one also has limitations based on its fundamental assumptions about the dynamics of human systems. The most rigorous of the existing models may apply only in narrowly defined theoretical contexts. The most imaginative, without benefit of disciplined research, may prove to be ineffective or even destructive in practice.

While these models of human interaction have served well enough in the past, their inherent weaknesses are beginning to show. They assume clear and distinct boundaries in space, time, and function, and our global economy transcends all bounds. They assume a low number of relevant variables and clear indicators of performance. The recent focus on systemic issues such as sustainability underscores the need to consider many factors at the same time, some of which are unpredictable or ambiguous. They assume linear cause and effect. Today massively complex information and resource networks contribute to nonlinear effects that cannot be ignored. As the world becomes more complex, the choices we have made to simplify our models seriously limit their reliability and usefulness.

---

[1] We will use the term *model* throughout to refer to "a simplified description, especially a mathematical one, of a system or process to assist calculations and predictions." (Oxford Dictionaries Online. (n.d.). *Oxford Dictionaries Online*. Retrieved from http://oxforddictionaries.com/.) We will characterize all systemic representations including qualitative and quantitative, positivistic and interpretive as models. We will make the distinction explicit when referring specifically to simulation, mathematical, conceptual, or computational models.

As challenges of humanity become more complex, the limitations of these models turn into fatal flaws. When individual, corporate, social, and political patterns are radically open to external influences, assumed boundaries of a model become irrelevant. As human systems at all levels are shaped by innumerable and constantly changing variables, model assumptions about a small number of dependent and independent variables are no longer valid. As our challenges involve more and faster feedback loops, model assumptions of linear cause and effect prove insufficient to capture the emergent dynamics of the system. In short, as human systems become more complex, our models—even the complicated ones—are not sufficient to inform either our research or our practice. Today's global challenges exceed the capacity of our historical models of human systems dynamics. Robust theory and effective practice demand a new generation of models and modeling techniques.

Even with all their flaws, the models of human interaction that currently exist provide insights to support historical analysis, current decision making, forecasting, and planning. In the same way that Ptolemy gave a "good enough" model of celestial movement, social and economic models of the twentieth century have been "good enough" to guide thinking and action across all levels of meaning making and action taking. Just as Copernicus introduced an alternative model to solve challenges that could not be solved under Ptolemy's geocentric worldview, we need a new model of human systems dynamics that will allow us to transcend the limitations of our past theory and practice to respond to uncertainty and radical emergence of our complex reality.

1.2 An Alternative to the Limitations of Current Models

Today, the inherent weaknesses of the existing models are increasingly apparent. Decision makers in all sectors and industries realize the limitations of the models and methods available to inform their action. Economists acknowledge that the conditions resulting in crisis and collapse are not represented in their econometric models. In spite of sophisticated technologies, intelligence communities have insufficient power to deal with the challenges of information collection, collaboration, and interpretation in the midst of unpredictable and complex networks of insurgents. Political upheavals, social movements, violent and nonviolent conflict generate phenomena that we urgently need to understand and influence. Around the world, institutions find themselves overwhelmed and without sufficient tools to see emerging patterns, understand their implications, and generate and select options for action to influence systemic patterns of health or sustainability.

Over the past two decades, research into nonlinear dynamics has revolutionized models and methods in a variety of physical science and mathematical disciplines. Techniques emerging from the study of nonlinear physical systems, such as nonlinear time series analysis and dynamical network theory, have been applied to social systems dynamics with some success. Research and practice indicate that even those models have limited utility in shaping effective theory and practice in complex social systems.

In this chapter, we will explore usefulness and limitations of some of the models of social interaction that have influenced research and practice in psychology and economics. We will also describe benefits and constraints of innovative methods that have emerged from nonlinear dynamics applied to social systems. We will explore these approaches in the contexts of the open, high dimension, and nonlinear patterns of today's complex human systems dynamics. We will introduce a conceptual model of the human systems dynamics based on a nonlinear paradigm of systemic interaction and emergent structuration. Based in both theory and practice, this conceptual model informs action while it assumes open, high dimension, and nonlinear dynamics of social systems at all scales. Finally, we will invite colleagues to engage with us to develop a computational model to quantify and test this emerging conceptual model.

Our experience is that this complexity-based conceptual model of human interaction resolves issues of previous models and helps individuals and groups see patterns in emergent systems, understand their implications in given contexts, and take intentional action to influence the patterns as they emerge. We speculate that this conceptual model might provide a strong theoretical grounding for a computational model to inform theory and practice, and that such a computational model would be robust enough to address our emerging challenges in complex human systems. After introducing the conceptual model and speculating about a possible computational implementation, we will propose a research agenda and invite colleagues to join us in creating a computational model that exceeds the benefits and resolves the risks of existing models of human systems dynamics and their applications in theory and practice.

2 Traditional Models of Social Interaction

"Essentially, all models are wrong, but some are useful." [1] Diverse fields in social sciences develop and apply mathematical and qualitative models and methods to represent human behavior. Each one has emerged from a specific discipline to respond to specific questions and inform certain kinds of decision making. Like any other model, each model applied to human interaction has its own inherent limitations.

2.1 Sources and Applications of Traditional Models

Models of human behavior have emerged from many different social sciences. Political science, education, arts, anthropology, industrial engineering, sociology, and an array of other fields focus on specific aspects of human systems and human behavior. Within each of these fields, a variety of conceptual and some computational models inform theory and practice. For the purposes of this paper, we will focus only on illustrative examples drawn from the fields of psychology and economics. Detailed analyses and critiques of even these models are beyond the scope of this paper. We intend only to acknowledge the widely accepted notion of the gap between human systems as they are experienced and as they are captured in current models.

Psychological models are based on a variety of theoretical frameworks and support diverse practical applications. For example, biological and cognitive models of time have emerged from psychological research and practice [2]. Models of mood disorders are diverse and emerge from a variety of theoretical frameworks [3]. Decision making is another application of psychological model making [4, 5]. One class of psychological decision models is particularly applied to intelligence efforts and the decision making and action in the intelligence community. Judgment and Decision Making (JDM), Analysis of Competing Hypotheses (ACH), Naturalistic Decision Making (NDM), foraging, and various group decision-making models [6] are all applied to the field of intelligence collection, analysis, and action.

These and many other models emerge from psychological research, and they can be applied with good purpose to enhance theory and practice when individuals and groups seek to see, understand, and influence change in social systems. As useful as they are, these psychological models are limited to specific contexts and challenges, and they are applicable at a limited number of human systems scales. Some focus on the individual (and occasionally a small group) level of organization in human systems. Other sets of models deal with patterns at the community, institution, or national scope. None of them is intended to speak simultaneously to all of the open, high dimension, and nonlinear patterns that emerge across the complex systems of human dynamics. In addition, most of these models are not amenable to computational modeling or other quantitative methods of inquiry. The lack of commensurability between qualitative and quantitative representation of dynamics fuels the on-going conflict between positivistic and interpretive epistemologies and research methodologies. The closer traditional models come to realistically describing open, high dimension, and nonlinear phenomena, the more difficult it becomes to represent their dynamics in mathematical or computational models. Simplicity and fidelity are constantly in tension in the whole range of models of psychological interaction. Where one succeeds, the other fails.

Economic models, from statistical analyses to rational choice theory and chaotic dynamics [7], have shaped individual, institutional, and market decision making for decades. The reliability and usefulness of such representations have been challenged for equally as long [8]. In spite of their acknowledged limitations, models of macroeconomic patterns of global interactions have become relatively common. They are used to influence decisions that affect global politics and commerce [9]. The reliability, robustness, and relevance of economic models can be characterized in many ways, but each model stands on its own foundation of assumptions and acceptable methods [10].

The insufficiency of current economic models is widely understood. In January of 2011, Ban ki-Moon, Secretary General of the UN proclaimed:

> It is easy to mouth the words "sustainable development", but to make it happen, we have to be prepared to make major changes—in our lifestyles, our economic models, our social organization and our political life. We have to connect the dots between climate change and what I might call here WEF—water, energy and food [11].

The systems we seek to sustain—physical, economic, social—are open, high dimension, and nonlinear. Models we use to represent those systems must be able to capture such complex dynamics. If we are to think simply and with fidelity about these systems, we must have new models that capture the complex dynamics of economic systems and their behavior. The current disconnect between micro- and macroeconomic models is one example of this challenge. Economic models isolate local action from the global patterns that capture consequences. The lack of integration also makes it difficult to incorporate emergent global patterns into local decision making. Global outcomes that depend on local action require new models for economic behavior that scale across levels of analysis and action while accounting for the massive complexity of nonlinear dynamics within every scale and among all scales. Current economic models are not able to satisfy any of these requirements.

We might add to this list of economic and psychological models ones from political science, sociology, management, organization development, education, and many other social sciences. All of these models are useful for their intended purposes, but none is robust enough to represent complex dynamics of human systems in ways that inform understanding or action in systems that exhibit complex, nonlinear, emergent behaviors. When we acknowledge that human systems are simultaneously open, high dimension and nonlinear, these models fall short in a variety of ways.

2.2 *Limitations of Traditional Models*

Traditional models serve many purposes, but they are not able to represent the complex dynamics of human systems as we experience them individually or in groups. In our action research across many kinds of social systems [12], we have found major limitations of the models emerging from assumptions of (1) a single level of analysis rather than massive interdependencies across scales; (2) closed boundaries rather than open interactions with emergent environmental landscapes; (3) low dimensionality rather than high and/or indeterminate number of relevant variables; (4) linear causality rather than nonlinear relationships and mutual causality; (5) random variability supporting statistical analysis rather than significant levels of ambient background correlation. While they make the problems more tractable, these assumptions also limit the usefulness of the models to address real dynamics of real problems.

Multi-level. Traditional models of human interaction tend to focus on a single scale of human activity. Like the disciplines from which they emerged, current models focus on one or at most two specific levels of social organization. One model might consider the state of the individual in relation to the group dynamics. Another might look at the firm in relation to a market. Yet another might consider political entities and their interactions in global patterns of behavior. None of the models supports a look across levels or at the interactions among multiple levels. None captures the scale-free pattern-forming processes that are common in complex adaptive systems. Even though we are keenly aware that global interactions can influence and be influenced by individual or group decision making, our models continue to represent one level in a way that is incommensurable with the levels above and below it. While some techniques, like traditional systems dynamics modeling, can include models within models to represent multiple levels of underlying dynamics; still, the ability to generalize insights or actions across scales is severely limited.

This limitation of being scale-bound is neither trivial nor merely theoretical. In many fields, analysis at the micro and macro scales are totally incommensurate, so critical information does not flow between local and global meaning making or action. Incidents of violent conflict demonstrate the effects of missing inter-scale communication. Individual peace makers interact with individuals in communities on the ground. They may use conflict resolution models and methods to quell emerging conflict between neighbors in a neighborhood. At the same time, the economic and geopolitical analyses may capture critical contextual cues that are not visible from the ground, but miss the messages that are local and specific to a particular hotspot. As a result, model-informed insights about both the local and global patterns are incomplete, and decision making and action taking in both contexts are constrained. Increasingly, social scientists refer to macro-, meso-, and micro-levels of interaction. While moving from one or two to three levels of analysis is definitely an improvement, such a conceptualization still misses the scale-free interdependence that is critical in the dynamics of complex adaptive systems.

Open. The second limitation of current models is that they require human systems to be bounded in space and/or time. Traditional models that represent human interaction are based on assumptions that limit the conceptual definition of a situation to make its problem space more tractable. Bounding conditions increase certainty, so limiting assumptions make it easier to manage the mathematics or theoretical descriptions. On the other hand, each assumption that limits the problem space makes it more difficult to correlate model behavior to real human behavior in the living system. Of course this is the purpose of the model—to represent the system in a simplified way. In the past, the trade-offs between bounded simplicity and real-world fidelity were manageable. Phenomena of social systems were simple enough that our finite, constrained representations were sufficient to inform theory and practice. As our local and global situations become more complex, however, it is increasingly difficult to support the delusion that our situations are as bounded as our models assume them to be. A robust and reliable model of human interaction must acknowledge and incorporate open system relationships if it is to support meaningful theory building, pragmatic decision making, and effective action.

While we know that all social systems are open to external influence, theoretical and mathematical models are seldom able to represent such open boundaries. Even the most closed examples of human institutions—prisons or fascist regimes for example—are subject to external influences. Many modeling methods constrain these unpredictable influences by artificially bounding a system in time or space. While such a compromise makes the mathematics more tractable, it limits the correspondence between behavior of the model and a real human system—whatever the scale.

High Dimension. High dimensionality is the third complexity of real human behavior that is difficult (if not impossible) to capture in traditional conceptual or mathematical models. Perforce, our models assume that any human decision or action depends on a finite number of relevant variables; while we know even the simplest decision in real life may be driven by a large and unpredictable number of parameters. Not only is the number of variables that influence human behavior high, they also change constantly. At one time, for multiple individuals or at different times for the same individual, different considerations will influence a particular decision or action. If this is true at the level of the individual, it is even more obvious for communities, institutions, or nation states.

This radical diversity of complex human systems is a major challenge to effective modeling. One drawback of any model is the distinction between the generalized, abstracted, perfect case represented in the model and the specific, embodied, particular example that occurs in reality. In order for a model to apply to diverse cases, assumptions must limit the amount of variability among the cases in the system. In so far as the variability is limited, the model fails to represent the particular case adequately.

Nonlinear and Non-random. Some models, such as those founded on rational choice theory, simply deny local variability in order to represent a consistent general case. Other models, including all of those based in traditional statistics, assume a random distribution of phenomena across a context. Beginning with the random distribution, modelers use statistical analyses to discern and characterize patterns of interaction and intention that might emerge over time. For example, when you assume random distribution, you can focus on average behavior as representative of the whole. Again, this strategy has been "good enough," but it breaks down in the class of systems considered to be complex adaptive. These systems generate system-wide patterns, so they do not begin from a state of normal distribution. When the goal is to model how individuals are influenced by each other, it is not "good enough" to imagine that their cultural or personal patterns are random to begin with. If human beings have free will and if they influence each other, which most model makers and users would like to believe, then we cannot assume an initial random distribution with no correlation among agents. Each individual case has the freedom to vary in unpredictable ways, and the assumption of zero natural correlation, which is required for statistical analysis, is no longer valid. When agent behavior is naturally and unpredictably correlated, as happens when many human beings are connected, their individual actions, driven by their free will, influence each other. As a result, we cannot assume a random distribution as a precondition for statistical analysis of overall system behavior, whether the system is a person, a group, institution, or community. We need a different way to conceptualize non-random, unpredictable differentiation that is common in our observations and experiences of human systems dynamics.

2.3 Current Options for Modeling

Limitations of modeling methods have been acknowledged for decades. Narrative is one modeling method that has successfully been used to represent open, high dimension, nonlinear, and locally variable phenomena [13]. Stories are powerful ways to represent reality in its own language, including its most complex characteristics. While the uses of narrative methods are becoming increasingly robust and rigorous [14] the dilemma of how to generalize or abstract narrative as a model of social interaction has not been resolved. Two computer-based narrative analysis methods are able to derive complex patterns of meaning from narrative data. CRAWDAD (www.crawdadtech.com) [15] uses the linguistic technique of centering resonance analysis to detect relations among noun phrases in a natural language sample and represent those relationships as a network of meaning [16]. Quantitative analysis of the network provides rich information about the patterns encoded in a narrative selection. Sensemaker (www.sensemaker-suite.com) [17] is another software-based narrative analysis process that transforms narrative into patterns of meaning. While both products create open, high dimension, and nonlinear models of narrative text, they share limitations of other complexity-inspired models which we will discuss later.

Historical models of human behavior have understandably compromised fidelity to make models more tractable and more generalizable. They focused on a single level of analysis; they bounded systems, focused on a small number of variables, assumed linear causality, avoided free will and interdependency by assuming random distribution of behavior. While such models serve specific purposes, they do not capture the complex dynamics that are relevant to decision making and action taking in the twenty-first century. In efforts to adapt models to match reality, many social scientists have embraced a variety of complexity-inspired methods. We will explore some of those methodologies next.

3 Complexity and Social Interaction

Traditional models of human interaction and human behavior have drawn from traditional scientific, linguistic, and mathematical models and methodologies. Since the mid-1970s, new and more complex analytical methods have emerged, and they have been applied to research and modeling of human systems [18]. These approaches break through some of the limitations of traditional approaches to modeling social systems because they deal explicitly with nonlinear causality. On the other hand, they fall prey to some of the traditional limitations while introducing some new limitations of their own. We will briefly introduce five modeling approaches that have been derived from nonlinear dynamical methods, describe how they support decision making and action taking in complex human systems, and explore their limitations as true and useful representations of complex dynamics of human systems for research and practice in emergent and uncertain environments.

3.1 Sources and Applications of Complexity-Inspired Models

Beginning in the mid-1970s and continuing to the current day, models, methods, and insights of the nonlinear dynamics in physical and mathematical systems have been applied to explore human systems dynamics. Scholars and practitioners have used these approaches more and less metaphorically to create simple models of complex human system behavior at a variety of scales. Five categories of models have been particularly useful, and we will describe them briefly here:

- Catastrophe theory;
- Dynamical network theory;
- Nonlinear time series modeling;
- Agent-based simulation modeling;
- Power law dynamics.

Catastrophe Theory. Renee Thom's catastrophe theory [19] emerged as one of the earliest quantitative models of complex dynamics. Not only did it deal with nonlinearity, but it also included ways to capture high-dimension interactions. It has been applied to a variety of social systems, including error and injury rates and growth of firms [20]. While catastrophe theory showed great promise in its ability to represent the dynamics of complex human interactions, it had limited practical use for a variety of reasons. First, the mathematical sophistication of the model made it complicated and difficult for practitioners to understand. In addition, its graphical representations of systems in more than three dimensions were impossible to see and even difficult for most people to imagine. So, while the qualitative explanations of Thom's work were powerful, and early interest in them was great, the quantitative applications proved too complicated to be useful for decision making and action taking. Within a few years, the promise of catastrophe theory as a definitive model of human interaction faded from most scholarly and practitioner applications.

Dynamical Network Theory. Dynamical network theory [21] has been used extensively as a powerful modeling method to explore market potential, social cohesion, and dissemination of information [22] and innovation [23]. In the past decade, a variety of software packages [24] have come on the market to simplify the methods of collecting and analyzing network-related data. Measures of network properties such as clustering, connectedness, density, and centrality have opened new ways to see and understand the patterns of social interaction and emergent social structures. Online social network sites have helped make such models familiar to the public and have accelerated the acceptance of network-based models of social systems. Stages of network evolution, from hub and spoke to scale-free structures, have informed an understanding of the development of social and computer networks over time. While this approach solves many of the issues of pre-complexity models, it is essentially descriptive, providing a snapshot of a current state without explanation of what came before or implications for options for action to influence the future.

Nonlinear Time Series Analysis. One of the earliest modeling methods from deterministic chaos involved a process of nonlinear time series analysis [25]. In this method, an extended time series is analyzed and plotted in phase space, looking not at change through time, but comparing the change in non-time variables from one point to the next across the entire time series. Such analysis allowed the researcher to characterize the nonlinear phenomenon as following the pattern of a random, point, periodic, or strange attractor pattern. Further analysis of the time series could reveal the dimensionality of the phenomenon by pointing to the number of key variables involved in the dynamics that shaped the pattern. In the early 90s many researchers used methods of nonlinear time series analysis, searching for strange attractors as evidence of deterministic chaotic dynamics in social systems.

Three challenges emerged in using this approach either for theory or practice development. First, the analytical method required a long and reliable time series as input data, and appropriate data was not often available from the systems under examination. Second, the mathematics required for the analysis were so complicated that they were not well understood by many researchers, so they were embedded in a variety of automated analysis tools. Lack of basic understanding of the underlying method led to a variety of errors in analysis and interpretation, including misinterpretation of results and unintended artifacts of the analytical methods themselves. The third challenge was embedded in practice. Even when strange attractor patterns were reliably discerned in time series data, the interpretation and meaning making based on the results were not clear or compelling. For these reasons and others, attractor pattern reconstruction as a modeling method to support decision making and action taking in human systems has been relegated to a small number of highly technical research applications.

Agent-Based Simulation Modeling. Agent-based computer simulation modeling has become a popular research method to demonstrate processes and outcomes of self-organizing dynamics of social systems [26] in decision science [27], financial markets [28], sociology [29], information and political science [30], conflict analysis [31], and a variety of other social science applications [32]. Given a set of initial conditions and agent characteristics, semi-autonomous agents in the model follow local rules, learn adaptive behaviors, and contribute to formation of system-wide patterns. These models can be used to visualize data about interactions, to test hypotheses regarding conditions and paths of self-organizing processes, and to teach about dynamics of complex change in human systems. The premier institution committed to the study of agent-based modeling and its influence in both physical and social sciences is the Santa Fe Institute (http://santafe.edu/). Researchers have used the method to model and explore a variety of dynamical systems and emergent phenomena. Though they demonstrate relationships between initial conditions and outcomes based on simple rules, the abstract and generalized structure of an agent-based simulation model limits its usefulness for decision making and action taking in real-world situations.

Self-Organized Criticality. Per Bak led a group that pioneered our fifth and final method of nonlinear modeling for human systems, the power law [33]. Power law dynamics, sometimes referred to as self-organized criticality, Pareto distribution, or Zipf's Law, have been used to describe major transition phenomena of markets and market development [34]. The idea was popularized as the "long tail" of internet-driven business models [35]. The power law has also been used to describe dynamics of population migration, violent conflict, and addiction. Because of its ability to capture inter-level dynamics and to account for discontinuous change, the power law has become one of the symbols of the changing dynamics of human systems. The power law is scale-free, so it works at every level of human system, from brain dynamics to world-wide conflict. The challenge is that there is currently no robust theoretical explanation of why these mathematical relationships emerge over time in complex systems. For this reason, the model can only be used to analyze transitions after the fact. It cannot inform prospective decision making and action.

All of these modeling methods have emerged from the study of nonlinear dynamics in physical and mathematical systems. They have been applied to human systems in an effort to develop models and methodologies that support better observation, understanding, and intentional action in all scales of social systems from intrapersonal reflections to international relations. Each one of these modeling techniques brings a special feature that helps the researcher or observant practitioner engage effectively and perform well in social systems that may be far-from-equilibrium and actively emergent. All of these models, however, share two characteristics that limit their usefulness in situations of practical decision making and action taking.

3.2 Limitations of Complexity-Inspired Models

Each of these models transcends some or all of the limitations of traditional methods of modeling human systems. Table 1 provides a brief summary of the five modeling methods and their relationship to the five limitations of the traditional models described above. Exceptions to these simple categorizations may certainly exist. Hybrid methods and models are emerging across many fields of study, but in general, these families of approaches share the assumptions described here. Networks and power law analyses deal with multiple levels of analysis and are able to represent massive interdependencies across scales. Networks, agent-based models, and power law distributions assume the possibility of open system boundaries. All of these methods except agent-based modeling assume high dimensional dynamics, and all of the methods account for both nonlinear and non-random pattern formation.

These models share a more realistic set of assumptions than previous modeling approaches, and they account for social systems' complex and nonlinear dynamics. They are far superior in representing the emergent dynamics of complex human interactions, but they still have two characteristics that limit application to formal and informal real-time decision making and action taking in social systems.

Retrospective. First, they are all inductive models, so they can only provide analysis in retrospect. None of them supports effective forecasting or even anticipation of future pattern formation. All of these methods draw data from the past to describe relationships and transformations that happened in the past. Time series analysis requires a long and detailed data set extracted from previous events. Network analysis captures past and current nodes and edges, but it does not help anticipate or recommend action for the future. Being a node in an information network can certainly inform decision making and empower action taking, but creating a network model from empirical data does not inform options for action or risk and benefit calculation except in very limited cases. Catastrophe theory bases model characteristics on existing data and its models represent historical dynamics. Agent-based modeling and power law dynamics both start with sets of assumptions, but the findings of the models are descriptions of historical systemic behavior as opposed to anticipation of future patterns of behavior. In essence, none of these models has the power to anticipate future patterns or to provide

Table. 1 Limiting assumptions for complexity-inspired models of human systems dynamics

Model characteristics	Catastrophe theory	Dynamical networks	Nonlinear time series analysis	Agent-based modeling	Power law dynamics
Multi-level		x			x
Open		x		x	x
High dimension	x	x	x		x
Nonlinear	x	x	x	x	x
Non-random	x	x	x	x	x

intelligence that can inform future action for decision makers who are engaged in self-organizing dynamics in real time.

On the one hand, this reliance on historical patterns is to be expected because complex dynamical systems are sensitive to initial conditions and, by nature, are unpredictable. You would not expect a model based on emergent dynamics to be predictive. On the other hand, if a model is to be of service to real decision makers in real situations, it must provide some level of intelligence about underlying dynamics and ways to intervene to influence an emergent future. An effective model also needs to provide information about the dynamics in any given moment to inform action that might shift those dynamics for future benefit. All of these complexity-inspired models are backward looking, and none of them proposes a causal mechanism, so none of them meets the requirement of support for prospective decision making in actual human systems.

Descriptive. Second, the models are descriptive rather than explanatory. A descriptive model represents the "symptoms" of dynamical self-organizing processes. These models help describe the behaviors that emerged over time in complex human systems. We can use them to determine whether the system patterns conformed to multi-dimensional manifolds (catastrophe theory); generated or broke connections (network theory); showed coherent behavior in phase space (nonlinear time series analysis); generated system-wide patterns from local interactions (agent-based simulations); or generated consistent ratios between numbers and sizes of transforming events (power law). Each of these models describes patterns that emerged among the components of a given system, but they do not capture explanations for those dynamics. An explanatory model, on the other hand, provides information about the underlying relationships that set conditions for observable behaviors to emerge. For example, the life cycle of the rhinovirus is an explanation for the common cold, while a runny nose is a true, but descriptive, symptom. Effective intervention depends on the explanation, and current complexity-inspired models of human behavior provide only descriptions.

We should note two possible exceptions to this rather radical observation. Per Bak and others who work with power law dynamics hypothesize a variety of explanatory theories [36]. Bak, himself, speculates that power law dynamics emerge from cycles of accumulation and release of tension among agents at various levels in a complex system. When enough tension accumulates in one scale of the system, structural changes take place in scales either above or below to release the tension and reach a more stable state. This mechanism is incorporated in the model we will propose shortly. The other notable possible exception is the work of June Holley regarding her network weaving approach [37]. Her work is derived from extensive experience in supporting development of entrepreneurial networks, but the connections between her practical advice and the structure of dynamical models is not always explicit. Her insights have also informed our emerging conceptual model as well as our practice in real human systems. With these slight exceptions, most models from the nonlinear array can, at best, describe the world as it has been. They cannot inform theoreticians or practitioners about conditions under which those patterns emerged or how they might be influenced.

Traditional models of physical and social dynamics provided explanations for phenomena in social systems, but they were of limited use because they drew from Newtonian mechanics, in which systems were closed, low dimensional, and linear. In those situations, the causality and explanations were clear, but not trustworthy when applied to complex human systems. In complex dynamics, where none of these limiting conditions persist, a new explanatory model is needed to inform human action inside emergent, complex adaptive systems.

Interpretation Not Action. Finally, because the complexity-inspired models are neither prospective nor explanatory, they cannot support decision making or action taking in a moment. They do represent systemic patterns of the past and can certainly support meaning making for individuals and groups who want to understand historical patterns. But if the goal is to create a computational model that informs wise forward-looking action, then it must be based on a conceptual model that provides prospective insight and explanation of self-organizing dynamics that help people see, understand, and influence patterns as they emerge in complex human systems.

4 Conceptual Model of Human Systems Dynamics

The challenge is to develop a computational model and modeling methodology that (1) represents the complex and unpredictable dynamics of human systems; (2) works at all scales from intrapersonal to global; (3) provides information about the possible futures of systemic behavior, even knowing that the future of complex systems cannot be predicted or controlled; (4) provides sufficient explanation of interactions within the system to inform options for action; (5) reveals meaningful patterns even in systems that are open, high dimension, and nonlinear; (6) represents the diversity of the parts and the coherence of the whole simultaneously; and (7) provides sufficient explanatory foundation to inform wise and responsible action, even under conditions of the greatest uncertainty. The computational model must be based on a conceptual model that rigorously meets all of these criteria, as well.

Social sciences are not the only contexts in which this challenge exists. Even across the physical and mathematical sciences, no single conceptual model of complex systems dynamics has proved to be general and robust enough to be universally accepted today. Fitness landscapes [38], dissipative structures [39], power law dynamics [40], synergetics [41], and dynamical networks [42] have all proven to be useful conceptualizations of complex dynamics, but none dominates across disciplines. Even in these contexts of physical and biological sciences (which are substantially less open, high dimension, and nonlinear than human systems), no model is accepted as a definitive representation of all complex adaptive phenomena. All accept limitations that compromise fidelity or specificity in favor of tractability.

We propose an alternative conceptual model of human systems dynamics that is derived from patterns common to a wide variety of physical science and mathematical models of complex adaptive systems. This emerging model also draws from philosophical foundations of perception and knowledge that are unique to the functioning of human systems. It is grounded in conscious and intentional action with real groups facing real challenges in organizations and communities [43]. In both theory and practice, this conceptual model captures the dynamics of human systems at all scales, in ways that inform decision making and action taking in complex and uncertain environments.

4.1 Conditions of Self-Organizing in Complex Systems

We accept the definition of *complex adaptive system (CAS)* as a collection of semi-autonomous agents that have the freedom to act in unpredictable ways, and their interactions generate self-organized patterns across the entire collection. As patterns form in the system, they constrain the options of the agents in subsequent phases as this self-organizing process continues [44]. We use an operational definition of *pattern* as similarities, differences, and connections that have meaning across space and/or time. Examples of CAS and the patterns they form are myriad in physical systems (e.g., whirlpools, heart rate variability, embryonic development, crystal formation), but we will focus here on such processes as they influence human systems. In the context of human experience, intrapersonal reflection and emotional and cognitive experiences generate self-organizing patterns for individuals [45]. (Some even argue that consciousness itself is self-organizing processes of a complex adaptive system [46], but that conversation is outside the scope of our current exploration.) Two or more people who form a coherent group for learning, work, or play function as agents in the complex adaptive system [47] of a group. Neighborhoods can be seen as CASs as well as agents that contribute to the patterns of an urban landscape. Firms self-organize from within and participate in emergent patterns of markets and industries. Provinces, nations, national allies all are examples of CASs in the realm of human systems.

Prospective. It is one thing simply to say that human systems self-organize. It is another to track self-organizing processes in retrospect through case study narratives, nonlinear analysis of time series, construction of network maps, and so on. Existing conceptual and computational models, as described above, are sufficient for such descriptive investigations. Analysis that supports proactive decision making in a complex human system, however, requires more. It requires an understanding of the conditions that influence the speed, path, and outcomes of self-organizing processes.

Constraints and Self-Organizing. Examination of diverse models of self-organizing processes in non-human complex systems revealed that self-organizing patterns were influenced by certain conditions. Any complex adaptive system only generates patterns when it is constrained in some way. The Belousov–Zhabotinsky reaction requires a containing vessel, a certain temperature threshold, and very particular chemical gradients. An ecosystem will be bounded in space and requires predator, prey, and reproductive interactions among organisms. Fractals require a nonlinear equation to act as a seed and the context of the complex number plane. A fitness landscape requires specific parameters that define fitness and feedback loops to determine survival on the landscape over time. A laser beam depends on both control and order parameters. A scale-free network has nodes and hubs with specified characteristics and criteria for connection. Without any constraints, regardless of the system or its substrata, no pattern self-organizes.

The same realization about the necessity of constraint in pattern formation came from personal and professional experience in social systems. Self-organizing processes in complex human systems, such as large group meetings or team performance, required some constraining conditions for patterns to emerge. Depending on those conditions, pattern formation was sometimes fast, direct, and clear. In other circumstances, the process of pattern formation was slow, wandering, and messy. Within a particular self-organizing process, the patterns were sometimes coherent and distinct, and sometimes they were distorted or ambiguous. Sometimes the emergent patterns were healthy or fit to purpose, and sometimes they were dysfunctional. Sometimes the self-organizing process moved quickly through exploratory stages and into exploitation. At other times the process got stuck in exploration and productive patterns emerged slowly, if at all. Patterns at the individual scale were sometimes in conflict with those at intrapersonal or group scales, and patterns in two separate parts of the system often contradicted each other.

These variations on the process of human systems emergence were not random. They were influenced by the context, and the context was formed by other self-organizing processes in other places and times. As one pattern emerged, it influenced the conditions that informed the speed, path, and outcomes of other patterns in the vicinity. This interdependency among patterns held whether the patterns were in scales above or below, or in remote parts of the system at the same scale. This mechanism of constraints influencing emergence of shifting, interdependent patterns held across environments, contexts, and levels of analysis.

Influencing Constraints. The question emerged for us: Would it be possible to identify parameters that influenced the variability of self-organizing processes, while acknowledging the open, high dimension, and nonlinear nature of self-organizing in human systems? If so, those parameters could be used to see self-organizing patterns in the moment, understand their potential for future pattern formation, and influence the conditions to nudge the emerging pattern toward desired outcomes.

These conditions that result from one self-organizing process and influence other self-organizing processes were the focus of our investigation and the foundations for the CDE Model for the conditions for self-organizing in human systems. CDE stands for the three fundamental, necessary and sufficient clusters of constraining conditions: container, difference, and exchange. The CDE Model is grounded in a multi-disciplinary study of nonlinear dynamics in a wide range of physical and mathematical sciences as well as action research in diverse human systems settings [48]. We will first introduce the model, then articulate ways in which it meets all of the criteria earlier defined for a conceptual model to inform theory and practice in complex human systems.

4.2 CDE Model for Conditions of Self-Organizing in Human Systems

The CDE Model of conditions for self-organizing in human systems is a qualitative conceptual model of a set of meta-variables that represent a wide variety of constraints that influence the speed, direction, and outcomes of self-organizing processes in human systems. The characteristics of these meta-variables and the nonlinear interdependencies among them form the foundation of a model of human systems dynamics that meets all of our criteria for a useful way to support seeing, understanding, and influencing the complex dynamics of social systems at all scales.

Container. Boundaries and boundary conditions have always been an integral concern of systems theory [49]. Open, high dimension, nonlinear conditions that are common in human systems challenge traditional understandings of system boundaries. Especially in complex human systems, practitioners and theoreticians challenge the traditional ways that systems are defined and system boundaries are represented. Boundaries of human systems must account for multiple, massively entangled levels of pattern formation, from personal insight to mob behavior. At the same time, human system boundaries have to account for the fact that "inside" and "outside" have distinct and significant meanings, and that those meanings change over time depending on perspective. To meet the needs of both flexibility and fidelity to a human system, the CDE Model must account for system boundaries that can be simultaneously open and closed, unitary and multiple, local and global.

In response to this need, the first meta-variable in the CDE Model highlights the boundaries of the focal system. This set is called *container* and is represented by "C". A container (C), of any self-organizing process includes any condition, or collection of parameters, that hold the agents of the CAS close enough together that they can interact and form patterns. A particular container may be physical, like a room or a mountain range. It may be conceptual, like a national identity, a stated purpose, or religious belief. It may be social, like an invitation to a party or an artistic performance. As these examples demonstrate, a container can be a bounding condition (fence), an attractive condition (magnet), or a combination of multiple mutual attractions (network).

In any real human system, there are innumerable containers at play simultaneously, and usually, the containers are massively entangled. Containers can be simply nested within each other, for example, a child is in a classroom, a classroom is in a building, a building in a district, and a district in an educational system. More often, however, the relevant containers are overlapping and interdependent. For example, the child is a member of a family, a scout troop, a baseball team, a cultural community, a gender group, a gang, and so on. Standing within all these boundaries simultaneously, the child is a participant in many different self-organizing complex adaptive systems. At every moment, the child both influences and is influenced by patterns emerging in any of these diverse containers.

C, in the CDE Model, is the meta-variable that encompasses all conditions that influence any particular complex adaptive system at any given point in time. Any particular emergent pattern of a CAS involves some subset of all possible Cs, and within a pattern, some Cs will be more and others will be less relevant to a particular purpose. For example, the child's elementary school class container may be irrelevant when he or she is playing first base, though it does not cease to exist. As intentional agents who see, understand, and influence patterns in self-organizing processes, we recognize the most relevant containers and focus our attention and action on them. At the same time, we understand that innumerable other containers exist and might become relevant at any moment. The simultaneous reality of one configuration of containers and the potential of another configuration of them allow us to deal with the system as if it were simultaneously bounded (to support computation and decision making) and unbounded (to represent the openness of observed social reality).

Containers manifest the self-organizing patterns at a specific point in time, but they also influence the potential for change in the pattern in the future by increasing or decreasing the degrees of freedom for all of the agents contained in the system. Larger or weaker containers reduce the pressure of constraint and increase the degrees of freedom for the agents to establish relations and to organize, so the process of pattern formation is likely to be slower and less articulated. Smaller or tighter containers, on the other hand, tend to increase constraints, decrease degrees of freedom, and so increase the likelihood of collision and the speed and clarity of the self-organizing process.

The correlation between container size and the speed and clarity of the self-organizing process serves as an explanation to support decision making and action taking by individuals and groups. It is significant to note, however, that the relationship is not simply causal. A particular change in the size of a container does not necessarily determine the effect on the pattern as a whole. The precise relationship between the container and the emergent pattern is not predictable in any given self-organizing situation because the influence of the container is mediated by the other two conditions for self-organizing—D (difference) and E (exchange).

Difference. Difference has also historically been an important factor in the ontology and epistemology of systems, and it is the second condition for self-organizing in the CDE Model. Even the ancients recognized the power of difference for causing change and making meaning [50]. Traditional physics considered difference as the key to all kinds of potential energy, and more recently, complexity scholars have explored the power of difference in understanding [51] and influencing systemic patterns [52]. The challenge is that in our open, high dimension, nonlinear human systems, we must deal with the fact that potential is embedded in many differences simultaneously, and that the relevance of a particular difference, and its influence on system-wide patterns, cannot be predicted and can shift without warning.

As a result, the CDE includes "D" as the second meta-variable to capture the myriad differences that influence change in a human system. A *difference* is any gradient or distinction that exists within any given container that bounds the complex adaptive system of focus. At a given moment, in any given human system, at any given scale, an indeterminate number of differences articulate the systemic pattern and hold the potential of the system to change.

In the system dynamics as captured in the CDE Model, relevant differences serve two functions. First, difference articulates a pattern as it emerges out of self-organizing interaction. Difference allows the pattern to be observed, analyzed, and influenced. Differences can be physical, emotional, social, political, financial, or any other dimension you can imagine. They may be subjective, objective, or normative. Regardless of the substantive manifestation of the difference, as agents interact, the interactions among their different characteristics manifest a pattern across the system. At any point in time or location in space, the pattern of a CAS consists of variation in one or another characteristic among the agents bounded by a given system container. For example, in a team, differences in expertise might contribute to the pattern of high performance. In a neighborhood, differences in household income might contribute to the architectural design for the whole. In a nation, differences in political assumptions and values shape the pattern of decision making and action for the government. Differences are relevant in a variety ways in diverse containers. For example, height can be a difference that makes a difference on a basketball team, but it might be irrelevant in an academic learning environment. Difference may refer to the degree of variation (more or less tall, more or less happy) or to the kind of variation (height or attitude). Discrete and large differences build clear patterns, and continuous differences or small ones contribute to fuzzy or ambiguous patterns.

The second function served by difference is to establish potential for change. Potential energy in physical systems is an example of the motive power of difference. Difference in height, spring tension, and heat are all examples of the ways in which potential for change is "stored" in differentiation between or among system-wide parameters. In human systems, the tensions created across differences also motivate action. The teacher knows more, and the student wants to learn. The racist is moved to violence. Platforms of political parties motivate voters. Gender, values, expertise, wealth, curiosity, expectations, age, power, faith commitments are just a few examples of differences that can hold the potential energy of a human system at any point in time. Any of these differences may shift a relevant pattern and result in new patterns self-organizing across a single human scale or between scales. For example, my level of confusion results from a difference within my own cognitive frame. It may cause me to bother my neighbor, and together we may interrupt the flow of an otherwise orderly class. A difference at one place or scale of a system motivates the local pattern to shift, and a shift in local pattern may result in shifts at a more global level.

Difference is such a powerful influence on change because it, too, represents levels of constraint (degrees of freedom). Large differences increase degrees of freedom and tend to motivate rapid or more turbulent change, while small differences constrain degrees of freedom and, therefore, convert more slowly. A system with few relevant differences (lower degrees of freedom, more constraint) will manifest a clear and coherent pattern, while one with many relevant differences (higher degrees of freedom, less constraint) may appear random and be stuck in an entropy trough.

Conversion of the potential energy of difference to the kinetic energy of change in a human system is not simply a causal process. At any moment, differences in multiple containers are influencing each other. For example, a social definition of political correctness, and my desire to make a good impression on my boss, may damp my action in regard to my political or racial bias. Even within the same container, multiple differences vie for dominance in the self-organizing process. For example, I love chocolate, and I am committed to weight control. Sometimes one difference sets the potential for action, sometimes the other does, and sometimes two differences are balanced, and the result is stasis.

The meta-variable of difference, both as demonstration of current pattern and motivation for change in the future, establishes another link in an explanatory pattern that can inform observation, meaning making, and action for people engaged in social systems. As long as any one difference or small set of differences is relevant at a given instant, the system may be manipulated as if it were simply a low dimensional phenomenon. Because "D" is a meta-variable, the CDE problem space is tractable enough for individuals to see (and groups to discuss) coherent mental models of a self-organizing social system. On the other hand, because an unlimited number of differences can be represented within any "C" in the CDE, the system as a whole can be understood to function simultaneously as a high- and low-dimension space. In this way, the model matches both the need for tractable representation and infinite variety of real experience.

While changes in "D" shift patterns in a system, they do not allow for total prediction or control. Multiple interacting differences influence a given pattern at a given moment, so an intentional change in one may be distorted or damped by any other. Even if differences are constrained to a small and manageable number, a change in the complex human system cannot be controlled. By itself, or even in tandem with defined containers, difference can influence direction but not predetermine outcome of a self-organizing process. The final set of meta-variables that simultaneously influence the emergent self-organizing patterns involves the connections across the system that allow difference to accumulate or dissipate.

Exchange. From feedback in traditional systems dynamics modeling [53] to the theory of constraints [54] and complex responsive processes [55] the idea of flow has been essential to conceptual and computational models of human systems. In the past decade, nonlinear relationships and feedback loops have been accommodated in models of human interaction, but they have usually been conceptualized in the context of low dimension and/or closed systems. When a nonlinear relationship is embedded in a quasi-bounded, high dimension space, it quickly becomes intractable. When coupled with high dimension and open boundaries, feedback has resulted in system patterns that are either disordered or radically subjective. These are critical modeling challenges because it is just such exchanges, in open and high dimension space, that are critical to understanding and action in human systems.

To meet this requirement, the final meta-variable in the CDE connects across the system to realize the potential stored in any of its differences. We call this meta-variable *exchange* and represent it by “E”. Exchange includes any transfer of information, energy, force, signal, material, or anything else between or among agents. It appears as flow from one part of the system to another. and it establishes relationships that are observed before, during, and after self-organizing processes in human systems.

While many kinds of exchanges influence pattern formation in human systems, one of the easiest to visualize is the flow of information during a conversation. Two people hold different views or expectations (D). When they come together for a particular purpose (C), they exchange (E) information, and if everything goes well, a coherent pattern of the whole emerges. The emergent pattern is not predictable. It may be increasing anger, distrust, frustration, and separation, or it may be shared mental models and harmonious friendship. The presence of the exchange cannot pre-determine the nature of the emergent pattern, but the absence of exchange will result in no shared pattern at all.

Exchanges are taking place in many different containers and across many different relevant differences simultaneously. An individual is thinking about one thing, her team is focusing a conversation on another, senior management is assessing the team’s performance against others, and other firms in the industry are seeking competitive intelligence. All of these Es are simultaneously influencing the emergent patterns at all scales of the system. A change resulting from one may well influence the efficacy of another. Exchanges that are invisible to one system participant may be quite powerful for another, and those that are global and formal are often less powerful than those that are local and informal.

Like containers and differences, exchange derives its influence on systemic behavior from increasing or decreasing degrees of freedom. The power of a given exchange is denoted by limiting degrees of freedom by the tightness of the connection it establishes across differences in a system. Tightness is determined by a variety of factors, including speed (time required to complete the connection), width (number of differences considered), strength (size of differences traversed). Tighter exchanges (increasing constraint and reducing degrees of freedom) tend to speed up self-organizing processes, and weaker ones (lower constraint, more degrees of freedom) slow it down. When exchanges are absent, no self-organizing change will occur at all.

Again, the “E” meta-variable can be used to assess and influence self-organizing patterns as they emerge. One might tighten exchanges to increase the speed and influence the outcomes of a self-organizing process, but the results of that intervention are unpredictable. The relationship is not simply causal. The variability of competing exchanges and the indeterminacy of containers and differences in the system at large make the future unknowable, even while exchanges can be manipulated with the intention of influencing the speed, path, or outcomes of a self-organizing process.

Interdependencies among C, D, E. The indeterminacy of the meta-variables of C, D, and E help capture the open, high dimension, and nonlinear nature of human systems. The relationships among the sets of meta-variables account for the emergent, self-organizing dynamics of those systems. Variability of the speed, direction, and outcome of a self-organizing process is influenced by the dynamic relationships among members of the same variable class within each of the meta-variables (C_1 to C_2, D_1 to D_2, or E_1 to E_2). The process is also influenced by the nonlinear relationships among the three collections of meta-variables (C_n to D_n to E_n). These interdependencies generate a variety of interesting consequences that increase the fidelity of the CDE Model to the lived reality of self-organizing processes in human systems. A shift in one difference begins a self-organizing shift in other differences, as well as in multiple exchanges within the same container and in related containers as well. For example, I observe an anomaly (D_1), this causes me to question (E_1) other observations (D_2) in the current experiment (C_1) as well as to challenge a protocol (E_2) that might be used by others on my team (C_2). Given the number of variables represented with each meta-variable and the interconnections among the meta-variables in any system moment, it is easy to understand how complex adaptive systems are sensitive to initial conditions. This complex interdependency among system conditions helps explain why complex adaptive systems can be unpredictable at the local and patterned at global scales.

Any feature of a social system that might influence a self-organizing pattern can be accounted for in this conceptual model. Even system features that would otherwise be seen as extraneous can be incorporated into the CDE portrait of any systemic pattern. The only adjustment that is required is to recognize the larger, relevant container that encompasses both what was originally "outside" with what was "inside." And, because the three conditions do not depend on time or distance, a CDE portrait is necessarily scale-free. For example, when we work with teachers in the throes of school reform, they focus primarily on their classrooms (C_1) and their students (C_2). When they perceive the requirement for high-stakes testing (E_1), they see it as an external force (C_3) that is being imposed on them and their students from the outside. Instead, we encourage the teachers to see high stakes testing as a difference that makes a difference (D) in their classrooms. By recognizing the test as a part of a classroom pattern, rather than an external imposition, new options for teaching and learning (E_2) emerge.

The three conditions (C, D, and E) are mutually determined, so changes in any one spontaneously result in changes in the other two conditions. It is possible that the broader universe of conditions holds a particular pattern (C or D or E) in place, so an adjustment may not be immediate or predictable. Ultimately, as the related conditions shift, more and more energy is required to resist the tendency of the system to adapt internally. For example, a corporate focus on profits (D_1) as a sole difference that makes a difference can pre-determine (E_1) policies and procedures (C_1) that contradict (D_2) individuals' (C_2) personal values (D_3). If exchanges are established that allow employees who share values to talk (E_2) with each other, they may amplify the values difference between them and their boss and, ultimately undermine profit-dominated patterns of behavior.

Difference and container have a special relationship in the context of complex adaptive systems. Within a container, a difference can denote a pattern and set conditions for self-organizing change. At the same time, if that difference is great enough, it may overwhelm exchanges that mediate the difference within the whole, until the system bifurcates along the fault lines formed by the difference. When this happens, the characteristic that functioned as a difference previously is transformed into a container, which bounds a new and somewhat autonomous system. In an opposite process, two containers may be connected by an exchange that is strong enough to elicit a shared pattern. In that case, the previous containers become mere differences that make a difference within a new emergent whole. Examples of both of these situations abound in real human systems. A team includes people of both genders, so gender is a difference that may or may not make a difference. A sexist joke or a harassing behavior can turn that difference into a container in which the men and the women face off against each other. At the same time, race and experience, which were also differences within the original container, may be invisible and equally distributed across the system, waiting for circumstances that transform them into features that contain rather than just differentiate an emerging pattern.

Recognizing the features that contain, differentiate, and connect across self-organizing patterns allows a conscious agent to observe the process of change as it occurs. Understanding the relationships among the conditions for self-organizing and seeing the pattern from a variety of perspectives allows the conscious agent to make meaning of social change as it emerges. Acknowledging options for action and anticipating the nonlinear consequences of a change allows the conscious agent to take intentional action to shift patterns in the course of self-organizing. Finally, continuing to observe the system patterns in formation allows the conscious agent to take adaptive action to amplify opportunity and mitigate risk in real time [56]. These three iterative problem solving steps—observe, understand, influence—engage the actor/observer with patterns as they emerge. They are the foundation of adaptive action and adaptive capacity for individuals, communities, and organizations.

4.3 Implications of the CDE Model

The CDE Model addresses and resolves many of the limitations of models of human systems that were informed by Newtonian and previous applications of complexity science, but it also introduces some challenges of its own as a foundation for either a conceptual or a computational model.

CDE Describes Social Phenomena. Like other complexity-inspired models, the CDE is able to address phenomena across levels of organization. As we stated before, none of the meta-variables necessarily includes either time or distance, so they are applicable on any scale of human system, as well as across scales in the same system at the same time.

The CDE Model represents human systems as simultaneously open and closed. While one might focus on a single configuration of C, D, and E at one moment, it is done with the consciousness that an infinite number of other containers, differences, and exchanges might become relevant at any moment.

CDE as a conceptual model also opens the space for a system to be both high and low dimension at the same time. Differences that make a difference at one moment may be supplanted by any one of a slew of other differences, all of which fit into the category of conditions represented by the meta-variable, D.

The CDE Model matches the nonlinear causality that is so familiar in human systems. The massive and mutual interdependency among the meta-variables reflects the massive interdependency among the conditions that shape the speed, direction, and outcomes of real self-organizing processes in the real world.

The dynamics of CDE are not predictable, but they are not random either. Changes in one condition or pattern are highly correlated to changes in others, both local and distant. On the other hand, CDE can account for random behavior when the conditions are under-constraining. Low constraint happens when the C is large, the Ds are many, and/or the Es are weak. In these situations, the coupling and correlations among the conditions for pattern formation are so low that system behaviors may be indistinguishable from randomness.

In addition to resolving the limitations of traditional models of social interaction, the CDE Model also resolves challenges unmet by other complexity-inspired models. The CDE captures the underlying dynamics of self-organizing processes by naming the categories of constraint that initiate and influence the shape of emerging patterns as well as the process of emergence. Because it explains underlying interrelationships, it can be used to inform intentional action to shift the conditions and, one can hope, influence the future path and outcomes. The causality is not absolute, and results are not predictable. The complex interactions among self-organizing conditions, both seen and unseen, make unintended consequences not just common but expected.

Finally, the CDE Model allows an observant actor to anticipate the future consequences of current action. In this way, the CDE allows a conscious actor to construct a conceptual model of an ethical dilemma. We can consider multiple options for action and imagine possible consequences, risks, and benefits for each option. When a choice is made, the consequences may or may not be as anticipated, but each cycle of adaptive action informs the next, so individuals and groups can learn and improve their capacity to adapt and influence patterns over time.

CDE Informs Practice. The CDE Model, as a conceptual representation of the complex dynamics of human systems, meets the challenges set out at the beginning of this paper. The CDE Model of the conditions for self-organizing in human systems (1) represents the complex and unpredictable dynamics of human systems; (2) works at all scales from intrapersonal to global; (3) provides information about future systemic behavior, even knowing that the future of complex systems cannot be predicted or controlled; (4) provides sufficient explanation of interactions within the system to inform options for action and anticipated possible outcomes; (5) reveals meaningful patterns even in systems that are open, high dimension, and nonlinear; (6) represents the diversity of the parts and the coherence of the whole simultaneously; (7) provides sufficient explanatory foundation to inform wise and responsible action, even under conditions of the greatest uncertainty.

As a conceptual model the CDE has been tested in a variety of contexts and under a wide range of conditions. It is currently informing theory and practice in education, school reform, program evaluation, occupational therapy, conflict resolution, organization development, management, leadership, team building, advocacy, public health, public policy advocacy and implementation at all levels of governance, human resource development, facilitation, diversity, ethics, and community development [57]. As a qualitative tool, it supports individuals and groups as they engage in adaptive action to see, understand, and influence the self-organizing patterns of complex human systems. The outstanding question is if and how the CDE Model might also form the foundation for a computational model of individual and collective human behavior.

5 Computational Model of Human Systems Dynamics

Over the years, we have experimented with a variety of computational models to represent the complex dynamics represented in the CDE Model. The most successful efforts at inductive model building and testing have focused on qualitative descriptions drawn from narrative or shared dialogue. We have also experimented with a variety of inductive and deductive mathematical models, including time series analysis, fuzzy logic, genetic algorithms, and dynamical networks. While all of these proved somewhat helpful, none was sufficient to inform decision making or action is real situations.

We continue to search, however, for a computational model that represents emergence of patterns of human systems dynamics that are open, high dimension, and nonlinear, like those represented conceptually by the CDE Model.

The computational or simulation model we propose will represent the C, D, and E as meta-variables that describe functional clusters of conditions that constrain and inform systemic patterns. At the most macroscopic level the three conditions are co-active collections of parameters that exist in the particular situation. A collection of "containing" characteristics holds the system together without generating an impermeable boundary. A collection of "differentiating" characteristics articulates the pattern and provides the energy and directionality for change to occur in the future. A collection of "exchanging" characteristics moves information, material, and energy around the system to release tension of difference in one physical or conceptual place and contribute to accumulating tension in another.

The model will make it possible for relevant Cs, Ds, and Es to be unique in any given situation and at any given time. Over time, even if the context remains constant, different Cs, Ds, and Es can increase or decrease in relevance to shift the pattern into self-organizing change and transform the potential energy stored in the pattern into actual systemic change.

In a robust computational model, a single feature might serve the function of container, difference, or exchange, depending on the context and the perspective of the system observer and the intention of the system analysis. In practice this happens often. I belong to a team (C). My team behavior is different from my other professional behavior (D). When I am in meetings, my team communications (E) are clear and effective. The computational model will need to acknowledge and incorporate this local specificity and global ambiguity.

The computational model would support nonlinear interactions among individual parameters as well as clusters of meta-variables representing the Cs, Ds, and Es, so that a change in one generates a change in the other where the tendency or direction of change can be anticipated, though the specific outcome is indeterminate. For example, I expect for the noise level (D_1) to decrease when I move a party to a larger space (C), but perhaps the guests will simply talk louder (E), and the overall volume (D_2) will remain unchanged.

An adequate model would support manipulation of a complex combination of constraints of the three conditions on each other or of particular conditions on others of its same category. A combination of conditions might influence the emergent patterns within the system and provide options for action for individuals and groups interacting with the system. In this way, it should be possible to model changes in multiple Cs, Ds, and Es either in series or in parallel. For example, if we simultaneously moved to the larger room (C) and encouraged people to dance (E), what would happen to the volume in the room (D)? What other conditions, perhaps not observed previously, would emerge as relevant to the pattern?

A model that met these criteria would represent human systems dynamics in a way that would be simultaneously flexible enough to match lived experience and elegant enough to support realistic processing and interpretation.

6 Conclusion

In this document, we have summarized and critiqued a variety of models designed to represent the dynamics of human systems. Some are based on traditional understanding of mathematical, physical, and social relationships. Others are based on nonlinear dynamics and principles of emergence in complex adaptive systems. All of these models have proven useful in specific applications in a variety of social science inquiries, but they all fall short of capturing the open, high dimension, nonlinear interaction of human systems dynamics in experience or of providing sufficient explanation to support prospective understanding and action.

While the CDE Model has proven to be a robust and useful interpretive conceptual model, its value to both research and practice in social sciences will be immeasurably enhanced when it can inform a computational model. Such a model will capture the nonlinear nature of the dynamics of human systems to inform the work of scholars and practitioners who face complex and emergent personal, professional, and political challenges. Scholars would use such a model to develop and test hypotheses about self-organizing dynamics at all scales of human interaction. Practitioners would use the model to inform decision making in real time and to develop individual and group capacity through training and instruction. Individuals and groups would build adaptive capacity to see, understand, and influence complex and unpredictable patterns as they emerge.

The network of Human Systems Dynamics scholar-practitioners invites you to engage with us in an inquiry that moves the CDE Model from the realm of conceptual and into the realm of computational modeling. Together we might create a model to help scholars and practitioners explore and develop adaptive capacity to leverage the emergent potential in our social systems today and into the unknowable future.

References

1. G.E.P. Box, N.R. Draper, *Empirical Model-building and Response Surfaces* (Wiley, New York, 1987), p. 424
2. J.E. Roeckelein, *Elsevier's Dictionary of Psychological Theories* (Elsevier, Amsterdam, 2006)
3. R. Morriss, Psychiatry **8**(3), 82–86 (2009)
4. Q. Miao, L. Liu, Soc. Behav. Personal. Int. J. **38**(3), 357–363 (2010)
5. M.K. Dhami, Psychol. Sci. **14**(2), 175–180 (2003)
6. B.J. Puvathingal, D.A. Hantula, Am. Psychol. **67**(3), 199–210 (2012)
7. L. Chen, G. Chen, Physica A: Stat. Mech. Appl. **374**(1), 349–358 (2007)
8. A. Gibbard, H.R. Varian, J. Philos. **75**(11), 664–677 (1978)
9. P.R. Mitchell, J.E. Sault, P.N. Smith, K.F. Wallis, Econ. Model. **15**(1), 1–48 (1998)
10. G.M. Lady, J. Econ. Dyn. Control **19**(3), 481–501 (1995)
11. Twentieth-Century Model 'A Global Suicide Pact', Secretary-General Tells World Economic Forum Session on Redefining Sustainable Development (2011), UN News Center. http://www.un.org/News/Press/docs/2011/sgsm13372.doc.htm. Accessed 28 Jan 2011
12. G.H. Eoyang, *Conditions for Self-Organizing in Human Systems*. Ph.D. Thesis. The Union Institute and University (2001)
13. C. Moraru, *Cosmodernism: American Narrative, Late Globalization, and the New Cultural Imaginary* (University of Michigan Press, Ann Arbor, 2011)
14. A. Bontea, *The Year's Work in Critical and Cultural Theory* (Oxford Jounals, Oxford, 2009), pp. 348–366
15. www.crawdadtech.com
16. T.Y. Choi, K.J. Dooley, J. Supply Chain Manag. **45**(3), p. 25 (2009)
17. www.sensemaker.com
18. M.S. Poole, *Organizational Change and Innovation Processes: Theory and Methods for Research* (Oxford University Press, New York, 2000)
19. R. Rosen, Bull. Math. Biol. **39**(5), 629–632 (1977)
20. J. Kaldasch, Physica A Stat. Mech. Appl. **391**(14), 3751–3769 (2012)
21. S.H. Strogatz, Nature 410.6825 268 (2001)
22. A. Krifa, M. Mendonca, R. Bin, N. Rao, C. Barakat, T. Turletti, K. Obraczka, ACM SIGCOMM Comput. Commun. Rev. **41**(4), 480 (2011)
23. X.H. Xiao, G.W. Ye, B. Wang, M.F. He, Physica A Stat. Mech. Appl. **388**(5), 775–779 (2009)

24. R. Toivonen, J. Onnela, J. Saramäki, J. Hyvönen, K. Kaski, Physica A Stat. Mech. Appl. **371**(2), 851–860 (2006)
25. S.J. Guastello, R.A. Gregson, *Nonlinear Dynamical Systems Analysis for the Behavioral Sciences Using Real Data* (CRC Press, Boca Raton, 2011)
26. J.M. Epstein, *Generative Social Science: Studies in Agent-Based Computational Modeling* (Princeton University Press, Princeton, 2006)
27. W.I. Sellers, R.A. Hill, B.S. Logan, Philos. Trans. Biol. Sci. **362**(1485), 1699–1710 (2007)
28. E. Samanidou, E. Zschischang, D. Stauffer, T. Lux, Rep. Prog. Phys. **70**(3), 409–450 (2007)
29. R.K. Sawyer, *Social Emergence: Societies as Complex Systems* (Cambridge University Press, Cambridge, 2005)
30. S. Rodriguez, V. Julián, J. Bajo, C. Carrascosa, V. Botti, J.M. Corchado, Eng. Appl. Artif. Intell. **24**(5), 895–910 (2011)
31. G.T. Jones, R. Hagtvedt, J. Bus. Ethics **77**(1), 85–97 (2007)
32. J.H. Miller, S.E. Page, *Complex Adaptive Systems: An Introduction to Computational Models of Social Life* (Princeton University Press, Princeton, 2007)
33. P. Bak, *How Nature Works: The Science of Self-Organized Criticality* (Copernicus, New York, 1996)
34. A.T. Stephen, O. Toubia, Soc. Netw. **31**(4), 262–270 (2009)
35. C. Anderson, *The Long Tail: Why the Future of Business is Selling Less of More* (Hyperion, New York, 2006)
36. M.J. Newman, Contemp. Phys. **46**(5), 323–351 (2005)
37. J.Holley, *Network Weaving Handbook*, http://www.networkweaver.com (2011)
38. S.A. Kauffman, *The Origins of Order: Self-Organization and Selection in Evolution* (Oxford University Press, New York, 1993)
39. I. Prigogine, I. Stengers, *The End of Certainty: Time, Chaos, and the New Laws of Nature* (Free Press, New York, 1997)
40. P. Bak, *How Nature Works: The Science of Self-Organized Criticality* (Copernicus, New York, 1996)
41. G. Hertel, J. Manag. Psychol. **26**(3), 176–184 (2011)
42. J.P. Bagrow, D. Wang, A.-L. Barabási, PLoS One **6**(3), e17680 (2011)
43. G.H. Eoyang, Conditions for Self-Organizing in Human Systems. Ph.D. Thesis. The Union Institute and University, (2001)
44. K. Dooley, A. Van de Ven, Organ. Sci. **10**(3), 358–372 (1999)
45. O. Gomes, Fractals **42**(2), 1206–1213 (2009)
46. J. Neisser, Conscious. Cogn. **21**(2), 681–690 (2012)
47. S. Mennin, Teach. Teach. Educ. **23**(3), 303–313 (2007)
48. G.H. Eoyang, *Voices from the Field: An Introduction to Human Systems Dynamics* (Human Systems Dynamics Institute Press, Circle Pines, 2003)
49. L.V. Bertalanffy, *General System Theory; Foundations, Development, Applications* (G. Braziller, New York, 1969)
50. Plato, *The Sophist: The Professor of Wisdom*, ed. by: E. Brann, P. Kalkavage, E Salem (The Focus Philosophical Library, Newburyport, 1996)
51. J. Kelso, D. Engstrøm, *The Complementary Nature* (MIT Press, Cambridge, 2006)
52. S.E. Page, *The Difference: How the Power of Diversity Creates Better Groups, Firms, Schools, and Societies* (Princeton University Press, Princeton, 2007)
53. D.H. Meadows, D. Wright, *Thinking in Systems: A Primer* (Earthscan, London, 2009)
54. E. Goldratt, *What is this Thing Called Theory of Constraints* (North River Press, Croton-on-Hudson, 1990)
55. R. Stacey, *Complex Responsive Processes in Organizations: Learning and Knowledge Creation* (Routledge, London, 2001)
56. G.H. Eoyang, R. Holladay, *Adaptive Action: Leveraging Uncertainty in Your Organization* (Stanford University Press, San Francisco, 2013)
57. HSD Institute: Directory of HSD Associates (2012), http://www.hsdinstitute.org/community/associates.html. Accessed 25 May 2012

Stochastic Complexity Analysis in Synthetic Biology

Natalja Strelkowa

Abstract In this chapter we present selected mathematical methods used for the design of synthetic gene networks. We showcase the method applications on the generalized repressilator model which simplest special cases—a genetic switch and a genetic oscillator—have been implemented in living bacterial cells. The concepts for switches and oscillators used in synthetic biology are originally from electrical engineering and were transferred to synthetic biology using mathematical analysis. However the straight transfer is not always possible due to low copy number stochasticity and time-limited cell environments. We explore the implications of inherent stochasticity and of nontrivial long lasting transients on gene network models revealing that these effects can lead to qualitatively different dynamics compared to steady states. Therefore, mathematical analysis of synthetic gene expression networks should not be limited to exploration of the steady states, but include the influence of stochasticity and transient dynamics.

Keywords Bifurcation analysis · Stochastic models · Transients · Synthetic biology

1 Introduction

Can we engineer biological systems in a predictable manner to perform new functions or to optimize already existing ones? And what are the underpinning principles behind the forward design of gene regulatory networks? These questions gave rise to Synthetic Biology—a rapidly advancing interdisciplinary science that aims to create new or redesign existing biological organisms [1–4]. Inspired in many cases by electronic elements, simple gene networks have been designed to

N. Strelkowa (✉)
Department of Bioengineering, Imperial College London, SW7 2Z London, UK
e-mail: natalja.strelkowa@googlemail.com

I. Zelinka et al. (eds.), *How Nature Works*, Emergence, Complexity and Computation 5, DOI: 10.1007/978-3-319-00254-5_8,
© Springer International Publishing Switzerland 2014

perform reproducible, low-level functions. The pioneering synthetic circuit implementations in 2000 include the genetic ring oscillator known as the repressilator [5] and the toggle switch [6].

Since then progress has been made in experimental implementation of engineered genetic transcription networks during the last years. Simple synthetic modules were used to create mammalian cell oscillators with integrated parts of natural cellular machinery [7] or cell-synchronized oscillators in bacterial populations [8]. Synthetic networks were also interfaced with cellular pathways to induce particular responses, as in the construct where the toggle switch was connected to the SOS pathway to induce DNA protection mechanisms in *E. coli* when exposed to UV light [9]. There are also examples for medically inspired designs as for instance selective invasion of cancer cells [10]. This experimental progress infuses the minds of mathematicians and engineers with new design ideas for more complex artificial gene networks, which would perform more sophisticated functions.

A successful design of an artificial genetic network is often preceded by mathematical modeling [11, 12], which usually includes abstraction of biochemical processes to differential equations. The abstraction steps include several assumptions on biochemical reactions as for instance that the laws of mass action are approximately valid or that the number of DNA copies in a cell is constant. After the abstraction a biochemical process can be coded in a mathematical model amenable to numerical solution. Very often inherent stochasticity of gene expression process strongly influences dynamic behavior of the network. Then the stochasticity can be explicitly included in the model by using either the Master equation formalism [13] for analytical calculations or Gillespie algorithm [14] for numerical evaluation.

In this chapter we review a gene expression model derivation and selected mathematical methods which are commonly used in the design of synthetic gene networks and show their application on the generalized repressilator model (uni-directionally coupled genetic repressors ordered in a ring). The generalized repressilator model has been extensively explored in the deterministic setting using nonlinear complexity analysis [15–18] and the special cases of two and three genes have already been implemented in living bacterial cells using synthetic biology tools [5, 6].

The model shows qualitatively different steady state characteristics if the rings consist of even number of repressors compared to rings with odd number of repressors [15]. Odd rings are oscillators in which stable oscillations emerge via Hopf bifurcations. The steady state of even rings is a dimerized up/down solution (i.e. they are switches), although even rings are also characterized by the presence of long-lived oscillatory transients that can dominate the observable dynamics in such systems [18].

In electrical engineering there are other well-studied uni-directionally coupled ring systems, which are currently in use for practical applications. Interestingly, the steady state feature that even numbered rings are switches and odd numbered rings are oscillations remains preserved for electrical as well as biological systems.

We show parallels between the generalized repressilator model and the well-established ring symmetric structures via bifurcation analysis and group theory and discuss how engineering concepts can be adapted to remain functional in noisy and time-limited cell environments.

2 Gene Expression in Bacteria

In the early 2000's, synthetic circuits for bacterial cells were implemented on plasmids using cloning techniques [19]. Plasmids enter a bacterial cell by transformation, i.e. synthesized DNA is taken up by the cell after a chemo-physical treatment (e.g. cell chilling in the presence of cations followed by a heat shock). Once the plasmid is inside the cell, the available transcription machinery of bacterial cell ensures the expression of the genes encoded on the plasmid. Any DNA sequence can be coded and synthesized on a plasmid creating great opportunity for design of new cell functionalities.

The engineering of synthetic networks requires a thorough understanding of natural gene expression in bacterial cells and their abstraction to mathematical models. For that we need to pinpoint biochemicals which control the gene expression process, e.g. are present in low or non-constant amounts. These process limiting biochemicals have to be contained in mathematical models either implicitly as constants or explicitly as dynamical variables. Time scales on which the reactions occur are also important for model derivation. Biochemicals which quickly reach the dynamical steady state do not need to be considered as dynamical variables, but can instead be approximated as constants defined by their steady state values. So, as describing the gene expression in the following we concentrate on the identification of limiting step biochemicals and will classify the reactions as fast and slow.

The production of synthetic proteins occurs in two steps: DNA transcription resulting in messenger RNA (mRNA) chains and mRNA translation leading to creation of proteins (see Fig. 1). As initiation of gene expression process a prokaryotic transcription initiation factor σ binds to RNA polymerase (RNAP) forming a holoenzyme (see Fig. 1a). The holoenzyme complexes are directed by σ-factors towards gene regulatory region so that DNA coding region is transcribed and its information is converted to mRNA—the basis for the engineered proteins. There are several different σ-factors in bacterial cells, but one factor is highly conserved across all bacteria and is responsible for expression of most indispensable genes during normal growth conditions. This σ-factor constantly binds and unbinds to the gene regulatory region located -35 to -10 basepairs upstream of the gene and the transcription can only start then the factor is bound to the DNA.

In order to take advantage of this natural transcription machinery the engineered DNA sequences must contain the upstream promoter region recognisable by the σ-factor. Then the abundantly present σ-factor will direct the holoenzyme to

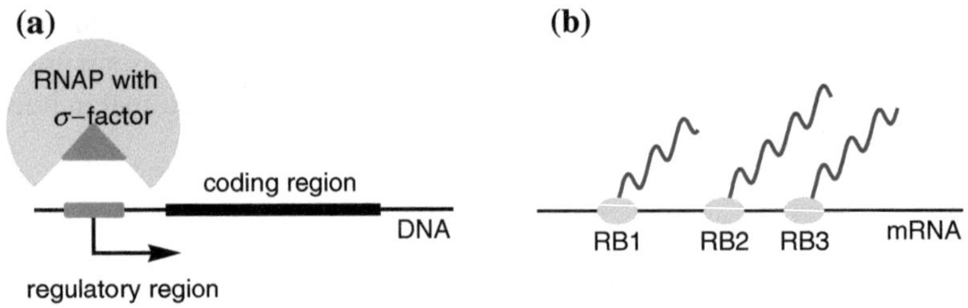

Fig. 1 Abstract view on gene expression in bacteria. **a** Transcription initiation. The structure of a typical bacterial gene consists of a regulatory region (*dark-grey filled rectangle*) followed by a coding region (*black filled rectangle*). The holoenzyme composed of RNA polymerase, (RNAP *light-grey filled circle*) and promoter recognising r-factor (*dark-grey inserted triangle*) is about to bind to the gene regulatory region. The transcription direction after binding of holoenzyme is indicated by a *black arrow*. **b** Translation in bacteria. Three ribosomes RB1, BR2, RB3 are translating one mRNA template into proteins, which are indicated with *dark-grey* curved lines. Since ribosomes are in abundance, protein production rate can be approximated as a linear first order reaction, which is proportional to the amount of mRNA. The rate constant for the translation reaction can be larger than one, because several ribosomes can simultaneously translate one template

the designed DNA sequence and make sure that the coding region is transcribed as if it was an indispensable gene in normal growth conditions. Both σ-factor and RNAP are abundantly present in the cells and are not limiting for gene expression. The binding and unbinding of the factor to the regulatory region of the DNA is classified as a fast reaction.

For an abstract mathematical model we need to identify biochemicals which control the transcription reaction, i.e. are limiting factors. A closer look at the involved biochemicals reveals that the number of DNA copies strongly influences the amount of mRNA. The number of mRNA transcripts is indeed directly proportional to the DNA copy number and the amount of mRNA is quite low (in the order of few tens) compared to other chemicals involved in the transcription process. There is a delay between the binding of the transcription activator and the appearance of the final mRNA transcript, because mRNA is slowly produced by chain elongation, i.e. biochemical letters are attached one by one to an already existing piece according to the DNA code.

The transcription of genes is regulated by activators (e.g. σ-factors) initiating the transcription and by inhibitors repressing the transcription. Inhibitors are other proteins which specifically bind to the regulatory region of the gene and hinder the transcription initiation. Activators and inhibitors constantly bind and unbind to the regulatory site of the gene and these reactions usually occur on a very fast time scale. The inhibition mechanism requires that the shape of the inhibitor protein is tailor-made for the associated regulatory region. The amount of mRNA will strongly depend on the amount of inhibitor proteins and this needs to be explicitly taken into account in a mathematical model. Indeed the pioneering implementations of the oscillator and the switch as synthetic networks are based on the inhibitory gene regulation [5, 6].

In summary, bacterial transcription majorly depends on the amount of regulatory proteins and the DNA copies in the cell. It involves biochemical reactions on two different time scales: the attachment and the detachment of the regulatory proteins is fast compared to the mRNA production in the elongation step. The overall number of mRNA transcripts is usually quite low, so that stochasticity due to low copy number needs to be considered in mathematical models.

The second step in the gene expression process is the translation of mRNA to protein sequences. Several natural biomolecules are involved in this reaction and they have to build a complex and attach to the mRNA, which was produced by the transcription reaction and contains the code for the designed protein. All natural biomolecules for the translation machinery are abundantly present in the cell, since they are involved in all translation processes and keep the cells alive. The only rate limiting factor of this reaction is the amount of mRNA containing the code for the engineered protein.

Similarly to mRNA, proteins are created by elongation of polypeptide chains (see Fig. 1b). Several copies are produced from a single mRNA chain at the same time, because several translation units can simultaneously attach to the same mRNA chain. So the overall number of proteins is usually larger than the number of mRNAs and is in the order of hundreds or thousands. Protein creation is a slow reaction occurring on similar time scales as the mRNA creation [11].

mRNAs and proteins produced in the transcription and translation reactions have limited lifetimes, which means they are destroyed regularly by cell internal processes. mRNA transcripts are degraded by ribonucleases. The exact process and which biomolecules are more important is still disputed [20, 21], but there are strong indications that there is only one initiation process. Once this process happened mRNA cannot be used as a protein transcript any longer and is degraded through a decay pathway. The decay molecules involved are present in approximately constant amounts in the cell cytoplasm and therefore mRNA degradation can be assumed as a first order chemical reaction. The amount of transcripts degraded per time is proportional to the mRNA copies and the proportionality constant is reciprocal to the mean life time of mRNA, which is $\approx$ 2–4 min in *E. coli* [22]. Proteins are longer lived than mRNAs with the average lifetime of 40–60 min [5, 23, 24]. Through similar degradation pathways as mRNAs the proteins are degraded by proteases and the degradation can be approximated as a single step biochemical reaction for the same reasons.

Summarising, we can identify four major reactions dictating the gene expression:

1. production of mRNA involving fast binding and unbinding of activators/repressor proteins and depending on the amount of DNA and the amount of activators/repressors
2. production of proteins mainly depending on the amount of mRNA
3. mRNA degradation mainly depending on the amount of mRNA
4. protein degradation mainly depending on the amount of proteins.

3 Abstraction of the Gene Expression Process and Model Derivation

3.1 Abstractions and Simplifications

The underlying gene expression process involves several biochemical steps, and some of them are more important for the qualitative and quantitative gene network dynamics than others. The aim of mathematical modeling is to take these important steps into account neglecting or approximating all other processes and provide an abstract model, which can be used for design development and evaluation of synthetic gene networks. The following is a summary of typical assumptions on biochemical processes often implicitly contained in mathematical models:

- **Laws of mass action.** Processes modeled by deterministic differential equations assume gas-like dynamics of mRNAs and proteins with characteristics such as free diffusion, random collision, and homogeneous mixture of molecules. If molecules involved in the model process have a relatively small size (smaller than 400 kDa) then this assumption holds in the first approximation. But we also have to keep in mind that bacterial cytoplasm is highly packed with proteins (340 mg/ml) [25] leading to molecular overcrowding and gel behavior of the cytoplasm. In particular, for large molecules the laws of mass action might not hold and overcrowding effects might need to be included in the model.
- **Synthetic genes are present in constant amounts per cell.** There are actually two ways how a synthetic genetic element can be incorporated into the bacteria. The first possibility is to encode the synthetic circuit onto a plasmid, which will be taken up by the cell. Plasmids do replicate inside the cell and this process must be controlled. For the classical examples of the toggle switch and the repressilator [5, 6] replication control ECol1 was encoded on the synthesized gene. A different way to encode synthetic elements is using expression cassettes, which would incorporate the artificial DNA directly into the bacterial genome. This procedure is more controlled and ensures the constancy of the gene copy numbers.
- **Engineered proteins bind inhibitory to the DNA regulatory regions without leakiness.** Basal transcription can easily be incorporated into the models as an additive term, however this effect should be negligible, if the engineered proteins are properly designed.
- **Binding and unbinding of the regulatory proteins and DNA is much faster than transcription and translation.** The dynamics of the intermediate complex consisting of DNA and regulatory protein occurs on much faster time scale than other reactions. Taking the time scale separation into account we obtain Michaelis–Menten-like kinetics for the transcription rate. Standard mathematical derivations are provided in detail in Sect. 3.2.

- **Protein and mRNA degradation are first order reactions.** Engineered bacterial proteins are degraded by simple mechanisms and we approximate their degradation as a first order reaction. But we have to keep in mind that there are other natural proteins, which degrade in several steps and in particular for eukaryote cells several steps are involved in folding/creation as well as degradation reactions [26]. Then the mathematical models need to account for these several steps using for instance an approximation.
- **Reaction constants are not time dependent.** Effectively we neglect cell growth and death and the change of the cell environment as the cells get older [21]. These effects would influence the reaction constants. A robustness analysis slightly perturbing the reaction rates around their reference values can be performed to ensure the network functionality under these circumstances.
- **All molecules are present in large amounts.** This constraint is only assumed for the deterministic description with ODEs. As it turns out stochastic fluctuations do have a qualitative effect on the overall dynamics and their role will be extensively discussed and showcased for our generic example the generalized repressilator model.
- **Cell division is a renormalization step.** One of the poorly investigated, but potentially very important effect on the synthetic circuit is the cell division, which for *E. coli* happens approximately every 15–60 min depending on the feeding conditions. The dynamical features such as switch toggle [6] or oscillation period [5, 8] take hours, which means they are spread over several cell generations. Formally we can assume that the cell division is strictly symmetric and the reactions are renormalizable (see for instance [27] Chap. 8 for definition of renormalization group). Effectively it means that the circuit dynamics do not depend on the system size, but scale with the system and their qualitative dynamics remain unchanged. These assumptions are often not fulfilled, because cell division and plasmid distribution between the daughter cells are not symmetric and it is not precisely clear how biochemical reactions scale with the cell size.

3.2 Derivation of a Deterministic Gene Expression Model

In previous sections biochemical reactions during the gene expression process as well as necessary abstractions lead to the conclusion that four major reactions, namely transcription of mRNA, degradation of mRNA, translation resulting in the protein production, and protein degradation, are the most important biochemical steps during the gene expression (see Sect. 2). Based on these insights we derive a deterministic ordinary differential equation model for gene expression involving these reactions.

We show the derivation for the repressed transcription, but a similar line of thoughts can be used to derive a model for transcription activators.

3.2.1 Repressed Transcription

We start by deriving a formula for mRNA transcription, which is inhibited in the presence of repressor proteins and is constant otherwise. Let the repressor proteins be denoted by P_r, the DNA copies D_0, the DNA regions not bound by the repressor D, and the DNA-proteins complex $DP_r \equiv X$. Note that the equality $D_0 = X + D$ holds.

A model for enzymatic repression with P_r can be described with a chemical reaction scheme [28, 29]:

$$D \xrightarrow{k_1} mRNA + D$$

$$D + P_r \xrightarrow{k_2} X$$

$$X \xrightarrow{k_3} D + P_r$$

Using the classical mass action kinetics the scheme translates to the following set of differential equations:

$$\begin{aligned}
\frac{d[D]}{dt} &= -k_2[D][P_r] + k_3[X] \\
\frac{d[X]}{dt} &= k_2[D][P_r] - k_3[X] \\
\frac{d[mRNA]}{dt} &= k_1[D]
\end{aligned}$$

using the brackets $[\cdot]$ to denote the concentration of biochemical molecules.

These equations contain chemical reactions with different rate scales: the mRNA molecules are produced at a very slow rate $k_2(\approx 0.015$ [30]) compared to the fast dynamics of the binding and unbinding of the repressor protein $k_1(\approx 0.17$ [30]) and $k_{-1}(\approx 10$ [30]). The three equations can be reduced to one under the quasi-stationarity assumption for the intermediate complex $d[X]/dt \approx 0$, which binding/unbinding dynamics are much faster compared to the mRNA production rate. With the quasi-stationarity assumption, we get rid of the variables $[X]$ and $[D]$ and describe the transcription inhibition only in terms of repressor protein concentration $[P_r]$.

$$\begin{aligned}
\frac{d[X]}{dt} &= 0 \\
k_2[D][P_r] &= k_3[X] \\
k_2([D_0] - [X])[P_r] &= k_3[X] \\
[X] &= \frac{k_2[D_0][P_r]}{[P_r]k_2 + k_3}
\end{aligned}$$

where the last line is the average complex concentration, which will be reached quickly due to fast binding and unbinding.

Then the time evolution of the mRNA can be written in terms of the repressor protein concentration $[P_r]$ as:

$$\begin{aligned}\frac{d[mRNA]}{dt} &= k_1[D] = k_1([D_0] - [X]) \\ &= k_1([D_0] - \frac{k_2[D_0][P_r]}{[P_r]k_2 + k_3}) \\ &= \frac{[D_0]k_1}{1 + [P_r]k_2/k_3}.\end{aligned}$$

The derivation is shown for monomers, i.e. only one protein binds to the regulatory region and represses the transcription. But bacterial genes are very often regulated not by a single, but by several proteins which cooperatively bind to a regulatory region of a gene. Their degree of cooperativity and the sequence in which the proteins can access the binding sites influences the exponent in the reduced equation for mRNA transcription:

$$\frac{d[mRNA]}{dt} = \frac{[D_0]k_1}{1 + [P_r]^h k_2/k_3} \equiv \frac{c_1}{1 + [P_r]^h}$$

The exponent h can be thought of as a fitting parameter and can be a positive real number. It is often referred to as the Hill coefficient and the repression functions are called Hill-functions [28, 29]. Experimentally it was found that $h = 1$ often explains the results worse than larger exponents, see for example [5], who used $h \approx 2$.

3.2.2 Degradation

We assume that mRNAs and proteins are degraded via first order reactions:

$$Y \xrightarrow{c_2} \emptyset$$

Here Y is either mRNA or protein molecules and c_2 denotes the according degradation constant. The change of molecule concentration $[Y]$ due to this reaction can be described with the ODE:

$$\frac{d[Y]}{dt} = -c_2[Y]$$

3.2.3 Translation

Translation of mRNA templates into proteins P can be thought of as an enzymatic production from the source, because all necessary components are present in the cell. The limiting step is the amount of mRNA and therefore the reaction constant is well approximated by a linear dependence on mRNA concentration $[mRNA]$:

$$\emptyset \xrightarrow{c_3[mRNA]} P.$$

The according concentration change per time due to this reaction is then

$$\frac{d[P]}{dt} = c_3[mRNA].$$

3.2.4 Full Deterministic Model

Putting together transcription dependent on repressor protein P_r, translation, and degradation reactions, we obtain a simplified deterministic model for repressed bacterial gene expression:

$$\begin{aligned} \frac{d[mRNA]}{dt} &= \frac{c_1}{K_m + [P_r]^m} - c_2[mRNA] \\ \frac{d[P]}{dt} &= c_3[mRNA] - c_4[P]. \end{aligned}$$

3.2.5 The Generalized Repressilator Model

The expression in Eq. 1 can be extended to N repressing genes wired in a ring and this network is then often referred to as generalized repressilator model:

$$\begin{aligned} \dot{m}_j &= f(p_{j-1}) - c_2 m_j \\ \dot{p}_j &= c_3 m_j - c_4 p_j, \end{aligned} \tag{1}$$

where $j \in 1, \ldots, N$, p_j and m_j describe the variables for proteins and mRNAs for each gene respecting the circular boundary conditions $p_0 \equiv p_N$, and $f(p)$ is a monotonically decreasing function acting as a repressor.

For reference, we provide literature-based parameter ranges which are taken as a reference point for numerical simulations of the generalized repressilator model:

- c_4 protein degradation rate is $\approx 0.06\,\text{min}^{-1}$ and corresponds to the protein half life of 10–60 min (see Ref. [5, 23, 24]).

- c_2 mRNA degradation rate is $\approx 0.12\,\text{min}^{-1}$, please note that the actual range is quite large 30 s to 50 min (see Ref. [22]).
- h Hill coefficient reflecting repressor protein cooperativity is ≈ 2 (see Ref. [5, 16]).
- $K_m \approx 1$ (see Ref. [16]).
- $c_1 \approx 1 - 10\,\text{min}^{-1}$ is calibrated to achieve the maximal values around a few tens.
- $c_3 \approx 0.16\,\text{min}^{-1}$ the translation efficiency is taken to be rather low. The value might be increased to 20 proteins per transcript [5].

A closer look reveals that the special case for $N = 2$ is a model for a genetic switch and for $N = 3$ for a genetic oscillator. Similar deterministic models have been used prior the implementation of the genetic switch and the genetic oscillator [5, 6]. Remarkably, a genetic switch and a genetic oscillator are indeed very similar since both networks contain repressing genes ordered in a ring topology (see Fig. 2a), but qualitatively they show completely different dynamics: the switch approaches a stable steady state with constant gene expression levels and the final state of the oscillator are periodically changing gene expression levels (see Fig. 2b).

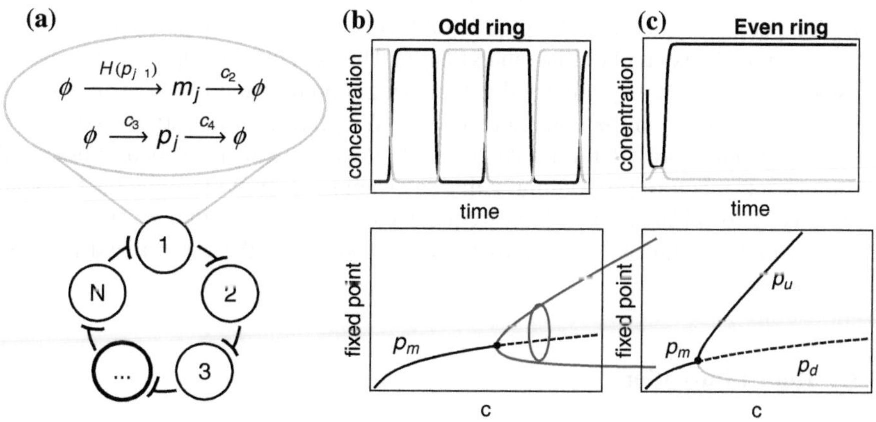

Fig. 2 Dynamical behavior of the generalized repressilator model. **a** Topology of the generalized repressilator: N genes in a cycle where each gene is repressed by the protein product of the preceding gene. Also shown is the reaction scheme underlying the dynamical system in Eq. 1 with production and degradation terms for the mRNA (m_j) and protein (p_j) of each gene. The repression of the production of mRNA is modeled by a Hill-type term $H(p_{j-1})$. **b** Typical time trace of the long-term deterministic dynamics of an odd ring and the associated bifurcation diagram. Odd rings converge to a globally attracting periodic solution after Hopf bifurcation (*black dot*). **c** Typical time trace of an even ring converging to fixed points and the bifurcation diagram. Even rings undergo a pitchfork bifurcation (*black dot*), leading to the emergence of two stable fixed points

Such qualitative differences in systems dynamics can be assessed during the design phase of synthetic circuits using the tools from complexity analysis. Complexity analysis systematically searches parameter spaces of differential equations returning parameter values at which qualitative changes in systems dynamics occur. These points are referred to as bifurcation points. In the following section we introduce needed background on two most common types of bifurcations—Hopf and pitchfork bifurcation—which both occur in the generalized repressilator model and lead to either oscillatory (Hopf) or switching (pitchfork) dynamics (see Fig. 2c).

3.3 Hopf and Pitchfork Bifurcations

Bifurcation analysis predicts qualitative change of the system behavior as one or several parameters change. A parameter value at which such qualitative change occurs is called bifurcation. The variation of a parameter can produce a change in the stability or the number of fixed points (points to which the system would converge as time goes to infinity $t \to \infty$). Typical questions answered by bifurcation analysis could be for example: for which parameter sets a model would express stable oscillatory behavior or how many fixed points exist for a given parameter set.

Bifurcation points can be found numerically using for example a continuation software packages such as AUTO [31] or MATCONT [32]. In rare special cases bifurcation analysis can be done analytically. The generalized repressilator model possesses high internal symmetry due to its ring topology and to a large extent the bifurcation analysis can be done analytically. We perform bifurcation analysis for the generalized repressilator model using both analytical as well as numerical techniques. But before proceeding with this example we provide some background on two types of bifurcation, which frequently occur in biochemical networks and also to a large extent explain the dynamical behavior of the generalized repressilator model.

3.3.1 Hopf Bifurcation

Stable oscillations in several biochemical networks arise through Hopf bifurcations [33]. Hopf bifurcation is a local bifurcation in which a stable fixed point looses its stability as a pair of complex conjugate eigenvalues of the linearization around the fixed point cross the imaginary axis of the complex plane [34]. Informally speaking, a stable system approaching a single constant steady state value becomes an oscillator as a parameter changes.

A formal definition of Hopf bifurcation for a system described by a set of differential equations $\dot{x} = f(x, \mu)$ is provided below:

Theorem 1 (Hopf Bifurcation: emergence of a periodic solution from a stationary solution [35]) *Suppose that the parameter dependent differential system*

$$\dot{x} = f(x, \mu)$$

$f : R^{m+1} \rightarrow R^m, f \in C^\infty$ *has an equilibrium* (x_0, μ_0) (i.e. $f(x_0, \mu_0) = 0$) *at which the following properties are satisfied:*

(H1) *The Jacobian* $D_x f(x_0, \mu_0)$ *has a simple pair of pure imaginary eigenvalues* $\pm i\omega$ *and there are no other eigenvalues with zero real part. Then (H1) implies that there is a smooth curve of equilibria* $(x(\mu), \mu)$ *with* $x(\mu_0) = x_0$. *There is pair of eigenvalues* $\lambda(\mu), \bar{\lambda}(\mu)$ *of the Jacobian* $D_x f(x(\mu), \mu)$ *along this curve with* $\lambda(\mu_0) = i\omega, \lambda(\mu_0) = -i\omega$ *at* $\mu = \mu_0$*which varies smoothly with the parameter. If, moreover, the following condition applies:*

(H2) $\frac{d}{d\mu}(Re\lambda(\mu))|_{\mu=\mu_0} \neq 0$,*then there exists a branch of periodic solutions bifurcating at* (x_0, λ_0).

The oscillations in the three gene repressilator are explained by Hopf bifurcations and in fact also in all generalized repressilator models with an odd number of genes in the ring.

3.4 A Condition for Switching Behavior

The dynamical feature, on which the genetic switch as implemented by [6] is based, is the pitchfork bifurcation as formally defined below:

Theorem 2 (Pitchfork Bifurcation [34]) *Suppose that the parameter dependent differential system is given by*

$$\dot{x} = f(x, \mu)$$

$f : R^{m+1} \rightarrow R^m$,$f \in C^\infty$ *where the following properties are satisfied:*

- *f is an odd function:* $-f(x, \mu) = f(-x, \mu)$
- *the first and second derivatives of f vanish, but not the third one:* $\frac{\partial f}{\partial x}(0, \mu_0) = 0$,$\frac{\partial^2 f}{\partial x^2}(0, \mu_0) = 0$, *and* $\frac{\partial^3 f}{\partial x^3}(0, \mu_0) \neq 0$.
- $\frac{\partial f}{\partial \mu} = 0$,$\frac{\partial^2 f}{\partial x \partial \mu}(0, \mu_0) \neq 0$

then there is a pitchfork bifurcation at $(0, \mu_0)$. *If the third derivative is positive the pitchfork bifurcation is sub-critical, if the third derivative is negative, it is super-critical.*

The qualitative behavior of the system before the bifurcation is that it converges to a single fixed point, in contrast after the bifurcation the system can converge to one of the two stable fixed points or one unstable fixed point depending on the initial conditions (IC). The pitchfork bifurcation induces a symmetry breaking in the system, which then leads to the switching behavior.

4 Numerical Bifurcation Analysis

On the abstract level the implemented genetic oscillator [5] and the switch [6] represent a ring of N identical elements repressing each other in a cyclic manner (see Fig. 2). Remarkably a system containing even number of elements is an inherent switch and a system containing an odd number of elements is an oscillator. These differences in the dynamical behavior can be explained by numerical bifurcation analysis. Widely used and well-established software solutions for numerical bifurcation analysis are AUTO [31] and MATCONT [32]. The software packages track solution branches of dynamical systems by solving algebraic equations and provide approximations for the parameter values at which bifurcations such as Hopf or pitchfork occur.

The bifurcation graphs for the repressilator model (see Fig. 3) show the steady state dynamics for odd element rings which become oscillators after the Hopf bifurcation. The steady state of even element rings are the two possible up/down solutions. Numerical bifurcation analysis reveals an interesting fact: in even and odd element rings we also find Hopf bifurcations branching from the unstable solution branch (marked in red dashed lines in Fig. 3). It turns out that these Hopf bifurcations play a crucial role for the dynamics of larger even element rings leading to long lasting but unstable oscillations [17, 18]. In [17] we proposed a design for a controllable genetic oscillator around these periodic solutions. A distinctive feature of the oscillator based on quasi-stable periodic solutions is that such oscillators are controllable through an outer regulation loop and can also be switched on and off on demand [17]. On the contrary, classical oscillators are based on stable periodic solutions, they oscillate autonomously and are not controllable.

5 Analytical Bifurcation Analysis of Generalized Repressilator and Unstable Periodic Orbits

For the generalized repressilator model given by the ODEs in Eq. 1 several aspects of the bifurcation analysis can be done analytically. Analytical calculations together with the symmetry arguments reveal a connection to other engineering concepts, as for instance the Duffing oscillators [36], magnetic field sensors based on unidirectionally coupled repressors in a ring [37] and also the oscillating behavior in the ring of quantum harmonic oscillators [27].

For analytical calculations in this chapter, a special case of a Hill function has been used with the Hill coefficient $h = 2$ in the repression function $f(x) = \frac{c_1}{1+x^2}$. However the outlined calculations are similar for any $f : U(\in R^+) \to U(\in R^+)$,

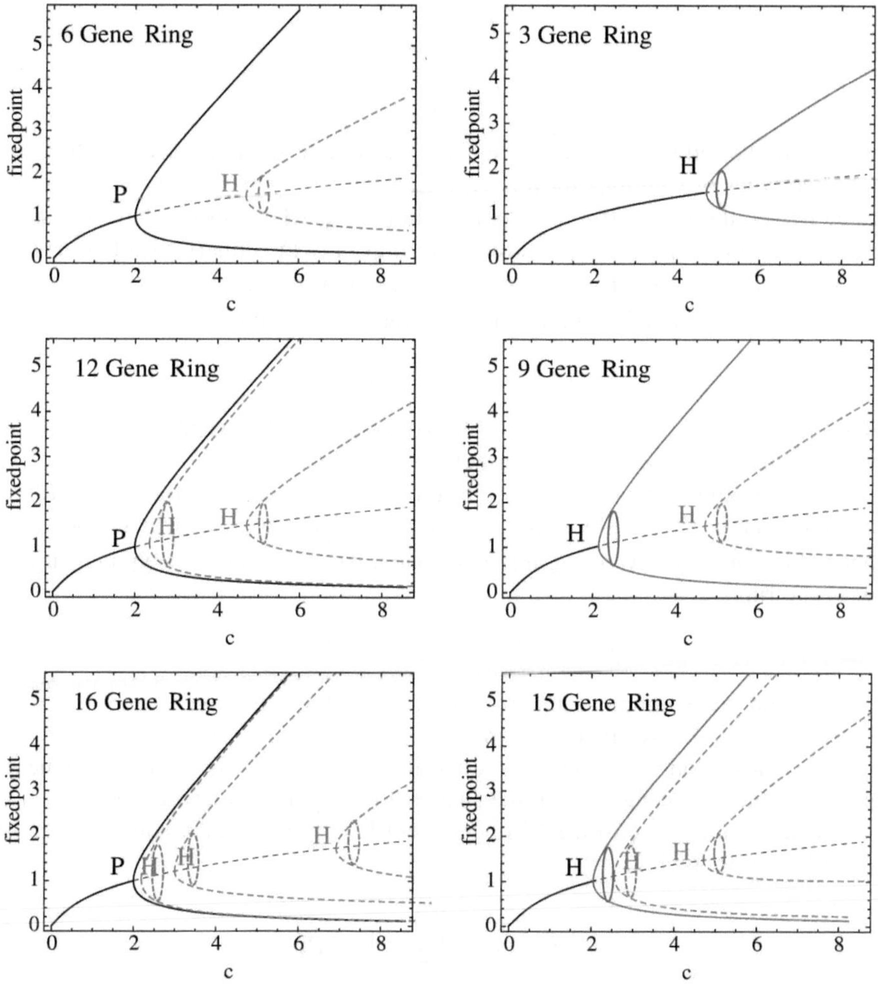

Fig. 3 Numerical bifurcation analysis for generalized repressilator model. Even rings undergo a pitchfork bifurcation, leading to the emergence of two stable fixed points. Odd rings undergo a Hopf bifurcation leading to the emergence of a limit cycle. Quasi-stable periodic solutions branch from the symmetric unstable solution leading to observable long lasting oscillations in even rings (indicated by *dashed lines*). In odd rings the unstable solutions also exist, but are not observable (i.e. if starting from random initial conditions) [17, 18]

which are bounded and monotonically decreasing in this region. The symmetric solution, where all proteins have the same concentration $p_j = p_{j-1}$ $\forall j$ is guaranteed for such $f(x)$. For a Hill-type repression function $f(x) = \frac{c_1}{1+x^h}$ a complete stationary solution can be formulated in terms of a recursion relation for the proteins:

$$p_j^* = \frac{c_1 c_3}{c_2 c_4}.$$
$$m_j^* = \frac{c_4}{c_3} p_j^*$$

In general, the full solution has to be obtained from nested fractions. But there are only two types of solutions for this symmetric system, which has been confirmed by numerics: the symmetric fixed point solution p_m, where all p_j's have the same concentration $p_j = p_{j+1} \equiv p_m$ and for even rings only the dimerized solutions $p_{u/d}$ where every other gene has the same concentration $p_j^* = p_{j+2}^* \neq p_{j+1}^*, \forall j$.

The symmetric fixed point solution p_m from the equation

$$p_m + p_m^{h+1} = c \tag{2}$$

exists for even and odd rings ($c \equiv \frac{c_1 c_3}{c_2 c_4}$ for convenience). It is stable for smaller c's, but eventually becomes unstable in both even and odd rings.

In even element rings, a pitchfork bifurcation occur at $c = 2$ for all N and this condition can be derived by requiring that $p_j^* = p_{j+2}^* \neq p_{j+1}^*, \forall j$, which essentially means tiling a larger even ring into gene pairs, where each pair is equivalent to a two gene ring.

$$p_1(1 + p_2^h) = c \,\&\, p_2(1 + p_1^h) = c$$

Positive and real solutions satisfying the two equations are given either by ($p_1 = p_u, p_2 = p_d$) or the symmetric alternative ($p_1 = p_d, p_2 = p_u$), where $p_u \approx c$ and $p_d \approx 0$. These conditions lead to dimerized solutions, where if one gene is expressed the following gene in the ring will be suppressed or vice verse.

For the specific case of the Hill coefficient $h = 2$, the explicit solutions can be given by the following:

$$\begin{aligned} p_j^* &= \frac{c}{2} + \sqrt{\frac{c^2}{4} - 1} \equiv p_u \\ p_{j+1}^* &= \frac{c}{2} - \sqrt{\frac{c^2}{4} - 1} \equiv p_d. \end{aligned} \tag{3}$$

Note that $p_u \to c$ and $p_d \to 0$ for large c. The emerged fixed points for larger c's correspond to two distinct dimerized states: one in which genes with odd indices are upregulated (p_u) while genes with even indices are downregulated (p_d); and another symmetric state where the genes with odd and even indices exchange their patterns of regulation. The pattern structure is similar to that of other dimerized degenerate solutions in classic models of unidirectionally ring-coupled elements [27, 36, 37].

It can also be shown analytically that the solution branches $p_{u/d}$ are stable (see [38] for details) by showing that the real parts of the Jacobian Eigenvalues are negative. We encourage the reader to reproduce the calculation by writing down the Jacobian of the system and evaluating it at the solution branches $p_{u/d}$ as in

Eq. 3. Eigenvalues of the Jacobian can then be found using the fact that the characteristic polynomial factorizes for ring symmetric structures. For more information on analytical bifurcation analysis please refer to [34, 38].

Our analytical calculations are in line with the numerics and confirm that after the bifurcation even element rings have two stable solutions, i.e. are bistable and behave like switches in the steady state.

5.1 Analytical Solutions and Cross-Discipline Transfer of Engineering Concepts

In numerical bifurcation analysis we saw the emergence of Hopf bifurcations from the symmetric middle branch solution p_m (see Fig. 3), there the first Hopf bifurcation for odd element rings leads to stable periodic solutions and all other bifurcations branch from unstable middle point solutions. The conditions for the emergence of those bifurcations ((H1) and (H2) in Theorem 1) can be calculated analytically. Detailed algebraic solution would go beyond the scope of this book chapter and the interested reader is referred to our previous publication [18].

Using these analytical calculations we obtain the number of possible Hopf bifurcations for rings with different number of elements and can explore the oscillating modes of the generalized repressilator model. The number of oscillating modes is directly proportional to the length of the ring N and depends on the parameter set and the repression function. For a biologically meaningful parameter set with a Hill coefficient $h = 2$ odd rings possess $N_{HB} = \left\lfloor \frac{1}{2}\frac{(N-1)}{2} \right\rfloor$ Hopf bifurcations and even rings possess $N_{HB} = \lfloor \frac{1}{2}(\frac{N}{2} - 1) \rfloor$ [18]. This result links the generalized repressilator model to other engineering concepts, which functionality is based on a ring symmetric topology of the elements.

The group theoretical approach predicts the existence of $N_{HB} = (N - 1)/2$ (odd rings) and $N_{HB} = (N/2 - 1)$ (even rings) possible solutions for strong enough element couplings in rings [39]. For the generalized repressilator the strength of element coupling is given by the form of the repressor function and the Hill exponent. We do not observe all possible solutions predicted by the standard group theory for the biologically meaningful parameter set and Hill suppression function with $h = 2$. These analytical results can be used for identification of possible oscillation modes and finally for the design of either autonomous or controlled synthetic oscillators.

Bifurcation analysis and group theory connect the generalized repressilator model to many other well-studied physical systems whose functionality is also based on discrete rotation symmetry (see for example [27, 36, 37]). The design of our special cases of the generalized repressilator model for N = 2 and N = 3 synthetic networks were indeed inspired by electrical switches and oscillators. Mathematical framework allows to make use of experience and insights from other engineering disciplines and aid cross-disciplinary transfer of ideas for functionalities which can be implemented in both physical and biological systems.

Specific properties of genetic networks have to be considered if transferring classical engineering concepts to synthetic biology. One such characteristic is that the gene expression comes with a non-negligible amount of noise due to low copy numbers of biochemicals involved in the transcription reaction. The number of DNA copies ($\mathbb{O}(1)$) or the number of mRNA molecules ($\mathbb{O}(10)$) involved in the process is low and stochastic fluctuations can lead to qualitatively different system dynamics even for the simplest genetic networks. In fact, the genetic switch (the generalized repressilator with $\mathrm{N} = 2$ genes) shows qualitatively different dynamics in stochastic low copy number regime compared to the deterministic equation. The inherent stochasticity also influences the oscillation properties of synthetic oscillators (e.g. the generalized repressilator with N=3 genes).

The time evolution of inherently stochastic systems are well described by differential equations for probability distributions. A convenient theoretical formalism for time dependent probability functions are Master equations and we will use them for modeling of network topologies taking the inherent gene expression stochasticity into account. We use the notation for this chapter introduced in a classical text book [13].

6 Stochastic Dynamics for Gene Regulatory Networks Using Master Equations

Molecular reactions involved in gene expression processes can be seen as discrete events randomly occurring in time. If the number of molecules involved is very high, the system can be well described with ordinary differential equations, but if the number of molecules is low, the randomness influences the system dynamics and the deterministic description is no longer appropriate.

To get an intuition about such processes imagine gas molecules, which are enclosed in a reservoir and react by collision. If the density of molecules is high, then several collision reactions will be happening per time period, then the fact that the reactions are discrete events becomes irrelevant for the description of the process, and the reaction rate can be well approximated as a deterministic rate law. A different situation occurs if the density of gas molecules in the reservoir is very low. Discrete collision events can be observed, which mathematically can be captured with the so called Master equation, describing time evolving probability distributions instead of the deterministic ordinary differential equations.

Instead of a real valued concentration vector in the deterministic formalism, we introduce the random variable $\mathbf{n}(t) \in \mathbb{N}^s$ describing the time dependent state of the system, which denotes the number of molecules of each biochemical s at the time t. A state $\mathbf{n}(t)$ can appear at a time t with the probability $P(\mathbf{n}, t)$, so that the evolution of the system will be given as evolution of probability function $P(\mathbf{n}, t)$.

The probability function is discrete in $\mathbf{n}$ the number of molecules, but continuous in time t. Hence, a general form of a Master equation:

$$d_t P(\mathbf{n}, t) = \sum_{\mathbf{n}'} W_{\mathbf{nn}'} P(\mathbf{n}', t) - W_{\mathbf{n}'\mathbf{n}} P(\mathbf{n}, t)$$

can be seen as a gain-loss equation for the probabilities of the system states. The transition probability from the state $\mathbf{n}'$ to the state $\mathbf{n}$ per unit time is given by the conditional probability $W_{\mathbf{nn}'}$. The first term of the equation is the gain to the state $\mathbf{n}$ by all other states and the second term of the equation is the loss of $\mathbf{n}$ caused by the transitions into the other states.

Describing molecular dynamics in this way we implicitly assume that the transition probabilities $W_{\mathbf{nn}'}$ only depend on current state of the system and are not affected by other previous states, i.e. the process history. The processes for which this kind of description in appropriate variables can be found are referred to as first order Markov processes [13]. The Markov property for successive times (i.e. $t_0 < (t_0 + \Delta t) < \cdots < (t - \Delta t) < t$) and homogeneous processes (dependent on time difference Δt, but not on time t itself) can be formulated as:

$$W(\mathbf{n}_t | \mathbf{n}_{t_0}; \mathbf{n}_{t_0+\Delta t}; \ldots; \mathbf{n}_{t-\Delta t}) = W(\mathbf{n}_t | \mathbf{n}_{t-\Delta t}) \tag{4}$$

The analysis of stochastic processes is well manageable if the Markov property is fulfilled, and therefore we will find an appropriate set of variables for the biochemical reactions involved in gene expression for which the Markov property is fulfilled.

6.1 Linear One-Step Birth-Death Processes

For biochemical reactions involved in gene regulation we consider a class of $W_{\mathbf{nn}'}$, where the transitions occur between the neighbouring states. So in a time period Δt a biochemical is either created or deleted. Such processes are called one-step birth-death processes (see Fig. 4). In order to describe those processes conveniently we introduce the creation $\mathbb{E}f(n) = f(n+1)$ and the annihilation operators $\mathbb{E}^{-1}f(n) = f(n-1)$ [13]. The schematically illustrated probability flows in Fig. 4 correspond to the reaction scheme:

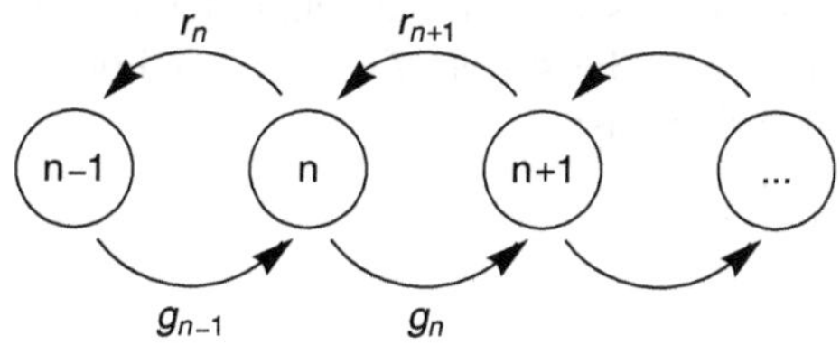

Fig. 4 One-step stochastic birth-death process illustration as in Ref. [13]

$$\emptyset \xrightarrow{g_x} x \xrightarrow{r_x} \emptyset$$

where the rate functions r_x, g_x are linear in x. Using the creation/annihilation operators the Master equation is then given by

$$\dot{p}_n(t) = (\mathbb{E} - 1)r_n p_n + (\mathbb{E}^{-1} - 1)g_n p_n \tag{5}$$

with $\mathbb{E}, \mathbb{E}^{-1}$ one dimensional step operators.

6.2 Multi-dimensional Creation and Annihilation Operators in the Master Equation for Biochemical Networks

For Master equations with several interacting biochemicals we introduce multi-dimensional step operators defined by

$$\begin{aligned} \mathbb{E}_i f(n_1, n_2, \ldots, n_i, \ldots) &= f(n_1, n_2, \ldots, n_i + 1, \ldots) \\ \mathbb{E}_i^{-1} f(n_1, n_2, \ldots, n_i, \ldots) &= f(n_1, n_2, \ldots, n_i - 1, \ldots) \end{aligned} \tag{6}$$

on the support $\mathbf{n} \in \mathbb{N}^s$. The main advantage of the formalism is that the probability distribution is discrete in the number of molecules. Low copy number reactions can be modeled as discrete transitions from one state to the next and conveniently described using the multi-dimensional creation and annihilation operators. Therefore, this formalism will be used for the stochastic modeling of gene expression.

6.3 Hill-Type Variable Reduction for Repressed Transcription

As for the deterministic example we consider the simplified reaction scheme for the repressed genetic transcription reaction:

$$\begin{aligned} D + P_r &\rightleftharpoons^{k_1}_{k_{-1}} X \\ D &\rightarrow_{k_2} D + mRNA. \end{aligned} \tag{7}$$

where D denotes DNA copies not occupied by the repressor protein P_r, X is the DNA-repressor complex and the conservation equation for the overall DNA present in the system $D_0 = D + X$ holds. Then the resulting Master equation can be written as:

$$
\begin{aligned}
d_t P(d,p,x,m;t) =& k_1(d+1)(p+1)P(d+1,p+1,x-1,m;t) \\
&+ k_{-1}(x+1)P(d-1,p-1,x+1,m;t) \\
&+ k_2 d P(d,p,x,m-1;t) \\
&- (k_1 dp + k_{-1}x + k_2 d)P(d,p,x,m;t)
\end{aligned} \tag{8}
$$

We assume that the number of repressor proteins is much larger than the number of DNA copies: $P_r \gg D_0$, so that the inhibition dynamics are almost unaffected by the binding/unbinding reactions and can be considered slow. For the slow stochastic variables P_r and $mRNA$, we assume intermediate values $P_r = \mathbf{p}^*, mRNA = \mathbf{m}^*$ and denote the fast dynamics probability distribution for the complex X, $p_n \equiv P(X = n)$[1].

The dynamics of the complex X happen on the considerably faster time scales compared to the mRNA production, because the binding and unbinding reaction rates of the inhibitor protein to the DNA are much larger compared to the transcription rate: $k_1, k_{-1} \gg k_2$. We are mainly interested in slow dynamics of the mRNA production and its dependence on the protein repressor dynamics, which also occurs on much slower rate than k_1 and k_{-1}. Therefore, we will decouple fast and slow time scales obtaining a reduced Master equation for the mRNA production, which will only depend on the dynamics of the inhibitor protein. We assume intermediate values for the slow variables $P_r = \mathbf{p}^*, mRNA = \mathbf{m}^*$ and write the fast dynamics probability distribution for the complex X:

$$
\begin{aligned}
d_t p_n =& k_1(d_0 - n - 1)\mathbf{p}^* p_{n-1} \\
&+ k_{-1}(n+1)p_{n+1} \\
&- (k_1(d_0 - n)\mathbf{p}^* + k_{-1}n)p_n \\
=& (\mathbb{E} - 1)r_n p_n + (\mathbb{E}^{-1} - 1)g_n p_n
\end{aligned}
$$

with $g_n = k_1 \mathbf{p}^*(d_0 - n)$ and $r_n = k_{-1}n$.

Following the same considerations as for a stochastic process confined between the two absorbing boundaries, we obtain an analytically solvable first order ODE for the mean of the complex X[56]

$$d_t {<}n{>} + (k_1\mathbf{p}^* + k_{-1}){<}n{>} = d_0 k_1 \mathbf{p}^*.$$

The solution is

$${<}n{>} = \frac{d_0 \mathbf{p}^* k_1}{\mathbf{p}^* k_1 + k_{-1}}(1 - e^{-(k_1\mathbf{p}^* + k_{-1})t}) - IC * e^{-(k_1\mathbf{p}^* + k_{-1})t}$$

[1] Note: $\mathbf{p}^*$ and $\mathbf{m}^*$ are stochastic variables, not mean values, which change slowly compared to the dynamics of the complex X. We assume that the time scale to reach the steady state of X is so fast, that the slow stochastic variables do not change during this period.

with IC denoting initial condition. The intermediate complex X will approach the steady state exponentially fast on the time scale, which is proportional to the fast reaction variables $\propto (k_1\mathbf{p}^* + k_{-1})$:

$$x_{st} = \lim_{t\to\infty} <n> = \frac{d_0\mathbf{p}^* k_1}{\mathbf{p}^* k_1 + k_{-1}} = \frac{d_0\mathbf{p}^*}{\mathbf{p}^* + k_{-1}/k_1}$$

leading to the reduced Master equation for the repressed transcription:

$$\begin{aligned} d_t p_m &\approx (d_0 - x_{st})k_2(\mathbb{E} - 1)p_m \\ &= \frac{d_0 k_2}{\mathbf{p}^* + k_{-1}/k_1}(\mathbb{E} - 1)p_m \end{aligned}$$

which again is in agreement with reduction result for the purely deterministic system.

Analogous calculations for the variance lead to the following result for the stationary coefficient of variation:

$$\lim_{t\to\infty} \frac{\sqrt{\text{var}\ (x)}}{<x>} = \sqrt{\frac{k_{-1}/k_1}{\mathbf{p}^* d_0}} \approx \sqrt{\frac{1}{\mathbf{p}^* d_0}}$$

This stationary variability of the intermediate complex x will be neglected in the reduced Master equation. Note that the approximation is better if the number of repressor proteins $\mathbf{p}^*$ and/or the number of DNA copies D_0 are much larger than one, so that the stationary coefficient of variation vanishes. Note also that in the deterministic description the coefficient of variation is implicitly assumed to be equal to zero.

It is worth mentioning that the elimination of the fast dynamics variables also affects the Markov property of the process description, in the sense that the production of the mRNA from the source is now directly dependent on the present amount of repressor proteins. In the Master equation (Eq. 8) before the elimination of the fast variable dynamics, the mRNA production was indirectly dependent on the repressor protein through several intermediate stochastic processes involved in the intermediate dynamics X. If gene expression model or this approximation capture the relevant dynamics of real gene expression process is still an active research topic [26, 40].

7 The Generalized Repressilator Model in the Master Equation Formalism

Similarly to the deterministic description we consider model transcription, translation and degradation of proteins and mRNA in the full Master equation resulting in 4 reactions per gene. We will use the creation (annihilation) $\mathbb{E}_{p_j}(\mathbb{E}_{p_j}^{-1})/\mathbb{E}_{m_j}(\mathbb{E}_{m_j}^{-1})$

for creation (annihilation) events of proteins and mRNA as introduced in the Sect. 6.2. Then the Master equation for the generalized repressilator amounts to:

$$\partial_t P(\mathbf{m},\mathbf{p};t) = \left[\frac{c_1}{p_{j-1}^2+1}(\mathbb{E}_{m_j}^{-1}-1) \right.$$
$$\left. + c_2(\mathbb{E}_{m_j}-1)m_j + c_3 \cdot m_j(\mathbb{E}_{p_j}^{-1}-1) + \cdot c_4(\mathbb{E}_{p_j}-1)p_j \right] P(\mathbf{m},\mathbf{p};t)$$

using as usual the index $j \in \{1,\ldots,N\}$ and respecting circular boundary conditions $p_0 \equiv p_N$.

7.1 Noise Qualitatively Changes Gene Network Dynamics

Inherent stochasticity can change the dynamics of genetic networks qualitatively. By qualitatively we mean here that the deterministic curve is not just slightly perturbed by the noise around its mean, but that low copy number dynamics shows features which are not observed in the deterministic description. To demonstrate that we compare the genetic switch model (generalized repressilator with $N = 2$) in the deterministic description and Master equation formalisms.

We take a closer look at the stationary state of the switch in the deterministic and stochastic cases. We learned in Sect. 5 that the stationary state of the switch in deterministic regime is given by two fixed points. Depending on the initial conditions the deterministic system will reach one of the two fixed points and once this has happened the state of the switch will remain unchanged.

In the stochastic regime the stationary state is given by a stationary probability distribution. The stationary distribution for the switch is bimodal with two hubs around the deterministic fixed points. The main difference between the deterministic and stochastic regime is that depending on the noise level there is a non-vanishing probability that the genetic switch can spontaneously reverse its state (see Fig. 5).

What we already see from this example is that the reliability and the functionality of genetic switches is strongly impaired in the low copy number regime. Clearly a switch which just spontaneously reverses its state if triggered by random noise is not a reliable bio-synthetic element and misses the design purpose. Low copy number dynamics should be addressed during the design phase by explicitly taking stochastic aspects into account and tuning the amount of noise. A useful framework for estimating the noise levels depending on the number of molecules involved in the reactions, i.e. the system size is given by the so called Ω-expansion.

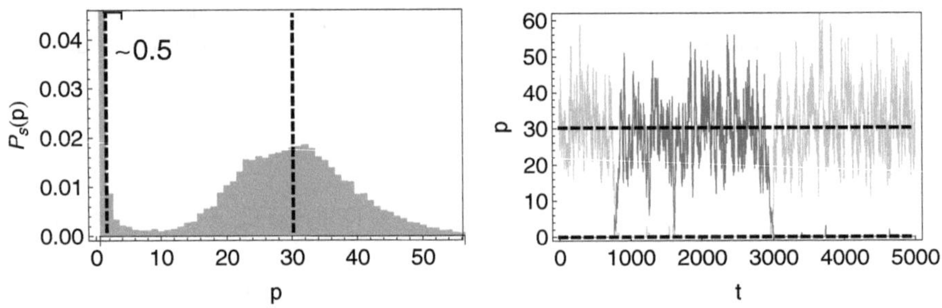

Fig. 5 Qualitative difference in the stochastic compared to the deterministic regime of a genetic switch. A numerical estimate of the stationary distribution for a genetic switch (*grey bars*) have been obtained using Dominated Coupling From The Past (DCFTP) introduced in Ref. [41]. As expected we obtain a bimodal distribution $P_s > pfi$ (*grey bars*) centered around deterministic fixed points marked by dashed *black lines*. The qualitative difference between the stochastic and the deterministic dynamics is that in the deterministic setting reaching the steady state is a final decision, the system dynamics will remain there for all times. In contrast for the stochastic setting there is a non-vanishing probability that switch spontaneously reverses its state. We show on the right a time trace of protein counts (p_1 in *dark-grey* and p_2 in *light-grey*) of the forward Gillespie simulation [14] containing two such reversal events

7.2 From Master Equations to ODEs and SDEs: Bridging the Gap with Ω-Expansion

Intuitively stochastic equations must go over into the deterministic as the number of molecules involved in biochemical reactions becomes high. The variable influencing how well deterministic dynamics approximate real system behavior is the number of molecules or the system size, which is usually denoted by Ω. Therefore, a systematic approach going from a fully stochastic to deterministic equation quantitatively estimating the errors for neglected terms is the application of Taylor expansion to the Master equation in the inverse of Ω [13].

As could be intuitively expected, the largest contribution to the Taylor expansion in terms of Ω-power is an ordinary differential equation, which describes the time evolution of the Master equation mean, i.e. the first moment.

The next order of the expansion contains equations for second moments, i.e. variances and covariances. It is also often referred to as "linear noise approximation" because the solution is determined by a Gaussian [13], which is fully characterized by the first and the second moments, therefore it is equivalent to an SDE:

$$x(t) = x_0 + \int_0^t b(x_s)ds + \int_0^t \sigma(x_s) \cdot d\omega_s \tag{9}$$

with $d\omega_s$ Gaussian white noise, b the drift term, and σ the diffusion term.

7.3 Reaction Rates Dependence on the System Size Ω for the Generalized Repressilator Model

The Master equation for the generalized repressilator model contains 3 linear reactions, which are the protein creation, the protein decay and the mRNA decay. The constants c_j in linear terms do not change with the size of the system Ω, because the biochemical rate is just proportional to the number of molecules in the system and if the size is increased, so will be their absolute number. The biochemical rate of linear reactions scales therefore automatically with the size of the system [13]. For the Hill-type transcription term scaling arguments and derivations for extensive and intrinsic variables result in the following transformation rule:

$$\frac{c_1}{1+p^2} \to \frac{c_1\Omega}{1+(\frac{p}{\Omega})^2}.$$

The Fig. 6 shows simulations of the stochastic system carried out with the Gillespie algorithm at different values of Ω. As expected if we increase the size of the system Ω the fluctuations disappear and the oscillations become regular.

Clearly, the oscillations and their quality are simultaneously affected by the parameter set and the system size Ω. But the parameter set itself also affects the system size Ω. In particular, our previously used bifurcation variable c_1 is directly proportional to the gene copy number D_0 which strongly affects the system size [38]. The gene copy number D_0 can be manipulated experimentally and used as a "biological knob" for induction of oscillations in synthetic repressilators implemented in living bacterial populations. The onset of stochastic oscillations in the generalized repressilator model needs to be explored in dependence of two variables, e.g. in the (Ω, c_1)-plane.

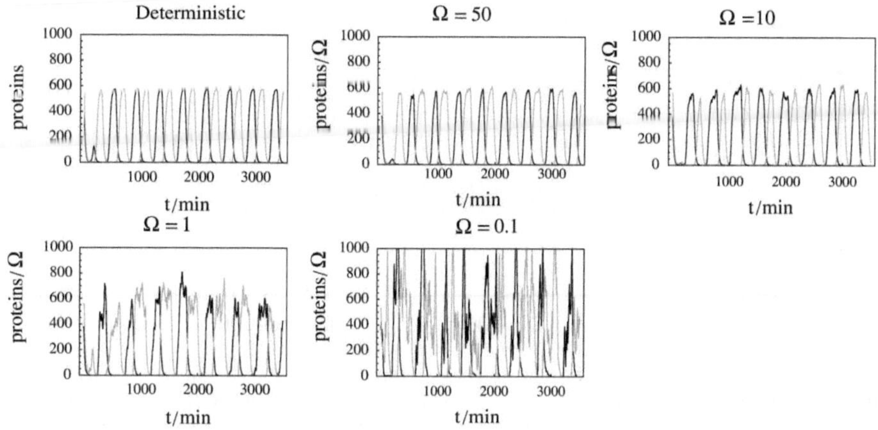

Fig. 6 From deterministic to stochastic regime in the generalized repressilator model using the dependence of reaction rates on the system size Ω

7.3.1 Onset of Oscillations in Stochastic Setting of Generalized Repressilator Model

The Fig. 7 illustrates snapshots in the (Ω, c_1)-plane around the onset of oscillations in the stochastic generalized repressilator model. The classical Gillespie algorithm [14] was used for numerical simulations. In the low copy number regime (small Ω), where the stochastic fluctuations are strong, the oscillating pattern are less distinct even after the deterministic bifurcation point (see Fig. 7a).

An interesting effect occurs in large-Ω regime right before the onset of oscillations in the deterministic model (see Fig. 7f, g). Due to the stochastic broadening of the Hopf bifurcation, the oscillating pattern appear in the generalized repressilator model already at smaller parameter value in the stochastic setting. Often it is an indication that microscopic fluctuations lead to emergence of coherent oscillations on the macroscopic level around the deterministic bifurcation point in the model [42].

7.3.2 Linear Noise Approximation Close to the Hopf Bifurcation

Following the numerical hint from the previous section we will investigate the premature oscillation onset shortly before the deterministic bifurcation point. We apply linear noise approximation in the fixed point phase close to the emergence of Hopf bifurcation (see [13] for procedure details).

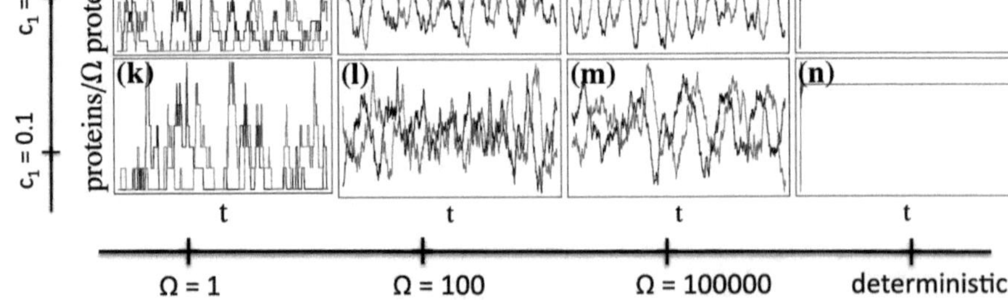

Fig. 7 Onset of oscillations in the stochastic repressilator model for $N = 3$ nearby the deterministic bifurcation point in the (Ω, c_1) -plane. In the deterministic regime the limit cycle oscillations start, then the bifurcation parameter c_1 reaches the value 0.21 so that the sub-figures **a–d** show the behavior of the system after the deterministic Hopf bifurcation and the sub-figures **e–n** before the bifurcation. Depending on the system size Ω oscillations appear already earlier in the stochastic version of the generalized repressilator model (**f, g**)

Denoting the system state $(\mathbf{m}, \mathbf{p}) \equiv \mathbf{x}$, the according SDE is

$$\mathbf{x}(t) = \mathbf{x}_0 + \int_0^t J \cdot \mathbf{x}\, ds + \int_0^t \sigma \cdot \mathbf{x}\, dW_s, \tag{10}$$

where J is the drift matrix equal to the deterministic Jacobian and σ is the diffusion matrix.

We use a standard procedure to determine if regular oscillations are contained in the dynamics of this SDE and calculate the autocorrelation function and its Fourier transform, the stationary power spectrum (see [43], Chap. 1.5.2). If the power spectrum is peaked around a frequency $\nu = \nu_{peak}$ then the system shows regular behavior with the frequency ν_{peak}.

For linear SDEs as in this case closed expression for the stationary power spectrum matrix $S(\nu)$ is known and it depends on the Jacobian J and the diffusion matrix σ (see the formula for multi-variable linear SDEs [43], Chap. 4.5.6):

$$S(\nu) = (J + i\nu)^{-1}\sigma \cdot \sigma^T (J - i\nu)^{-1}. \tag{11}$$

The power spectra are obtained from the diagonal contributions on $S(\nu)$ and shown for values in the fixed point phase close to the bifurcation point in Fig. 8. The power spectrum peak can be made divergent to ∞ around the resonance values. The linear noise approximation suggests that in all odd numbered rings microscopic fluctuations lead to coherent oscillations on the macroscopic level close to the Hopf bifurcation point. The Fig. 8 illustrates this effect for the ring N = 3.

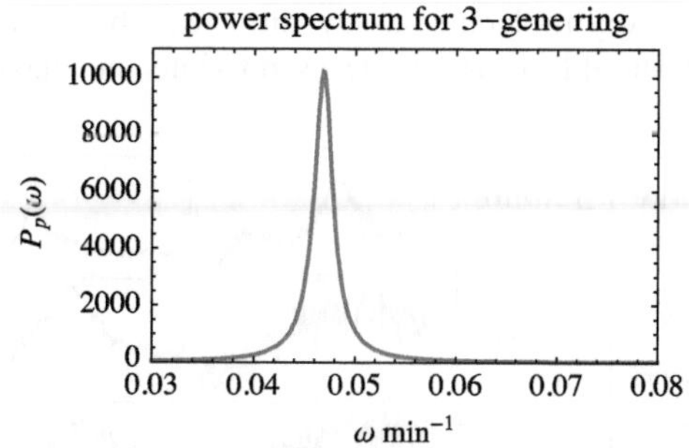

Fig. 8 Power spectra obtained from the linearization around Hopf bifurcations leading to stable periodic orbits in the generalized repressilator model. In the stochastic regime the oscillations emerge already for smaller parameter values, e.g. prior to deterministic Hopf bifurcation. According to the linearized model, we expect coherent stochastic oscillations to appear already at $c_1 = 0.162$ for the stable limit cycle in $N = 3$. The frequency peak can be made divergent around these parameter values, which physically means that stochastic fluctuations occurring on the microscopic level lead to coherent oscillations on the macroscopic level [42]. Direct Gillespie simulations around these values confirmed the numerical hints of the linearized spectra.

7.3.3 The Quality of Stochastic Oscillations in the (Ω, c_1)-plane.

In previous sections we saw that the regularity or the quality of stochastic oscillations in the repressilator model is affected by the system size Ω and the bifurcation parameter c_1. For engineering of synthetic oscillators in living bacterial cells it can be interesting to quantitatively explore the quality of oscillations depending on both parameters. Therefore, we navigate the (Ω, c_1) -plane for the special case $N = 3$ applying the following quantitative measure [44]:

$$Q(\nu_{peak}) = \frac{\nu_{peak} S(\nu_{peak})}{\int S(\nu) d\nu} \tag{12}$$

In the non-oscillating phase ($c_1 = 0.1$) the power spectrum is flat, and the quality score assumes its lowest values (see Fig. 9). The resonance oscillations prior the Hopf bifurcation point are more prominent in larger systems and a steady increase of the quality score is observed in the second row of the Fig. 9 for $c_1 = 0.162$. In the regime long after the bifurcation ($c_1 = 1.6$) the quality of oscillations according to this score strongly depends on the system size Ω. In particular, in small systems the quality of oscillations according to this score does not become better even if moving far away from the bifurcation point.

8 Discussion

8.1 Mathematical Models and New Functionalities of Network Topologies

The design of synthetic gene networks is usually based on a simplified parametric model including only key molecules as dynamic variables. Mathematical models for gene expression should be as simple as possible, but have enough detail on

	$\Omega = 1$	$\Omega = 100$	$\Omega = 100.000$
$c_1 = 1.6$	$2.4 \cdot 10^{-3}$	$8.0 \cdot 10^{-2}$	$1.3 \cdot 10^{-1}$
$c_1 = 0.162$	$1.0 \cdot 10^{-3}$	$2.0 \cdot 10^{-3}$	$3.0 \cdot 10^{-3}$
$c_1 = 0.1$	$1.4 \cdot 10^{-4}$	$1.4 \cdot 10^{-3}$	$1.3 \cdot 10^{-3}$

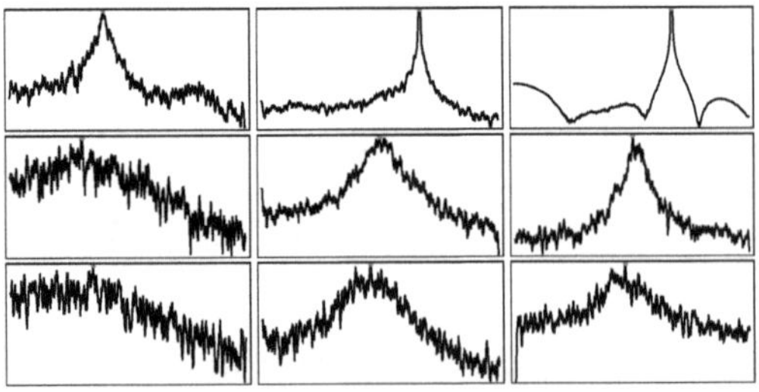

Fig. 9 Snapshots of the oscillation quality in the (Ω, c_1) plane. The quality score defined in Eq. 12 has been applied for the repressilator model with $N = 3$ genes in the non-oscillating phase ($c_1 = 0.1$), at the coherent oscillations prior the Hopf bifurcation point ($c_1 = 0.162$) and further away from the bifurcation point ($c_1 = 1.6$) for the different system sizes $\Omega \in \{1, 100, 100.000\}$. The power spectra have been obtained from Gillespie simulations, are normalized and shown on the logarithmic scale to provide an intuition for oscillation quality score shown in the table.

dynamic description for key molecules. The biochemical processes which are of main interest should be treated as model variables and molecules which are not of interest and/or can be assumed constant should be absorbed into the model parameters.

Further simplifications can be made by time scale analysis and approximation of fast reactions by their steady state values and their absorption into model constants. For gene expression in bacteria we identified mRNA and proteins as model variables and replaced a fast dynamics variable by its steady state value.

One of the key issues by performing this sort of abstraction steps is that the resulting mathematical description remains at least approximately Markovian, because non-Markovian process description are much more difficult to simulate or analyze and they are therefore less suitable as design aid in synthetic biology.

Once a gene expression model is created numerical or analytical bifurcation analysis can be applied to investigate, which qualitative dynamics can be expected for the proposed network topology and different parameter sets. We use the generalized repressilator model as an example to illustrate how bifurcation analysis can reveal different functionalities of network topology depending on parameter set.

In weak repression regime generalized repressilator model has a simple steady state dynamics: it converges to an equilibrium state where all genes have the same expression. For strong enough repression term bifurcation analysis predicts qualitatively different steady state dynamics for even or odd numbered rings. Odd numbered rings for strong enough repressor term undergo Hopf bifurcation and become oscillators. Even numbered rings for strong enough repressor term undergo pitchfork bifurcation and become switches. Indeed bifurcation analysis was an essential part of design phase prior the implementation of the first synthetic elements—the synthetic switch by Ref. [6] and the synthetic oscillator by Ref. [5]—in real biological cells.

Further bifurcation analysis of generalized repressilator model for longer even rings ($N \geq 4$) revealed the existence of quasi-stable periodic solutions in which longer even numbered rings often get trapped in long transient oscillations before reaching the switch-like steady state. Based on these oscillation modes we have recently proposed a design for a controllable oscillator with an outer feedback loop for oscillation induction and maintenance [17]. This example shows how bifurcation analysis resulted in an idea for a new engineering application of gene network topology.

8.2 Transferable Concepts

Another source for engineering ideas and functionalities for synthetic gene networks are classical electrical engineering. It provides ideas on how to design parts out of low-level functions as for example repression or activation. The seminal contributions from 2000 by Ref. [5, 6] for synthetic switch and oscillator have also been build as electrical circuits prior their implementation in bacterial cells. Several other achievements in synthetic biology have been adapted from classical engineering including cascades, logical gates, other designs for oscillators and spatial patterning (see [45]). Bifurcation analysis and other mathematical concepts helped to transfer the concept from electrical part design to synthetic biology.

In general, mathematical analysis and group theory as demonstrated here with ring symmetries can help identify classical engineering or physics concepts, that can be transferable to synthetic biology. Structures with N repressors ordered in a ring are an example that topology-based designs are potential candidates for portable concepts across several disciplines: odd numbered rings oscillate in natural biological systems [46], as an electro-magnetic device [37], and also as a synthetic gene expression network [5]. Abstract ideas such as for example group theory and bifurcation analysis help to identify transferable designs, which can function as synthetic networks inside bacterial cells. Topology-based concepts not relying on special functional form or a small parameter region have the best chances to function in several disciplines and are therefore worth considering as an adaptable concept.

There are fundamental differences between the design requirements for classical and genetic engineering. Classical designs were invented for static environments where electrical or mechanical parts usually quickly reach their predicted steady states, but synthetic networks have to be designed for operation in time-limited, fluctuating cell environments. It means that the focus of mathematical analysis for the parts design has to be shifted from the routinely performed deterministic analysis of steady state towards the analysis of the transients and noise influence which takes the system size into account.

Nontrivial transient behavior can arise in genetic networks if the network topology for instance includes several feedback loops. A prominent example of such behavior is our generalized repressilator model with longer even rings which if starting from random initial conditions often exhibits long lasting transient oscillations even though the steady state is stable and switch-like [18]. For the genetic networks, in particular if their topology becomes complex and contains several feedback loops, the dynamics that is observable and that matters for the cells can be transients and not steady states and this has to be taken into account during the design of synthetic gene networks.

8.3 Noise

The origins of noise in genetically identical populations are attributed to inherent stochastic nature of biochemical reactions [47, 48]. Therefore gene expression noise is unavoidable and is observed in all living organisms. A prominent example for the gene expression noise manifestation in complex organisms are human or animal twins, which are genetically identical but show different phenotypic features, e.g. coat/finger print pattern or personality [47].

Indeed, naturally occurring noise in biology is far from being a nuisance on the contrary necessary noise seems to be an evolutionary strategy keeping genetically identical, but phenotypically diverse populations. This strategy has been reported in several microbial populations [49–51] and most likely it enhances survival chances in case of sudden changes in environmental conditions. A well-characterized example is a soil bacterial population *Bacillus subtilis*, which determines the cell fate using stochastic switch between "dormant" and "competent" states [52].

Noise in engineered organisms does not have to be a nuisance. The pursuit should be to create designs for synthetic circuits, which take advantage of noise. Noisy low copy number regime has also the advantage of small effective system size requiring overall less heterologous proteins needed to be produced by the cells and therefore resulting in smaller cell burdens. Previous synthetic constructs [5, 6] have been designed to operate in large number regimes mainly to showcase the proof of concept with clear signals, but for larger genetic networks the cell burden becomes too high if all nodes are operated in the full capacity deterministic regime.

Still only few design ideas are available which explicitly take advantage of the noise in gene networks [53]. One of the reasons is that we still require better techniques for characterisation of noise distributions. The development of in vivo real-time measurements of mRNA and protein concentrations with single cell and single molecule resolution is ongoing research in this direction [54]. From this research we expect to obtain knowledge on transcriptional noise distributions in real cells. We can then integrate this knowledge into models for synthetic gene network designs using stochastic complexity analysis, as for example the (Ω, c_1)-plane framework introduced in this chapter.

8.4 (Ω, c_1)-Plane Framework for Analysis of Noise

We have shown that low copy number noise influences both the steady state of engineered genetic network as well as the onset of bifurcations. In the generalized repressilator model bifurcation analysis required navigation on the (Ω, c_1)-plane, where Ω is the system size and c_1 the deterministic bifurcation parameter directly proportional to gene copy number D_0 [38, 55]. Both parameters have influenced the bifurcation onset of the repressilators. The gene copy number D_0 can be seen as a biochemical "knob" for the generalized repressilator model, which is controlled by the experimentator and triggers the emergence of bifurcations in synthetic realizations of the model inside bacterial cells.

In the associated deterministic model the gene copy number increase would in principle just reproduce the bifurcation diagrams [17]. However, in the stochastic setting the increase of gene copy numbers has an effect on two model parameters: D_0 strongly affects the bifurcation parameter c_1 (direct proportionality) as well as the system size Ω and the noise level. The biochemical knob D_0 can be seen as a diagonal navigator on the (Ω, c_1)-plane.

Synthetic gene networks in living bacterial cells should be implemented in sufficiently large Ω regimes in order to achieve reliable functionality. For synthetic gene networks there will always be a trade off between the operational quality and the amount of metabolic burden due to high levels of expressed heterologous proteins. The (Ω, c_1)-plane approach can be used for search of operation regimes with sufficient functional quality while achieving the lowest possible metabolic burden.

Classical parts design can serve as sources for ideas in gene network engineering. Bifurcation analysis and other mathematical techniques help to create models for gene expression and transfer known engineering concepts as for example switches and oscillators as synthetic networks into bacterial cells. However, these concepts need rethinking and adaptation taking into account transcriptional noise due to low copy numbers and time-limited bacterial cell environments. The model analysis during the design phase should therefore not be confined to steady state analysis but also include the analysis of transients and stochasticity, which can lead to qualitatively different dynamics.

References

1. E. Andrianantoandro, S. Basu, D.K. Karig, R. Weiss, Synthetic biology: new engineering rules for an emerging discipline. Mol. Sys. Bio. **2**, 0028 (2006)
2. M. Heinemann, S. Panke, Synthetic biology–putting engineering into biology. Bioinformatics **22**, 2790–2799 (2006)
3. A.S. Khalil, J.J. Collins, Synthetic biology: applications come of age. Nat. Rev. Genet. **11**, 367–379 (2010)
4. A. Cheng, T.K. Lu, Synthetic biology: an emerging engineering discipline. Annu. Rev. Biomed. Eng. **14** (2012)
5. M.B. Elowitz, S. Leibler, A synthetic oscillatory network of transcriptional regulators. Nature **403**, 335–338 (2000)
6. T. Gardner, C.R. Cantor, J.J. Collins, Construction of a genetic toggle switch in escherichia coli. Nature **403**, 339–342 (2000)
7. M. Tigges, T.T. Marquez-Lago, J. Stelling, M. Fussenegger, A tunable synthetic mammalian oscillator. Nature **457**, 309–312 (2009)
8. J. Stricker, S. Cookson, M.R. Bennett, W.H. Mather, L.S. Tsimring et al., A fast, robust and tunable synthetic gene oscillator. Nature **456**, 516–519 (2008)
9. H. Kobayashi, M. Kaern, M. Araki, K. Chung, T.S. Gardner et al., Programmable cells: interfacing natural and engineered gene networks. Proc. Natl. Acad. Sci. USA **101**, 8414–8419 (2004)
10. J. Anderson, E. Clarke, A. Arkin, C. Voigt, Environmentally controlled invasion of cancer cells by engineered bacteria. J. Mol. Biol. **355**, 619–627 (2005)
11. G. Karlebach, R. Shamir, Modelling and analysis of gene regulatory networks. Nat. Rev. Mol. Cell. Biol. **9**, 770–780 (2008)
12. A. Raj, A. van Oudenaarden, Nature, nurture, or chance: Stochastic gene expression and its consequences. Cell **135**, 216–226 (2008)

13. N.G. Van Kampen, *Stochastic Processes in Physics and Chemistry*, 3rd edn. (Elsevier, Amsterdam, 2007)
14. D.T. Gillespie, Exact stochastic simulation of coupled chemical reactions. J. Chem. Phys. **8**, 2340–2361 (1977)
15. H. Smith, Oscillations and multiple steady states in a cyclic gene model with repression. J. Math. Biol. **25**, 169–190 (1987)
16. S. Müller, J. Hofbauer, L. Endler, C. Flamm, S. Widder et al., A generalized model of the repressilator. J. Math. Biol. **53**, 905–937 (2006)
17. N. Strelkowa, M. Barahona, Switchable genetic oscillator operating in quasi-stable mode. J. Roy. Soc. Interface **7**, 1071–1082 (2010)
18. N. Strelkowa, M. Barahona, Transient dynamics around unstable periodic orbits in the generalized repressilator model. Chaos: An Interdisc. J. Nonlinear Sci. **21**, 023104 (2011)
19. F.M. Ausubel, *Current Protocols in Molecular Biology* (Wiley, New York, 1987)
20. H. Celesnik, A. Deana, J.G. Belasco, Initiation of RNA decay in escherichia coli by 5′ pyrophosphate removal. Mol. Cell **27**, 79–90 (2007)
21. J. Richards, T. Sundermeier, A. Svetlanov, A.W. Karzai, Quality control of bacterial mRNA decoding and decay. Biochim. Biophys. Acta **1779**, 574–582 (2008)
22. J.G. Belasco, J.G. Brawerman, *Control of Messenger RNA Stability* (Academic Press, San Diego, 1993)
23. S. Mizusawa, S. Gottesman, Protein degradation in escherichia coli: the lon gene controls the stability of sula protein. Proc. Natl. Acad. Sci. USA **80**, 358–362 (1983)
24. C.V. Rao, A.P. Arkin, Stochastic chemical kinetics and the quasi-steady-state assumption: Application to the gillespie algorithm. J. Chem. Phys. **118**, 4999–5009 (2003)
25. W. El-Sharoud, *Bacterial Physiology: A Molecular Approach* (Springer, Berlin Heidelberg, 2008)
26. J.M. Pedraza, J. Paulsson, Effects of molecular memory and bursting on fluctuations in gene expression. Science **319**, 339–343 (2008)
27. A. Altland, B. Simons, *Condensed Matter Field Theory*, 1st edn. (Cambridge University Press, New York, 2006)
28. A.V. Hill, The combinations of haemoglobin with oxygen and with carbon monoxide. I. J. Physiol. **40**, 4–7 (1910)
29. J. Weiss, The Hill equation revisited: uses and misuses. FASEB J. **11**, 835–841 (1997)
30. A.M. Kierzek, Stocks: Stochastic kinetic simulations of biochemical systems with gillespie algorithm. Bioinformatics **18**, 470–481 (2002)
31. E. Doedel, Auto 07. software for continuation and bifurcation problems in ordinary differential equations. download—http://indycsconcordiaca/auto/ (2007)
32. A. Dhooge, W. Govaerts, Y.A. Kuznetsov, Matcont: A matlab package for numerical bifurcation analysis of odes. ACM Trans. Math. Softw. **29**, 141–164 (2003)
33. O. Purcell, N.J. Savery, C.S. Grierson, M. di Bernardo, A comparative analysis of synthetic genetic oscillators. J. Roy. Soc. Interface **7**, 1503–1524 (2010)
34. S.H. Strogatz, *Nonlinear Dynamics and Chaos* (Westview Press, Cambridge, 1994)
35. E. Hopf, Bifurcation of a periodic solution from a stationary solution of a system of differential equations. Berlin mathematische physics klasse, Sachsischen Akademic der Wissenschaften Leipzig **94**, 3–32 (1942)
36. P. Perlikowski, S. Yanchuk, M. Wolfrum, A. Stefanski, P. Mosiolek et al., Routes to complex dynamics in a ring of unidirectionally coupled systems. Chaos **20**, 013111 (2010)
37. J.F. Lindner, A.R. Bulsara, One-way coupling enables noise-mediated spatiotemporal patterns in media of otherwise quiescent multistable elements. Phys. Rev. E. **74**, 020105 (2006)
38. N. Strelkowa, Stochastic analysis of nonlinear dynamics and feedback control for gene regulatory networks with applications to synthetic biology, Ph.D. thesis, Imperial College London (2011)
39. M. Golubitsky, I. Stewart, D. Schaeffer, *Singularities and Groups in Bifurcation Theory*, Volume II. Applied mathematical sciences, vol. 69. (Springer, Berlin, 1988)

40. T.B. Kepler, T.C. Elston, Stochasticity in transcriptional regulation: Origins, consequences, and mathematical representations. Biophys. J. **81**, 3116–3136 (2001)
41. M. Hemberg, M. Barahona, A dominated coupling from the past algorithm for the stochastic simulation of networks of biochemical reactions. BMC Syst. Biol. **2** (2008)
42. R.P. Boland, T. Galla, A.J. McKane, How limit cycles and quasi-cycles are related in systems with intrinsic noise. J. Stat. Mech: Theory Exp., P09001 (2008)
43. C.W. Gardiner, *Handbook of Stochastic Methods*, 3rd edn. (Springer, 2004)
44. S. Risau-Gusman, G. Abramson, Bounding the quality of stochastic oscillations in population models. Eur. Phys. J. B **60**, 515–520 (2007)
45. P.E.M. Purnick, R. Weiss, The second wave of synthetic biology: from modules to systems. Nature **10**, 410–422 (2009)
46. A. Takamatsu, R. Tanaka, H. Yamada, T. Nakagaki, T. Fujii et al., Spatiotemporal symmetry in rings of coupled biological oscillators of physarum plasmodial slime mold. Phys. Rev. Lett. **87**, 078102 (2001)
47. J.M. Raser, E.K. O'Shea, Noise in gene expression: Origins, consequences, and control. Science **309**, 2010–2013 (2005)
48. T.J. Perkins, P.S. Swain, Strategies for cellular decision-making. Mol. Syst. Biol. **5**, 326 (2009)
49. R.J. Johnston, J. Desplan, C. Desplan, Stochastic mechanisms of cell fate specification that yield random or robust outcomes. Annu. Rev. Cell. Dev. Biol. **26**, 16.1–16.31 (2010)
50. M. Acar, J.T. Mettetal, A. van Oudenaarden, Stochastic switching as a survival strategy in fluctuating environments. Nat. Genet. **40**, 471–475 (2008)
51. G.M. Süel, J. Garcia-Ojalvo, L.M. Liberman, M.B. Elowitz, An excitable gene regulatory circuit induces transient cellular differentiation. Nature **440**, 545–550 (2006)
52. H. Maamar, A. Raj, D. Dubnau, Noise in gene expression determines cell fate in Bacillus subtilis. Science **317**, 526–529 (2007)
53. A. Eldar, M.B. Elowitz, Functional roles for noise in genetic circuits. Nature **467**, 167–173 (2010)
54. L. Cai, N. Friedman, X.S. Xie, Stochastic protein expression in individual cells at the single molecule level. Nature **440**, 358–362 (2006)
55. N. Strelkowa, M. Barahona, *Stochastic Oscillatory Dynamics of Generalized Repressilators*, ed. by T. Simos, G. Psihoyios, C. Tsitouras, Z. Anastassi. Numerical Analysis and Applied Mathematics ICNAAM 2012, number 1479 in AIP Conference Proceedings. (American Institute of Physics, New York, 2012), pp. 662–666
56 N.S. Goel, N. Richter-Dyn, Goel74, Stochastic Models in Biology, Academic press New York, San Francisco, London,215-223 (1974)

Automatic Computation of Crossing Point Numbers Within Orthogonal Interpolation Line-Graphs

Victor J. Law, Feidhlim T. O'Neill and Denis P. Dowling

Abstract The recording of atmospheric pressure plasmas (APP) electro-acoustic emission data has been developed as a plasma metrology tool over the last couple of years. In this work low moment analysis of acoustic time-series data is examined for structure complexity (in terms crossing points) within 2- and 3-dimensional time-series line graph datasets for the purpose of plasma control. A theoretical analysis of the structural complexity analysis is given, and the embedding algorithms obtained are implemented using LabVIEW™ for the purpose of real-time plasma control. The software uses a threshold segmentation process to map high contrast images of the line-graphs into binary (red and black) images that maps particles (red) directly to the number of crossing points within the cluster. It is found that single and bimodal cluster systems can be analyzed within the pixelation error limit. The approach taken here is generic and may be transferred to other (non-acoustic) datasets.

Keywords Plasma · Acoustic · Threshold segmentation · Cluster analysis

1 Introduction

With the increased use of atmospheric pressure plasma (APP) systems in industrial applications such as surface activation of polymers prior to bonding within the automotive [1] and the aerospace industry [2, 3] there is a growing recognition for the need for the development of non-invasive in situ process monitoring techniques. One such metrology is based on plasma acoustic emission. The study of acoustic emission has a long history dating back to W Duddell's and V Paulson's

V. J. Law (✉) · F. T. O'Neill · D. P. Dowling
Surface Engineering Group, School of Mechanical and Materials Engineering, University College Dublin, Belfield, Dublin 4, Ireland
e-mail: Victor.law@ucd.ie

I. Zelinka et al. (eds.), *How Nature Works*, Emergence, Complexity and Computation 5, DOI: 10.1007/978-3-319-00254-5_9,
© Springer International Publishing Switzerland 2014

ionized gas sound production experiments at the turn of 19–20th century when radio technology was in its infancy [4, 5]. In more recent times (since 1990s) optical-acoustic studies of plasma welding process [6], acoustic metrology of anomalous arc discharge in plasma processing equipment [7], plasma and laser welding [8–10], and plasma anodizing [11] have been reported. More recently plasma acoustic emission metrology has been demonstrated on industrial scale APP reel-to-reel systems [12, 13] and APPJ systems that use compressed helium gas [14] or compressed air [15–17] as the working gas. In addition acoustic emission has been reported in a helium linear-field jet [18]. Arguably one of the drivers of this metrology is it relative low cost of implementation as compared to optical emission spectroscopy and electrical measurement [19–21].

This chapter explores acoustic signatures that are characteristic of the process environment between the PlasmaTreat Open-air^{TM} atmospheric pressure plasma jet [15, 16, 20–22] and a temperature sensitive surface. In particular we look at the low entropy plasma jet flowing afterglow region where the process temperature is around $\sim$ 300–400 K. This temperature region has a characteristic plasma-surface gap distance of 5–60 mm. This region is also of interest because the low temperatures reached enable thermally sensitive materials to be processed: for example, the activation of polymers and carbon epoxy composites prior to adhesive bonding, the therapeutic treatment of open wounds, and cell treatment.

Whereas operating the APPJ below a gap of 5 mm produces a high entropy state (1700 $\pm$ 100 K) due to the proximity of the plasma arc and therefore thermally damages (burns) these materials.

Today's atmospheric plasma pressure metrology uses advanced digital time- and frequency-domain instrumentation linked to multivariate analysis tools [12–19, 22, 23] to capture the interaction between the plasma and treated surface. To examine the low entropy state signals we bring together the emerging Low-order moment (Mean, Standard Deviation (SD) and Skewness (Sk)) analysis of acoustic time-series data [16] with structural complexity techniques in the mathematical analysis of random polygons that are composed of freely jointed segments of equal length (equilateral), and knot systems that display morphological organization [24–27]. One of the emphases of these polygon and knot system studies was the development and understanding of the structural complexity of polymer chains in solvents; in particular the determination of the knot type and topological (minimum) crossing point numbers within polymer chains.

The aim of this work is to extend Ricca's crossing point algebra techniques [27] to Cartesian interpolation line-graphs, where the graphs are formed from high contrast (black and white) regions. In addition the crossing point technique is extended to the 3 orthogonal planes to extract both the average crossing point which is defined as the average crossing point number over all three planes. This last embodiment within our approach is in-line with previous reported work [26] where it has been observed that average crossing point number correlates well with the experimentally observed speed of electrophoretic migration of knotted DNA molecules of the same size but of various knot types, and with relaxation dynamics of modelled knotted polymers.

The approach taken here greatly simplifies the choices of viewing angle by locking the analysis to 3 orthogonal planes. Then using this restricted viewing angle, threshold segmentation software is constructed and employed to automatically process and extract the crossing point number from the interpolation line-graphs. In this work the 3 orthogonal planes are: XY: μ-SD, XZ: μ-Sk, and XY: SD-Sk.

To describe this new non-parametric cluster analysis process, the rest of the chapter is divided into three sections. Section 2 provides a theoretic analysis of crossings point numbers in relation to interpolation line-graphs and an outcome interpretation of in terms stochastic and deterministic events. The LabVIEW^{TM} 2011 programing is described in Sect. 3, and crossing point number analysis of single and double clusters is presented in Sect. 4. Section 5 provides a discussion of this work in the context of process control where stochastic and deterministic effects need to be identified. The algorithms described in this chapter were implemented using LabVIEW^{TM}, however, it is worth noting that they could be implemented in many other software packages including Matlab$^{®}$. A 4 page extended abstract briefly describing this work (Interdisciplinary Symposium on Complex Systems, Kos, Greece 2012) can be found in Ref. [28].

2 Analysis

The time-series acoustic data used in this work was first published by Law VJ et al. [16]. A total 26 datasets containing 300 sequentially sampled data points is used here: 13 dataset for 19 kHz and 13 datasets for 25 kHz plasma electrical drive frequency. Each dataset is then reduced to their μ, SD and Sk datum points. Figure 1 depicts the result of this reduction process. These Low-order moment planes describe the temporal-spatial heterogeneous APPJ process from 5 mm nozzle to surface distance (arc) to 60 mm far flowing afterglow.

Inspection of the six Low-order shape-space line-graphs reveals that the high entropy arc datum points tend to be separated from the flowing afterglow datum points. In addition as the axis parameter changes from μ to Sk the low-entropy part of each cluster becomes more complex: moving from a near-linear shape entanglement cluster to cyclic or torus-like. This mapping transition [29] will be discussed later with the analysis of the crossing point numbers in Sect. 2.1. For now it is sufficient to understand that these datasets illustrate a specific plasma processing case however this analysis approach may be used on other datasets that describe different phenomenological processes.

To extract meaningful information from these six line-graphs, a modified form of the falling stick model as described by Ricca [27] is used. Ricca's falling stick model is mathematically defined in Eq. 1, (where C_{max} is the maximum number of crossing points, N is the number of sticks and $N-1$ is the embedded delay). The model allows the sticks to fall on to each other, or fall away from each other, and with the ends of the sticks being allowed to cross and extend above the previous

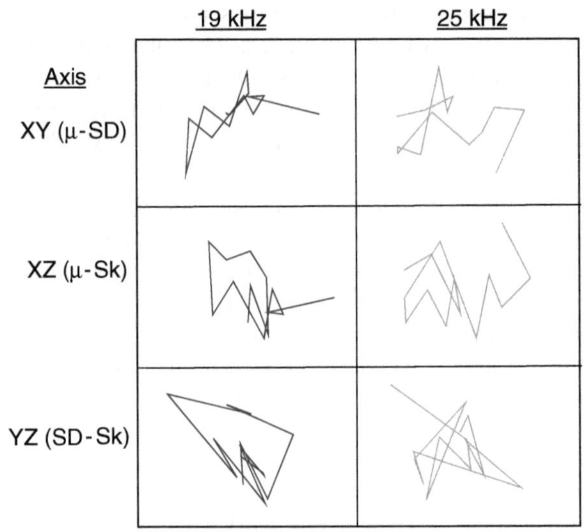

Fig. 1 Low-order orthogonal interpolation line-graphs of APP data

fallen stick. In this process the sticks are allowed to exhibit morphology. In this scenario, morphology means that an observer can trace a stick going under another stick and emerge on the other side of the upper stick due to surface shading and texture.

$$C_{max} \sim \frac{N}{2}(N-1) \tag{1}$$

Equation 1 is then re-analysed for the interpolation line-graph as shown in Fig. 1. In this case, the length of each interpolation line is not fixed at some equilateral value, but is variable in length and extends between and up to their associated twin datum points. Furthermore the interpolation lines cannot extend in length beyond the initial and following datum point, that is to say the lines are contiguous. The direction of each interpolation line is allowed to be variable so that they express the underlying physical process that they are to characterise. There is no morphology in this process as the interpolation lines are totally formed from one colour (black) seating upon a white background and when lines cross another there is no layer information.

A reinterpretation of the falling stick model with morphological sticks overlaying each other is schematically shown in Fig. 2a, and the interpolation line-graph model is shown in Fig. 2b. In Figs. 2a and 2b, $N = 3$. From this simple representation it can be seen that 3 crossing points for the stick model are produced and 1 crossing point for the interpolation line-graph model is produced. In addition it is important to note that morphology does not take any part in the deduction of this outcome.

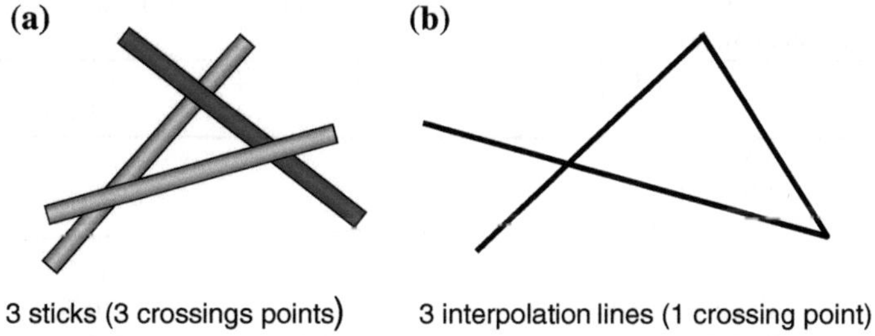

Fig. 2 **a**, **b**. Schematic representation of falling stick model, including morphology (**a**) and, line interpolation model (**b**)

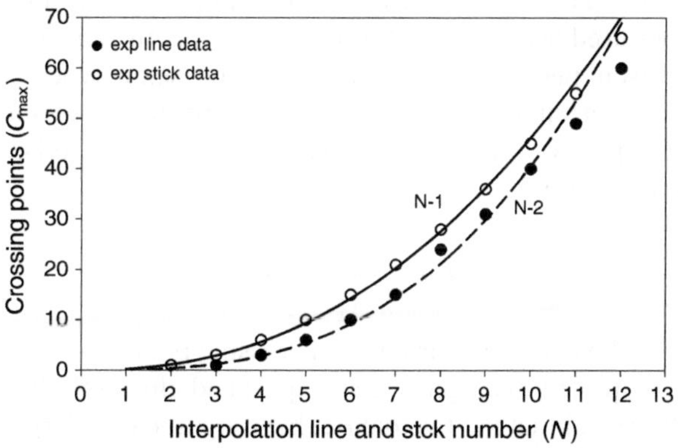

Fig. 3 Experimental and embedding delay models for stick and interpolation line-graph scenarios

Figure 3 shows the experimental observations of both models and the analysis derived from Eqs. 1 and 2. Note in both equations the approximation sign is used because the fitted curve of the models departs from experimental observations above $N = 11$. In this new analysis, Eq. 1 has now been transposed into Eq. 2 with an embedding delay of $N{-}2$.

$$C_{max} \sim \frac{N}{2}(N-2) \tag{2}$$

Given the knowledge that the sum of the crossing points within each orthogonal interpolation line-graph can be computed and the average crossing point $\left(\overline{C_{\perp}}\right)$for all three planes can also be computed by adding all the crossings points and then dividing the sum by 3 to yield the final value a simple procedure may be mathematically expressed as shown in Eq. 3.

$$C \approx \overline{C_{\perp}} \approx \left(\frac{C_{xy} + C_{xz} + C_{yz}}{3}\right) \tag{3}$$

Table 1 Crossing point numbers for interpolation line-graphs in Fig. 1 and their orthogonal average and efficiency index

1	$N = 12$ and $C_{max} = 60$	Crossing point # for 19 kHz	Crossing point # for 25 kHz
2	XY: μ-SD	4	4
3	XZ: μ-Sk	5	4
4	YZ: SD-Sk	6	11
5	Average	5	6.3
6	Efficiency index (%)	8.3	10.5

The structural complexity information within $(\overline{C_\perp})$ may also be normalised to C_{max} for a given number N within the datasets to provide an efficiency index (I_{eff}). This procedure is mathematically expressed in Eq. 4, where for the 12 interpolation line-graphs used here $C_{max} = 60$. The index may then be used for transfer or make a comparison between other plasma of the same type, between similar plasma processes (cleaning, activation and/or deposition), formed by different plasma reactor types, or indeed other datasets that characterise different phenomenological processes.

$$I_{eff} \approx \frac{\overline{C_\perp}}{C_{max}} 100(\%) \tag{4}$$

Table 1 shows the calculated orthogonal crossing points for the 6 datasets given in Fig. 1 along with their averaged value (Eq. 3) and efficiently index (Eq. 4). The tabulated results reveal that the two dataset have significantly larger crossing point values in the YZ: SD-Sk plane (row 4) as compared to the mean and standard deviation values. The differences are reflected in the average crossing point (row 5) and the efficiency index (row 6). For example efficiency index for 19 kHz and 25 kHz data sets are 8.3 and 10.5, respectively. Some of the possible origins of these differences are explore in Sect. 2.1.

2.1 Stochastic Thermal Noise Comparison

In this section the efficiency index is examined with respect to the interpolation line-graphs in Fig. 1, and to random (stochastic) electrical noise measurements. To test the efficiency index, the electrical thermal (room temperature: 293–298 K) noise from two 1 MΩ input terminals of a 200 MHz bandwidth analogue-to-digital digitizer are measured and presented. Under these open-circuit conditions the sampled dc voltage values are expected to have a Gaussian probability function over time and be symmetrical around the zero voltage crossing point. The standard deviation will also be approximately equal to the route mean square of sampled datum points [30].

For the purpose of this analysis, each channel measurement is repeated 3 times and the sampled datum points (14 points per channel) are plotted against each

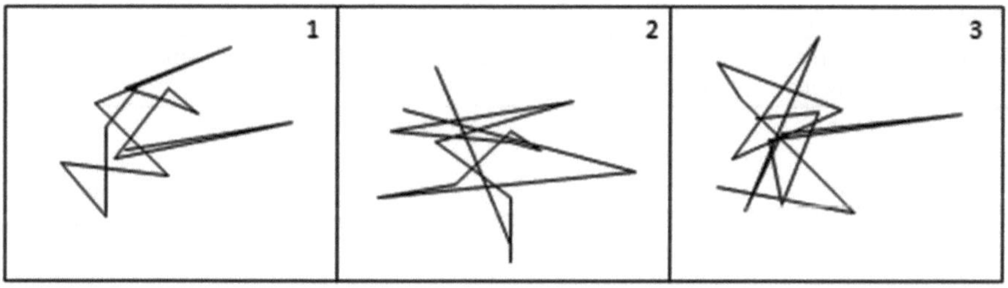

Fig. 4 Cartesian space representation of the three noise measurements

other to generate a series of Cartesian graph plots. The results of these plots are shown in Fig. 4. The crossings point of each graph is then computed and the average taken. This procedure therefore enables a direct comparison to be made with the crossings points of the plasma datasets and a general classification of the efficiency index to be made in relation to a statistics family of values.

Figure 4 (1, 2, & 3) shows the results of the stochastic noise measurements in Cartesian space. Channel 1 is plotted on the horizontal axis and channel 2 is plotted along the vertical axis and the individual consecutive datum points are connected with interpolation lines. In these three representations the main feature of note is that the interpolation lines form a single cluster, or entanglement, with no deterministic clustering such as linear or cycle shapes as observed in corresponding datasets (see Fig. 2). The salient statistical description of these plots (standard deviation of each channel measurement, the mean of the channel measurements plus the voltage (peak–peak) per channel) is shown in Table 2.

The tabulated results reveal a SD mean value of 0.000507 for channel 1 and a SD mean value of 0.009288 for channel 2. In voltage amplitude terms, these values are approximately 33 % of their respective noise voltage amplitudes.

The orthogonal analysis results of the three datasets are tabulated in Table 3, where the computed orthogonal averaged crossing point and efficiency index are given for each datum point. These results reveal that the averaged crossing point increase from 0.33 to 13.33 for $N = 3$ to $N = 13$ and their corresponding

Table 2 Electrical noise per channel measurements

Measurement	Standard deviation	Voltage (peak-peak)
Channel 1		
1	0.000485	0.18027–0.17840
2	0.000579	0.18024–0.17822
3	0.000457	0.18027–0.17895
Mean	0.000507	
Channel 2		
1	0.006997	0.07156–0.04495
2	0.011276	0.07845–0.03794
3	0.009595	0.07356–0.04236
Mean	0.009288	

Table 3 Orthogonal analysis

Data point	N	Crossing points	Crossing points	Crossing points	Average crossing point	Clmax	Efficiency index
1	0	0	0	0	0	0	0
2	1	0	0	0	0	0	0
3	2	0	0	0	0	0	0
4	3	0	1	0	0.33	1	33.33
5	4	1	2	0	1	3	33.33
6	5	2	3	1	2	6	33.33
7	6	2	4	3	3	10	30
8	7	2	6	7	5	15	33.33
9	8	4	6	10	6.66	24	27.75
10	9	4	6	12	7.33	31	23.64
11	10	4	9	14	9	40	22.5
12	11	6	9	18	11	49	22.44
13	12	7	11	22	13.33	60	22.20
14	13	8	16	24	16	72	22.22

efficiency index decreases from 33.33 % to 22.22 %. These values equate to an average crossing point efficiency index range of 27.64 % ± 5.68.

As we are visualizing the Cartesian shape-space of Fig. 4 in pixels (picture elements) rather than the real-world voltage dimension it is possible to characterize the efficiency index using non-parametric cluster analysis.

Let us first consider 14 datum points forming a continuous straight line in Cartesian space with coordinates: a,b_1, a,b_2, a,b_3.... For this example the gradient of the line and intercept are mathematically represented by the straight line equation: $y = mx + b$. Therefore the efficiency index = 0, and by definition there cannot be any crossing point. Hence there must be a strong deterministic process behind the formation of the line. Now consider the 14 datum having a crossing point efficiency index of 100 %. This upper (or maximum) boundary scenario is again described by a linear deterministic model. Between these boundary limits, random (or stochastic), processes can operate. Classification the crossing point efficiency index can therefore help in providing an estimation of deterministic effect within time series datasets. For example, the stochastic electrical noise generated by the unit provides an efficiency index of between 33.33 to 20 %. Outside this band deterministic effect may be expected to be observed and their effect becoming stronger as the boundary limit is reached.

Figure 5 displays this analysis in graph form, where the abscissa axis represents the line interpolation number and the crossing point number is plotted along the ordinate axis. In this figure the black datum points represent the upper boundary limit as described by the embedding delay of (N–2) of Eq. 2. The lower, or straight line equation, boundary is defined by the zero value on the ordinate. Between these two limit boundaries, stochastic crossing point behavior is

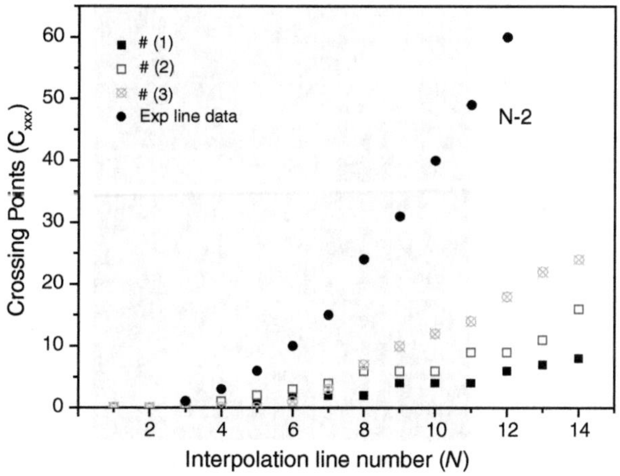

Fig. 5 Interpolation line-graph C_{max} values and stochastic data

located. The estimated location of the stochastic behavior is outlined by the black square and open square datum points as measured by the stochastic noise measurement.

2.2 *Bimodal Clusters*

So far we have considered a single cluster of data points. However in the real world of plasma processing datasets sometimes contain double [3, 12, 13, 16] or multiple [31] clusters over the supposedly stable processing time period. This is particular true of parallel-plate reel-to reel plasma processing system where two or more plasma reactors are electrically driven from one electrical power source. Typically these atmospheric reactors are used to plasma treat polymers and composites. The bimodal behavior arises from two contrasting electrical effects: firstly, the sequentially nature of the process in that the treated material adds to the electrical impedance of each reactor in turn as the material passes through the reactors or, secondly mode jumping where the treated material drastically alters the plasma chemistry and hence the plasma electric impedance.

Figure 6 shows a principle component analysis (PCA) loading plot of one such bimodal behavior as measured on the Dow Corning® SE-1100 AP4 helium-atmospheric reel-to-reel system [16]. In this example the plasma electrical parameters (current and voltage) of the dual reactor are measured with the process starting at the bottom right and progressing to the top-left with each datum point added at are rate of 1 point per 0.5 s over a total process time period of 2 min. In this example note how the datum points jump 8 times between the two clusters states within the total time period.

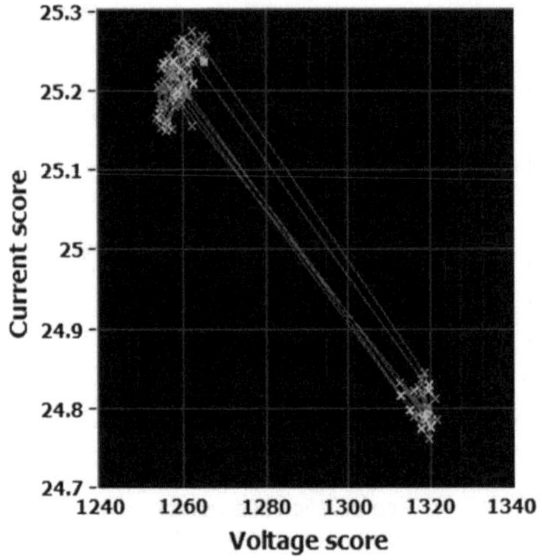

Fig. 6 PCA loading plot of helium plasma driven at 500 W as a function time [16]

This bimodal cluster behavior was originally analyzed for cluster separation in Euclidian co-ordinate space rather than angle measurement and local threshold variation within the Euclidian space. However in this work, the application of crossing point number analyses requires we treat each cluster as a separate identity where the interpolation lines projecting out of the clusters are terminated in empty space between the clusters. An example of this new LabVIEW Software treatment of bimodal behavior is described in Sect. 4.1 and shown in Figs. 10 and 11.

3 LabVIEW Programing

This section describes the non-parametric cluster analysis (NPCA) using LabVIEW™ 2011 [National Instruments Ltd] and the Vision builder A1 (3.5) application software.[1] An earlier vision of the software using LabVIEW 8.2 with Vision builder has been used for non-parametric cluster analyses of the space separation between data clusters mapped onto a Cartesian graph. The mode classification results of which are published in Chaotic Systems: Theory and Applications, 2009 [12].

The new software uses a threshold segmentation process to map any high contrast image. In this case the low-order moment interpolation line-graphs of

[1] The complete program may be obtained from the Author (victor.law@ucd.ie). In addition information on National Instruments LabVIEW software can also be found at http://www.ni.com/
.

each of the three orthogonal planes are mapped. For ease of portability the software is installed in a Windows environment on a Dell laptop computer.

The LabVIEW software packages form a complete programming language based entirely on a graphical user interface. Within the software a sub-program called a virtual instrument (VI) creates graphical constructs equivalent to for-loops, do-while loops, case structures, and subroutines (called sub-VIs). These components are placed on a panel (block diagram) which is ultimately hidden from the operator. The data flow through the VIs is controlled by connecting "wires" between objects, somewhat like constructing an electrical circuit. The graphical user interface or "Front panel" is created from a flexible set of predefined graphical displays and control features. The programme is designed to accept and read an 8-bit Bitmap or JPEG image of the line-graph and exports its 0–255 pixel bit depth greyscale component into the Vision builder software. The image read VI is located outside of the while loop so when the program is operated the image VI opens and stays open to allow interactive processing of the image within the while-loop. Within the while loop the vision builder software contains a suite of hierarchical particle measurement VIs tools. The parameters of the VIs are user preset to look for bright objects within a local adaptive threshold which has a lower and upper boundary limit. The local threshold operation defines the graphs interpolation lines in terms of pixel depth (typically between 226 and 290 bit depth): pixels with a bit depth below 226 and outside the crossing point areas are reset to a background pixel value of $I = 0$ (black); pixels not reset $I = 0$ and completely surrounded by the interpolation lines are regarded as holes and are set to a new pixel value of $I = 1$ (red). Hence the pixel surround by three interpolation lines form a red particle, the result of which produces a black-and-red binary image that represents the number of crossing points (in terms of particles).

Figure 7 shows the complete source code at the block diagram level. For this publication the code is turned through 90 degree to fit the book format, otherwise the code is be to read from left to right with the create and read image VIs and final error handling VI outside of all the while-loop.

For a simple image containing one or two crossing points the structural complexity information can be readily visualised and identified by the human eye. However to remove any possible counting error an automatic counting algorithm is used to count and display the number of particles that are formed by the interpolation lines. In addition, for each individual particle within the region-of-interest their areas calculated, logged and subtracted from the interactive region to produce a percentage particle area value. This process also allows further analysis to be performed at a later date and provides the means of pixelation error correction, see Eq. 5.

Finally the operator can generate a report containing the non-parametric cluster analysis information which is time-stamped in units of hours, minutes and seconds and exported as comma separated file (*.csv).

It should be noted that the digital image processing is pixel based rather than using real-world units (i.e. mm, cm, etc.…). There is however a limitation to the program functionality which is found in the quantization process of the pixelated

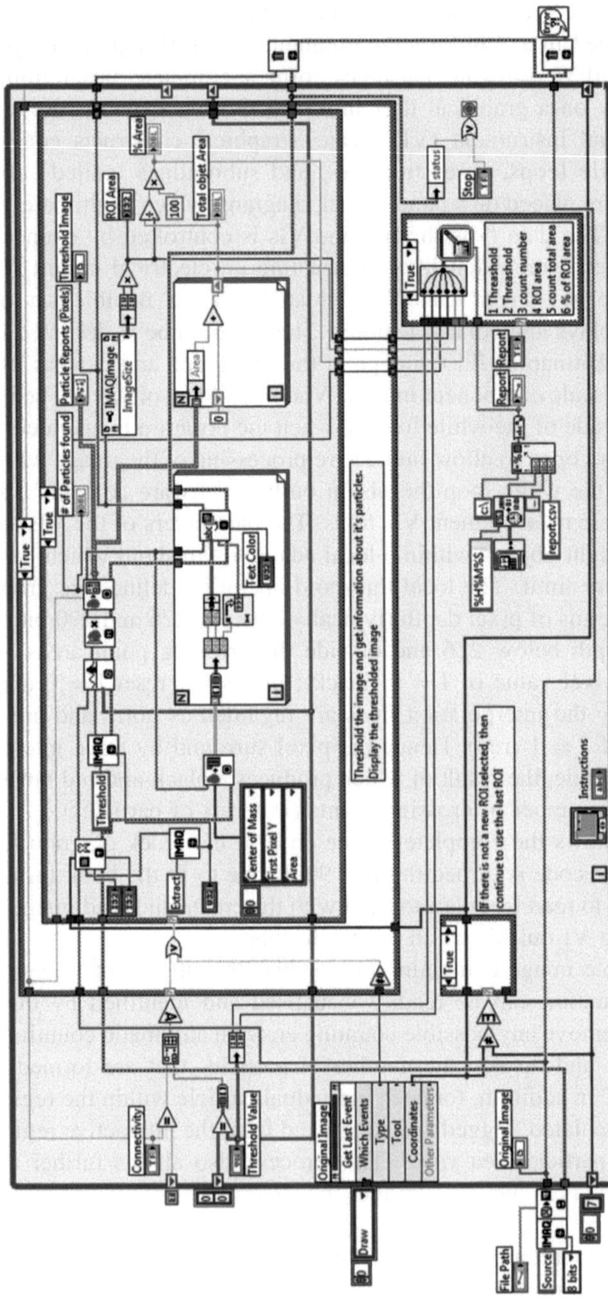

Fig. 7 Block diagram of completed LabVIEW^{TM} software programme

line widths. In this case, the LabVIEW boundary limit is defined by the $640 \times 480 = 307{,}200$ pixels or 0.3 megapixels arranged in regular two-dimensional grid field format of the read image VI; and as long as all of the line graph information is within these quantization limits the program will function as designed. The interpolation line-graph image fidelity is also a limiting factor as they are formed by pixels, and where too few pixels produce a jagged line which may be degraded further if the original image was in the JPEG format which has undergone compression by 10:1. Figure 11 provides a good example of pixel limitation where a 195 x 191 pixel format is used in the original image. Under these restricted conditions the program is operating but the jagged line edge starts to present a problem to the counting algorithm.

For the reader who may not be familiar with LabVIEW block diagrams a simplified block diagram of dataflow through the Vision build VIs is shown in Fig. 8. In this figure all arrays, Boolean (true or false) operators, data displays, data collection and data reporting have been removed for clarity. The diagram shows that the main vision VIs are nested within two binary (True or False) case structures: the false state of these case structures handle all the error commands: the errors being no file to read or no change in the threshold values. These case structures operate within while loop, located outside of which are the 'create' and 'read' file command Vis, and the error handling VI.

An example of the front panel of the software program is shown in Fig. 8. In this example the Cartesian shape-space image form Fig. 1 is displayed on the left-hand-side of the panel and, the automatically reconstructed binary image (black and red) is displayed on the right-hand-side. Using a threshold window pixel value matrix of minimum = 226 and maximum = 296 it can be visually observed (red coloured triangles) the number of particle analysis outcome is 4. Note, the visual comparison and digital count directly compares to the number of crossing point in the original image.

The non-parametric cluster analysis information is displayed in numerical panels on the front panel, and includes the number of particles, total area of particles and the percentage area of the particles within the region of interest.

4 LabVIEW Results

This section presents the LabVIEW^{TM} software processing two images of different pixel formats, 462 by 362 pixels 195 by 191 pixels, respectively. The two images are of a single cluster and cut away of a single cluster taken from a bimodal cluster system. These two examples show how the clusters are counted and also how pixel resolution affects the crossing point counting process. The software front panel of these two clusters are shown in Figs. 8 and 10, respectively. In addition, Sect. 4.2 describes how the number particles, hence the number crossing points, is altered by interpolation line pixel resolution.

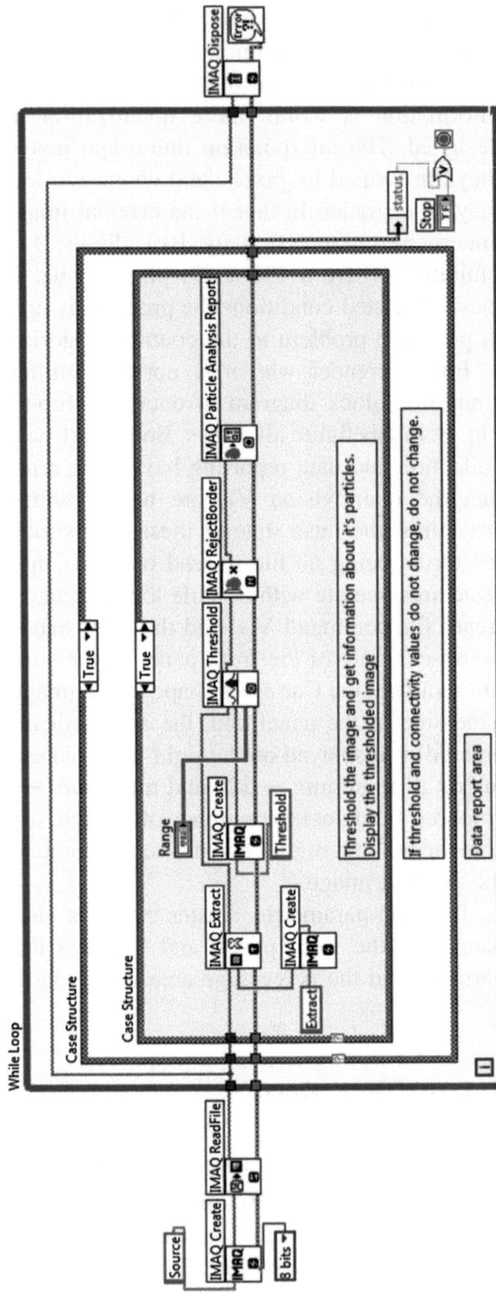

Fig. 8 Simplified block diagram showing the dataflow through the main VIs

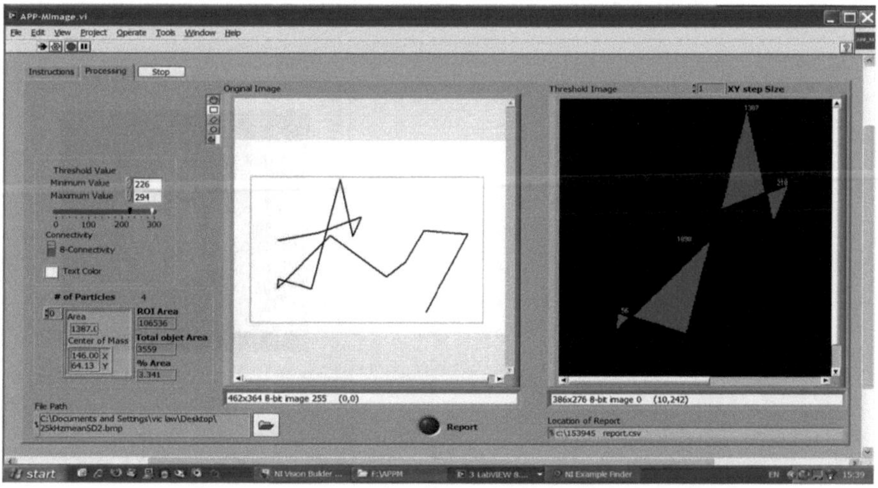

Fig. 9 LabVIEW™ software front panel showing the analysis of a single cluster

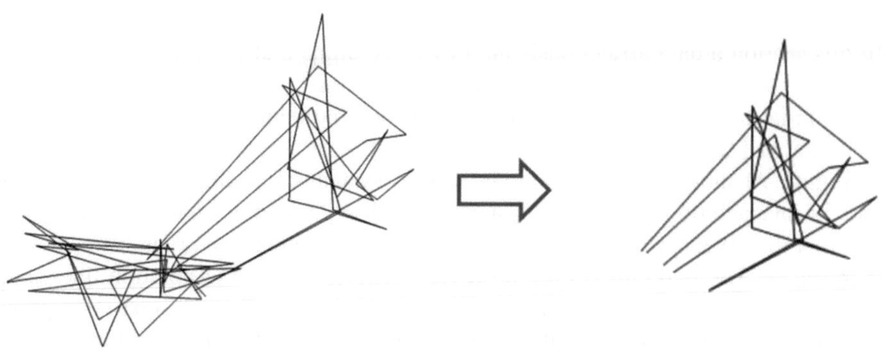

Fig. 10 Bimodal cluster system (*left*) and single cut away cluster (*right*)

4.1 Single Cluster

Figure 9 displays the front panel analysing a 462 by 362 pixel image of 25 kHz (μ-SD) line graph image as depicted Fig. 1. The left image is the original image and the right image is the reconstructed binary image. In this example it can be seen that the original image has 4 crossings points that form a total of 4 particles. The automatic reconstructed binary (back and red) image clearly depicts this visual analysis. In this case there is no need for further analysis.

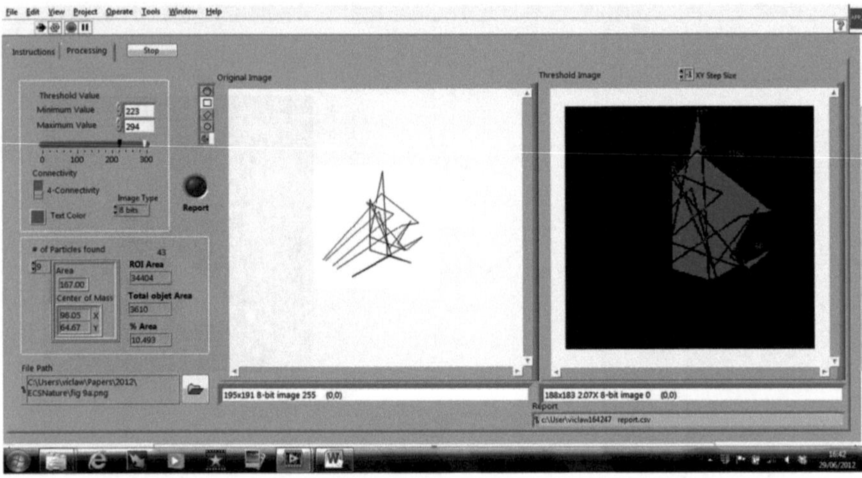

Fig. 11 LabVIEW^{TM} software front panel showing cut away cluster analysis

4.2 Single Cluster Analysis Within a Bimodal Cluster System

In this section a data array consisting of two columns that form a bimodal cluster system is considered. These data cluster formations generally described plasma system that has two time dependant states which momentary jump from one state to the other [12].

Figure 10 shows a typical bimodal cluster system consisting of 53 data points and 52 interpolation lines. The clusters are well separated as characterised by the fact that the pixel space between the two clusters is of the same order as the cluster themselves. The two clusters are however connected by 6 interpolation lines. For purpose of this work the right hand cluster is cut away at the mid separation point and analysed for the number of crossing points. The most striking feature of note for the cut away cluster is that the six connecting interpolation lines are no long connected at both ends, but rather the lines are terminated in free space. That is the lines are hanging in free space.

The software analysis of the cut away cluster is shown in Fig. 11. In this case a 195 by 191 pixel image is used along with the similar threshold values (min = 223 and max = 295). The first feature of note is that the interpolation lines that are hanging in free space do not interfere with the crossing point algorithm. However the number of crossing points in the original image is difficult to visualize: whereas, the reconstructed binary image yields 43 particles and hence 43 crossing points. On closer visual inspection of the original image the number of crossing points is 31 and an interpolation line pixel width of 1–2 pixels.

Table 4 lists all 43 particles and their particle size in terms of pixels for the reconstructed image. The table shows there are 31 particles having a pixel value of 3 or greater and 12 having 2 or less pixels (7 particles have 1 pixel and 4 particles

Table 4 Particle # and area within the reconstructed binary image of Fig. 10

Particle #	Area	Particle #	Area	Particle #	Area
1	117	15	177	29	161
2	205	16	1	30	2
3	1150	17	30	31	77
4	18	18	31	32	35
5	2	19	6	33	60
6	1	20	2	34	370
7	3	21	37	35	1
8	9	22	202	36	59
9	10	23	16	37	1
10	167	24	268	38	4
11	71	25	1	39	7
12	78	26	29	40	1
13	27	27	1	41	95
14	2	28	44	42	31
				43	1

have 2 pixels). The discrepancy of 12 between the original and the reconstructed binary image appears to be due to pixelation of the interpolation lines. This is the cause for the misinterpretation of image data, that is to say that the counting algorithm over counts the number of particles.

To correct the counting error and thus make the count process more accurate a secondary analysis procedure is required, and one that is based on pixelation error, that is a change in grey scale of individual elements within a group of pixels that go to make-up the types of objects (lines and particles) in the binary image. One way of doing this is to manually calculate the total number of particles ($Part_{max}$)

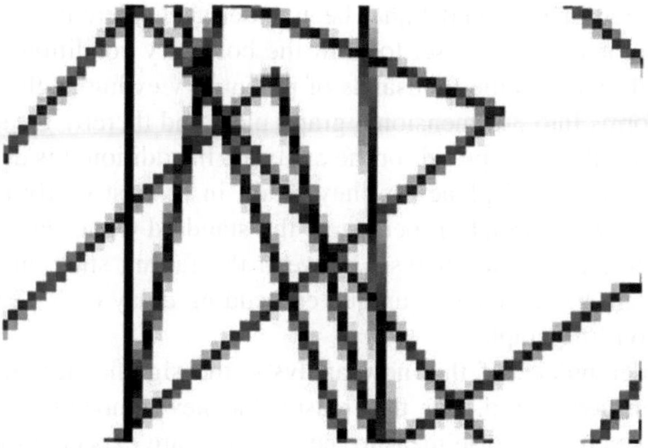

Fig. 12 Exploded view of the pixelated lines of the original cut away cluster image in Fig. 10. The pixelated image shows the smallest particle has two pixels and the line width is constructed from 2 *grey scale* picture elements

then subtract this value from this value the number of particles that have pixelation error ($Part_{error}$) so generating a good estimation of the true number of particles ($Part_{true}$). Here $Part_{error}$ is defined as the number of particles that have a similar number of pixels to the grey scale pixels that construct the width of the interpolation lines. This procedure is mathematically expressed in Eq. 5. With this pixelation correction an accurate estimation of particles within the cluster is achieved, and so leading to a good estimation of the number crossing points.

$$Part_{true} \approx Part_{\max} - Part_{error} \tag{5}$$

The validation of the value of $Part_{error}$ is simply performed by enlarging the cut away cluster line detail and counting the pixel and inspecting both the line and particle pixelation. Figure 12 shows an exploded view of the cluster. In this view the characteristic grey scale (black through to grey and on to white) mosaic features. Note how the smallest particle is made up from 2 pixels and the lines width are constructed from 2 grey scale pixels. This parity now defines the pixel error.

Incorporating this pixelation correction procedure into the software, the original number of crossing points (43) now reduces to 31. Using this additional procedure, the software can now be used to map out the crossing point number for each orthogonal plane of a single and multiples clusters.

5 Discussion and Outlook

This chapter has described the amalgamation of low-order moment analysis with 3-dimensional structural complexity techniques applied to time-series datasets. In this case the time-series data describes spatial-temporal acoustic emission from an air atmospheric pressure plasma jet. A comparison between the morphology falling stick problem and orthogonal (XY: μ-SD, XZ: μ-Sk, and XY: SD-Sk) 3-dimensional interpolated line-graph data has been undertaken. Here the normal family of the three orthogonal planes is set to limit the boundary conditions. Although this approach greatly reduces the thousands of potential viewing angles, the outcomes directly transforms into 3-dimensional graph plots and thereby greatly reduces the computing power that is required for the analysis. In addition it is desirable to have only the three orthogonal planes as they retain in at least partly a mathematical tractability and some formal properties of the standard Cartesian projection.

A comparative algebraic analysis between the falling stick and the Cartesian graphing problem has revealed a unique embedding delay of $N-2$ (see Eq. 2) for the interpolation line-graph.

For the wider uptake of this new analysis, the significance of the efficiency index has been considered. For this reason the new embedded delay has been tested against stochastic noise measurements which are expected to have a known time independent Gaussian probability function. The analysis indicates that the efficiency index (Eq. 4) may be used as a classification system to separate out stochastic process from deterministic effects within a dataset. For example: a low (0–20 %) or high index score (33–100 %) may indicate a deterministic process occurring within the dataset; an index between 22–33 %, in conjunction with single cluster, would indicate stochastic processes are most likely to be occurring within the dataset.

With this new insight we turn back to the plasma datasets which have efficiency indexes of 12.5 % at 19 kHz and 21.6 % at 25 kHz, respectively. These values may initially suggest a deterministic effect within the process and indeed the linear and open cyclic space-shape (Fig. 1) of the dataset gives support to this idea. However the 25 kHz SD-Sk dataset does exhibits a single cluster with an associated efficiency index is 21.6 %. Both these criteria would be sufficient for the dataset to be classified as having a stochastic behaviour if it was not for the factor the 3rd order of moment (SD, or, Skewness) needs careful interpretation as a zero SD does not imply that the mean = median, as in the case of discrete distributions which are multimodal [16]. Given this corollary the SD-Sk does point to a known hot process processing window where the entropy of the arc dominates Plasma-Treat afterglow. Clearly this processing window needs to be investigated further.

As the new embedded delay algorithm describes geometric features in 3 dimensional orthogonal Cartesian space the algorithm has been incorporated using National Instruments LabVIEW software which employs a graphical user interface. The software programing has been designed to sample high contrast (black and white) digital images with minimal user interactive control settings, and automatically compute the crossing point number within a single cluster.

In the case of bimodal cluster systems it has been demonstrate the two clusters can be analysed separately. The cut away cluster has crossover interpolation lines that hang in free space, these however do not interfere with the counting algorithm has they have are one ended. By extension this process can be employed in multiple cluster systems.

Poor preparation of the original image in terms of interpolation line and particle pixel resolution (pixelation) may require additional processing to achieve the correct crossing point number (particle count). Experimental observations indicate that loss of line definition occurs when the line thickness falls below 2 pixels. For images that contain 40 lines and above 265 by 265 pixel array within a $640 \times 480 = 307{,}200$ pixels format is required. In the present work this line definition is not limited by the number of lines. An approximation of the correct crossing point number (particles) can be obtained by rejecting particles which contain a similar number of grey scale picture elements that construct the width of the interpolation lines. Experimentally this is found to be approximately 2 pixels and defined graphically using the software user interface.

Finally it has been found that the graphical user interface software approach developed here agrees well with the manual inspection of the test plasma datasets. This agreement indicates that the software may be applicable to analyse other related dataset that describe different phenomenal processes. This approach may therefore constitute a new non-parametric cluster analysis tool.

5.1 Future Morphology Software

This work has focused on high contrast black and white 8-bit grayscale images of interpolated line-graphs that contain no perspective information. To advance this work forward to enable processing images that contain perspective information, i.e. that delineate lines and features that overlay each other to form entanglements of physical features [27, 32], a move away from simple binary images that contain no graduation of edge definition is required. As morphology information is generally considered to be generated by visual perspective, two methods of producing morphology may be considered. Firstly, a graded grey scale may be used to define edge detail. Secondly, use colour images and separate out the colour planes into various sets of primary components, such as RGB (Red, Green, and Blue) where each colour plane becomes an 8-bit image that can be processed like a grey scale image. Such software does exist within the suit of National Instruments vision software and is being investigated by the author and shall be reported on.

Acknowledgments We are grateful to acknowledge and thank Professor Ricca RL (University of Milan) for first suggesting to use crossing point number as a means of cluster analysis of our data. In addition this material is based upon works that were part supported by Science Foundation Ireland under Grant No.08/SRC/I1411.

References

1. R. Suchentrunk, H.J. Fuesser, G. Staudigl, D. Jonke, M. Meyer, Plasma surface engineering–innovative processes and coating systems for high-quality products. Surf. Coat. Technol. **112**(1–3), 351–357 (1999)
2. J.K. Kim, H.S. Kim, D.G. Lee, Adhesion characteristics of carbon/epoxy composites treated with low- and atmospheric pressure plasmas. J. Adhesion Sci. Technol. **17**(13), 1751–1771 (2003)
3. V.J. Law, A. Ramamoorthy, D.P. Dowling, Real-time process monitoring during the plasma treatment of carbon weave composite materials. JMSE **1**(2B), 164–169 (2011)
4. V. JY, Music in electric arcs: an english physicist, with shunt circuit and keyboard, made them play tunes. The New York Times, 28 April, 7 (1901)
5. A.N. Goldsmith, Modulation of the Poulsen arc. In *Radio Telephony* http://oz6gh.byethost33.com/poulsenarc.htm (1918)
6. D.F. Farsona, K.R. Kim, Generation of optical and acoustic emissions in laser weld plumes. J. Appl. Phys. **85**(3), 1329–1336 (1999)
7. M. Yasaka, M. Takeshita, R. Miyagawa, Detection of supersonic waves emitted from anomalous arc discharge in plasma processing equipment. Jpn. J. Appl. Phys. **39**, L1268–L1288 (2000)
8. Y. Wang, P. Zhao, Noncontact acoustic analysis monitoring of plasma arc welding. Press. Vessel. Pip. **78**, 43–47 (2001)
9. S. Szymanski, J. Hoffman, J. Kurzyna, Plasma plume oscillations during welding of thin metal sheets with a Co2 laser. J. Phys. D Appl. Phys. **34**, 189–199 (2001)
10. E. Saad, H. Wang, R. Kovacevic, Classification of molten pool modes in variable polarity plasma arc welding based on acoustic signature. J. Mater. Proc. Technol. **174**, 127–136 (2006)

11. M. Boinet, S. Verdier, S. Maximovitch, F. Dalard, Application of acoustic emission technique for in situ study of plasma anodizing. NDT&E Int. **37**, 213–219 (2004)
12. V.J. Law, J. Tynan, G. Byrne, D.P. Dowling, S. Daniels, The application of multivariate analysis tools for non-invasive performance analysis of atmospheric pressure plasma. in *Chaotic Systems: Theory and Applications*, ed. by C.H. Skadas, I. Dimotikalis (World Scientific Publishing, 2010), pp. 147–154
13. J. Tynan, V.J. Law, P. Ward, A.M. Hynes, J. Cullen, G. Byrne, D.P. Dowling, S. Daniels, Comparison of pilot and industrial scale atmospheric pressure glow discharge systems including a novel electro-acoustic technique for process monitoring. Plasma. Sources. Sci. Technol. **19**, 015015 (2010)
14. V.J. Law, C.E. Nwankire, D.P. Dowling, S. Daniels, Acoustic emission within an atmospheric helium discharge jet. in *Chaos Theory: Modeling, Simulation and Applications*, ed. by C.H. Skiadas, I. Dimotikalis, S. Skiadas (World Scientific Publishing, 2011), pp. 255–264
15. V.J. Law, F.T. O'Neill, D.P. Dowling, Evaluation of the sensitivity of electro-acoustic measurements for process monitoring and control of an atmospheric pressure plasma jet system. Plasma. Sources. Sci. Technol. **20**(3), 035024 (2011)
16. V.J. Law, N.T. O'Neill , D.P. Dowling, Atmospheric pressure plasma acoustic moment analysis. Complex Syst. **20**(2), 181–193 (2011) (Complex systems, Publications, Inc)
17. V.J. Law, D.P. Dowling, J.L. Walsh, F. Iza, N.B. Janson, M.G. Kong, Decoding of atmospheric pressure plasma emission signals for process control. CMSIM**1**, 69–76 (2011). ISSN 2441 0503
18. N. O'Conner, S. Daniels, Passive acoustic diagnostics of an atmospheric pressure linear field jet including analysis in the time-frequency domain. J Appl. Phys. Lett. **110**, 013308 (2011)
19. C.E. Nwankire, V.J. Law, A. Nindrayog, B. Twomey, K. Niemi, V. Milosavljević, W.G. Graham, D.P. Dowling, Electrical, thermal and optical diagnostics of an atmospheric plasma jet system. Plasma Chem. Plasma Process. **30**(5), 537–552 (2010)
20. D.P. Dowling, F.T. O'Neill, S.J. Langlais, V.J. Law, Influence of dc pulsed atmospheric pressure plasma jet processing conditions on polymer activation. Plasma Process Polym. **8**(8), 718–727 (2011)
21. D.P. Dowling, F.T. O'Neill, V. Milosavljevic, V.J. Law, DC pulsed atmospheric-pressure plasma jet image information. IEEE Trans. Plasma Sci. **39**(11), 236–237 (2011)
22. J. Pulpytel, V. Kumar, P. Peng, V. Micheli, N. Laidani, F. Arefi-Khonsari, Deposition of organosilicon coatings by a non-equilibrium atmospheric pressure plasma jet: design, analysis and macroscopic scaling law of the process. Plasma Process Polym. **8**(7), 664–675 (2011)
23. J.L. Walsh, F. Iza, N.B. Janson, V.J. Law, M.G. Kong, Three distinct modes in a cold atmospheric pressure plasma jet. J. Phys. D Appl. Phys. **43**, 075201 (2010)
24. K.C. Millett, E.J. Rawdon, Energy, ropelength, and other physical aspects of equilateral knots. J. Comput. Phys. **186**, 426–456 (2003)
25. A. Dobay, J. Dubochet, K.C. Millett, P.-E. Sottas, A. Stasiak, Scaling behavior of random knots. Proc. Natl. Acad. Sci. **100**(10), 5611–5615 (2003)
26. Y. Diaoyx, A. Dobayzk, R.B. Kusner, K.C. Millett, A. Stasiakz, The Average crossing number of equilateral random polygons. J. Phys. A. Math. Gen. **36**, 11561 (2003)
27. R.L. Ricca, On simple energy-complexity relations for filament tangles and networks. Complex Syst. **20**(3), 195–204 (2012) (Complex systems, Publications, Inc)
28. V.J. Law, F.T. O'Neill, D.P. Dowlin, 3-dimensional (orthogonal) structural complexity of time-series data using low-order moment analysis. *AIP Conference Proceedings. 10th International Conference on Numerical Analysis and Applied Mathematics*, vol. 1479, (2012), pp. 670–673
29. D.G. Aronson, M.A. Chory, G.R. Hall, R.P. McGehee, Bifurcations from an invariant circle for two-parameter families of maps of the plane: a computer-assisted study. Commun. Math. Phys. **83**, 303–354 (1982)

30. J.B. Hagen, *Radio Frequency Electronics: Circuits & Applications* (Cambridge University Press, Cambridge, 1996), pp. 311–336
31. V.J. Law, N. Macgearailt, Visualization of a dual frequency plasma etch process. MST **18** (3), 645–649 (2007)
32. C.F. Barenghi, R.L. Ricca, D.C. Samuels, How tangled is a tangle. Physica D **157**, 197–206 (2001)

Computational Tactic to Retrieve a Complex Seismic Structure of the Hydrocarbon Model

Tatyana A. Smaglichenko, Maria K. Sayankina and Alexander V. Smaglichenko

Abstract Passive seismic is a direction of human activity, due to which the prospecting of hydrocarbons can be performed by producing a minimal amount of drilling. The synthetic data are presented as arrival times of waves from regional events. Seismic rays pass through the thin layers that compose 2-D hydrocarbon model. The conventional view views mathematical methods as weakly reliable in a detection of hydrocarbons. Nevertheless we demonstrate that numerical reconstruction of seismic velocities in the gas-saturated reservoir can be accurate if the tactic of solution includes a statistical analysis of different subsets of observations and a selection of computational techniques, which resolve a complexity of the structure. Unconventional computing is in that we first make an assumption about geophysical properties of a subject, and only then the appropriate calculation scheme is chosen with following checking of parameters of a resolution. The basic techniques involve the relaxation scheme of the gradient descent method or CSSA to retrieve the large-size structure beneath the gas reservoir and the modification of Gaussian elimination that is effective to overcome the problem of an uncertain error of seismic observations.

Keywords Hydrocarbon model · Passive seismic data · Methods of transmission tomography

T. A. Smaglichenko (✉) · M. K. Sayankina
Research Oil and Gas Institute of Russian Academy of Sciences,
Moscow, Russia
e-mail: t.a.smaglichenko@gmail.com

M. K. Sayankina
e-mail: msayankina@gmail.com

A. V. Smaglichenko
Schmidt Institute of Physics of the Earth of Russian Academy of Sciences,
Moscow, Russia
e-mail: losaeylin@mail.ru

I. Zelinka et al. (eds.), *How Nature Works*, Emergence, Complexity and Computation 5, DOI: 10.1007/978-3-319-00254-5_10,
© Springer International Publishing Switzerland 2014

1 Introduction

The data of seismic waves passing through the Earth from their natural sources to the survey's receivers mainly include the information about inhomogeneous deep structure. Purpose of any inversion technique is accurately as possible to reconstruct images of real geological bodies located in depth in order to perform their analysis as well as to see processes that are connected with them and occur in the invisible Earth interior. Complexity of seismic structure is among the factors that significantly influence on the inversion effectiveness. Natural structures have a small- or large-size and combine both these characteristics. For instance the large-size structure having a strong low velocity of P-waves often designates a motion of a hot mantle substance or a mantle plume. Such anomalous spot has been found in ocean beneath central Iceland [23]. Another clear low-velocity anomaly characterizes the area of an extinct volcano Eifel (Germany) [15]. Strong low-velocity zone neighboring with high velocities has been found within the rift zone in the Ruwenzori mountains (Africa) [4]. These large-size structures are examples of models that can not be easily retrieved by tomographic methods.

Small-size structures are determined by means of zones of low and high velocities that alternate with each other. Normally they can be found in the fault zones, wherein extremally low velocity can be inside in the fault plane when high velocities characterize rocks in a vicinity of this plane. Structures can be changed with time and be transformed into different hierarchies. The example of unstable geodynamic state is the Western Nagano fault area (Central Japan) that appeared after the 1984 Nagano-Seibu earthquake having a magnitude 6.8. During many decades numerous aftershocks followed the main shock and then they led to complex fault patterns [3], which are characterized by the structure simultaneously having small- and large-size velocity anomalies [20]. Inversion methods are conventionally applied are not able easily resolve such structures. In this study we focus our attention on the part of Earth, which near a surface (till a depth of 0.54 km) and will investigate the complex model that has a meaning for a mining.

It is a long time since seismic exploration methods use borehole data and effectively find horizons of hydrocarbon deposits in the subsurface. However it is not always possible to determine details of a shape of the complex hydrocarbon reservoir that contains water- and gas- saturated layers and surrounding rocks. This opportunity can provide seismic tomography, which works with the data of seismic waves passing through the inhomogeneous Earth from natural sources to the survey's receivers. Due the tomography application one can particularly obtain the wave propagation velocities and to reconstruct physical properties of rocks in the reservoir. This allows predict the hydrocarbon potential for the being investigated region, to claim a presence of hydrocarbons at the given place and only then perform drilling.

In order to illuminate the Earth beneath the reservoir traditional exploration methods use active sources, i.e. explosions. The seismic tomography can also process information from passive seismic sources i.e. earthquakes. The advantage of such approach is ability to utilize the data in more wide frequency range than under typical hydrocarbon exploration [5]. Moreover under passive seismic there is no intervention to an environment, and the geodynamic balance of our planet is not disturbed. Nevertheless tomographic computations are based on limits that include the linearized formulation of the problem (see the next section), a choice of the appropriate blocks (cells) in accordance with a morphology of hydrocarbon reservoir (see Sect. 3.1) and finally hierarchical properties of the geophysical medium (different sizes of the structure), which are widely considered in this work.

The purpose of this study is to demonstrate the computational tactic of numerical reconstruction of the hydrocarbon model in conditions of passive-source seismic data. As a synthetic example of this model we choose so called "dipping thin-bed reservoir model" that has been considered in the work [14]. The proposed algorithm combines methods of transmission tomography: the Modification of Gaussian Elimination (MGE) that was recently developed [19, 20]; the Consecutive Subtraction of Selected Anomalies method (CSSA), which corresponds to the innovative scheme of the method descent gradient [18] and traditional inversion techniques as possible part of the algorithm.

2 Passive Source Seismic Data

The transmission seismic tomography has been considered among methods for hydrocarbon detection in a subsurface region using passive source seismic data (item 22 of [5]). Seismic data consist of P –wave arrival times that are observed from earthquakes (passive sources) and registered by seismometers. Assume that for each event we have initial guesses about epicenter, focal depth and origin time T^0_{source}. For the seismic ray (source-seismometer) a non-linear function of the observed time T^{obs} has been approximated by seismologists Aki and Lee in the work [1]:

$$T^{obs} = T^{cal} + \frac{\partial T}{\partial x}\Delta x + \frac{\partial T}{\partial y}\Delta y + \frac{\partial T}{\partial z}\Delta z + \Delta T_{source} + \sum\nolimits_k T^{(k)} F_k + E \qquad (1)$$

where T^{cal} is the calculated time based on an initial velocity model V_0, Δx, Δy, Δz, ΔT_{source} are unknown corrections to the source parameter, $\sum_k T^{(k)} F_k$ is unknown correction to the seismic velocity parameter. The last correction is expressed as sum of multiplications of unknown fractional perturbations of velocity slowness $F_k = \left(V^{-1} - V_0^{-1}\right)/V_0^{-1}$ and travel times $T^{(k)}$ that can be calculated with respect to the initial model V_0 for blocks (cells) of a medium, through which the given ray passed. E represents a sum of the approximation error and error in observations. Actually the non-linear function of the earthquake wave travel time is generalized function of the variables x, y, z. Decade prior to the research [1] mathematicians Lavrentiev and Romanov [7, 8] approximated this function and proposed the linearized formulation of an inversion problem for the case of two variables x, z that can be presented in the following form:

$$T^{obs} \approx T^{cal} - \int((V(x,z)/V_0^2))dx, dz \qquad (2)$$

where an integral is taken along a curve representing the seismic ray trajectory based on an initial model V_0. Note, that the formulation of (2) is valid under the next condition is satisfied:

$$abs\left(\frac{V - V_0}{V_0}\right)\Bigg) \ll 1 \qquad (2.1)$$

Later, Romanov [16] showed that the three-dimensional case can be resolved using a set of planar tasks. Thus a transition from the nonlinear to linear formulation having the integral form (2) or the differential form (1) is justified if the model $V\,(x,y,z)$ meets the condition of (2.1).

Let us stress that the merit of Aki and Lee is in that they did formulation of the velocity model via blocks (cells). On the one hand, this leads to adding of some small error into the last item in the Eq. (1). However on the other hand, it gives a simple way to a solution of the problem when we have deal with incomplete data sets, which are normally in all fields of practice of seismic research.

Suppose, P-wave arrivals that were registered in the prospecting area of hydrocarbons form the vector **b** of observation data with components of $T^{obs}-T^{cal}$. A dimension of this vector is equal to the number of seismic rays. Respectively to the equation of (1) parameters $\Delta x, \Delta y, \Delta z, \Delta T_{source}$ define the vector **y**, the dimension of which is in correspondence with the number of sources and parameters F_k determine the vector **x**, components of which are velocity perturbations that deduce changes in density and porosity of the geophysical medium. A dimension of the vector **x** is equal to the number of blocks (cells). Thus, for all observation data the tomography problem can be written as the solution of the next system of linear equations:

$$\mathbf{Ax} + \mathbf{Hy} = \mathbf{b} + \boldsymbol{\varepsilon} \qquad (3)$$

where **b** is the vector of data of passive-source seismic survey; **A** is the matrix, whose elements are calculated via lengths of seismic rays that penetrate from passive natural sources to a surface, wherein seismometers are located; **x** is the vector of values of velocity perturbations at different locations in the subsurface region; **H** is the matrix, whose elements are determined via seismic rays that are distributed from each source; **y** is the vector of the source location corrections; **ε** is the vector of errors.

Sources from regional earthquakes can be used as passive sources. Hypocenter of an earthquake can be very far away from a subsurface region, wherein the hydrocarbon exploration is performed. A high frequency of a strong earthquake will be decreased due to attenuation of seismic waves passing through the inhomogeneous medium from the source of earthquake to the exploration area and seismometers can record this event as the low frequency signal. In this case the arrival time of the signal should be compared with times of seismic waves arriving from the earthquake to permanent stations, which are located in the epicentral region. Thus the information from low frequency signals is the main part of the data set. If seismicity of the district is high then high frequency signals will form additional data that could be very essential to improve a resolution.

A complexity of the velocity model **x** that can be determined due to the equation of (3) is in a presence of an additional unknown vector **y**, which corresponds to the source parameters. Computational practice shows that traditional methods converge very slowly when two parameters (velocity and hypocenter locations) are involved to the inversion process of large systems of (3). Moreover the system is "burdened" by zero elements of matrices A and H that lead to inadequate results. In order to overcome these difficulties the probabilistic approach has been proposed in [20]. If an error in the source coordinates is in limits of pre-assumed accuracy then by using this approach it is possible to show that the problem of (3) can be simplified in the next form:

$$\mathbf{Ax} = \mathbf{b} + \hat{\boldsymbol{\varepsilon}} \tag{4}$$

where $\hat{\boldsymbol{\varepsilon}}$ is the vector of errors that involve errors in data and errors in a location of sources.

Let us remark that the system (4) should be also solved when we have deal with explosions (active sources). Even for such simple system as (4) the next complexity issue appears. In accordance with seismic experiment as a rule we should solve a strongly over-determined system. From a numerical point of view there is infinite number of solution that satisfies to the Eq. (4) with different small values of the calculation error. In accordance with the least-squares we search for minimal value. But it often happens that the computational result is not consistent with the condition of (2.1), under which the linearized formulation is valid.

Frequently the obtained values of the vector **x** are nonrealistic in order to characterize the seismic velocity structure. Seismologists try to overcome this problem by resorting to various tricks. They simply cut the data set by selecting respectively small values of the vector **b**. However this way leads to the situation when two images showing quite different physical properties characterize the same district, and the contradiction is in that the calculation error is small enough for these images.

The following complexity is in the uncertain structure of the error of $\hat{\boldsymbol{\varepsilon}}$. Let us point that errors in seismic data can be in wide range of different values. The error of each seismic ray or consequently the value of each component of the vector $\hat{\boldsymbol{\varepsilon}}$ in the system of (4) depends on accuracy of measurements, choice of the initial model, linear properties of the seismic model and can not be accurately estimated [12]. From computational point of view we do everything in order to get the correct solution, for which the unique minimum of the least square functional is reached. However on the other hand, everything is done in order to reach a consistency of this "correct" solution with observations having a chaotic behavior of the error.

It is known that in the applied mathematics Tikhonov's regularization is used to get the solution that is stable with respect to the data error. However when we add the smoothing operator to the least squares norm we suppose that errors are small. In the case of seismic data errors can not be under a control of a researcher. Therefore we apply the GME technique [19] and sub-divide the initial data set into a number of sub-sets, which can be reliably resolved with the point of view of a uniqueness of the sub-system solution. The selection of a stable value among all outcomes of sub-systems provides a protection of the final result from the influence of an uncertain error.

3 Hydrocarbon Model

3.1 Synthetic Example of Reservoir

The hydrocarbon model or "the thin-bed model" has been presented in [14] as the 2-D model that consists of layers: water- and gas- saturated reservoirs, surface rock, encasing shale, which are above and beneath reservoirs. Combination of large-size structures is the morphological manifestation of the hydrocarbon reservoir. In Sects. 4 and 5 we will demonstrate that there is a difficulty in the reconstruction of large-size structures if we apply conventional methods. Therefore in this work we propose the using of other inversion schemes. A complexity of the hydrocarbon model is the closeness of the values of seismic velocities for water- and gas- saturated reservoirs. This can lead to the registration of similar seismic signals from waves that were passed through these reservoirs and the same shale. From numerical point of view the vector-rows of the extended matrix A will linearly dependent. Then the system of (4) becomes underdetermined. This means that we have no sufficient number of seismic rays in order to resolve this system at whole. However if we follow the basic idea of GME that is in the division of the initial system into sub-systems then the chance appears in order to invert the homogeneous vector-rows in order to reconstruct the part of the model (see below Sect. 6) and then reconstruct other parts step-by-step.

"The thin-bed model" is consistent with the generally accepted scheme of hydrocarbon deposits (Fig. 1), which was described in many schoolbooks for example in [21]. Suppose a multiple drilling and explosions (active seismic sources) are in an using to create seismic waves then it is difficult to define the segments of seismometer data, which could correspond to the amplitude-frequency characteristics of P- waves passing through the reservoir. Because of a small thickness of oil and gas layers, the travel times, which are components of the vector **b** in the equation of (4) are almost equal for different types of seismic waves. Passive seismic is based on a registration of P-waves from regional events. These waves are created by nature. They pass through the reservoir and their amplitudes and frequencies can be clearly determined in the seismometer records

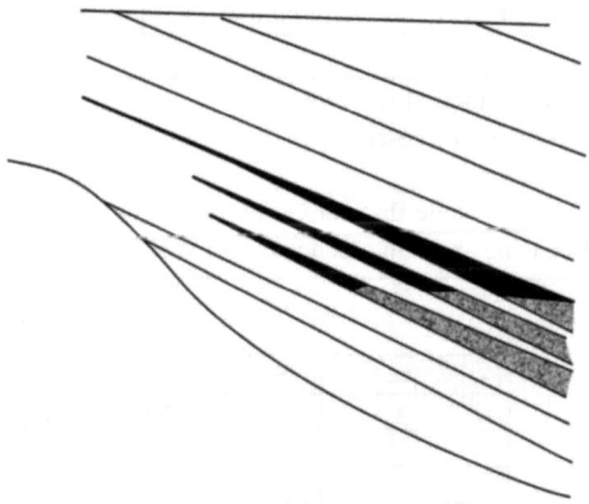

Fig. 1 A simplified scheme of hydrocarbons. Gas- (water-) saturated reservoir is colored in *black* (in *gray* with *dots*). Different kinds of shale are delineated by *lines*

at a surface. As told above the arrival times of waves should be additionally agreed with the information from permanent seismic stations in the region. Thus passive seismic provides more wide opportunity for transmission tomography than the traditional seismic exploration.

Let us guess the initial velocity $V_0 = 2.53$ km/s that corresponds to a visible layer at a surface. Table 1 illustrates velocities in layers of "the thin-bed model" and respectively velocity perturbations relatively to the initial velocity.

We defined the grid-size 0.112×0.067 km (Fig. 2). This size has been selected based on two conditions. First, physical conditions of propagation of seismic waves through inhomogeneous blocks must be met. Therefore the block-size has been obtained by calculating the Fresnel zone radius and assuming that regional events with low frequency will participate in the given experiment. Second, blocks should be located in accordance with the geometry of reservoirs.

Figure 3 shows the distribution of seismic velocities in blocks in accordance with Table 1.

3.2 Seismic Experiment

In this study we will conduct our experiment for seismic rays of P-waves transmitted through the Earth to a surface, wherein they are registered by seismometers. The distance between seismometers is 112 m (Fig. 4). This separation is optimal in order to register body waves (P-waves) as well as surface waves (S-waves). In real research the bringing of information about surface waves can significantly improve a reliability of our results. S-waves travel along the Earth surface and their using can well illuminate velocities in homogeneous layers that close to a surface. This information can be incorporated to the inversion process of the transmission tomography.

Seismic rays form a rectangular matrix **A** of dimension 264×88, whose elements are determined by a length of rays in blocks. Synthetic data $\mathbf{b}^{synth}$ were

Table 1 Velocity perturbations of synthetic model

	V_p	$(Vp-V_0)/V_0$
Water-saturated reservoir	3.049	0.21
Gas-saturated reservoir	2.789	0.1
Surface rock	2.53	0.0
Encasing shale that above reservoirs	2.643	0.044
Encasing shale that beneath reservoirs	2.695	0.065

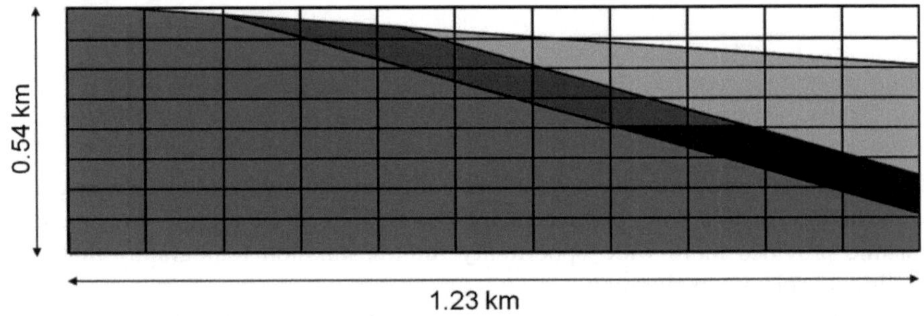

Fig. 2 Water- and gas- saturated reservoirs, surface rock, encasing shale that are above and beneath reservoirs are colored in *black*, *dark-grey*, *white*, *light-gray* and *gray*, respectively Geometry of colors is in agreement with "the thin-bed model" that is given in [14]. The grid illustrates the configuration of blocks of the seismic medium

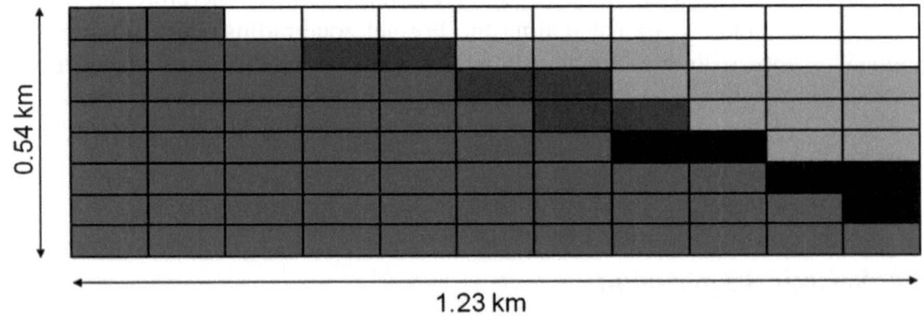

Fig. 3 Water- and gas- saturated reservoirs, surface rock, encasing shale that are above and beneath reservoirs correspond to blocks, which are colored in *black*, *dark-grey*, *white*, *light-gray* and *gray*, respectively

calculated taking into account velocity perturbations that were assigned to each block in accordance with the "true" seismic model (see Table 1, Fig. 3). Thus our task is to solve the system

$$\mathbf{Ax} = \mathbf{b}^{synth}, \tag{5}$$

and to compare components of vector x with values of the "true" model.

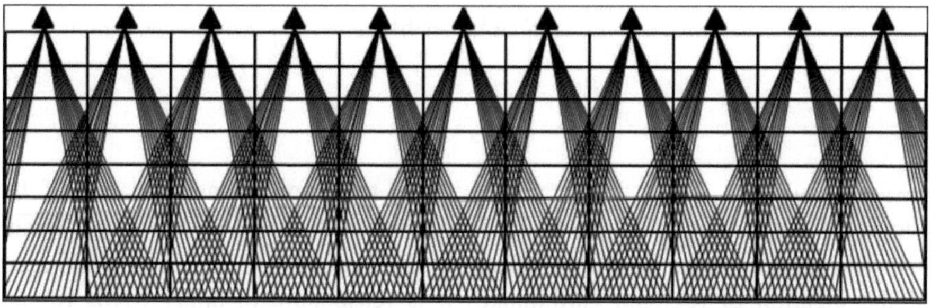

Fig. 4 Geometry of experiment. Seismometers are denoted by *triangles*. The distance between measurement points is 112 m

4 The Modification of Gaussian Eliminations

The MGE technique has been developed to protect the system from various errors, which distort seismic images and lead to an ambiguity of tomography results. The sources of error in data can include inaccurate measurements, incorrect reading from seismograms, inappropriate choice of the model parameters. From a numerical point of view the presence of recurrence formulas can create the source of round-off error, which leads to the lost of adequacy of the solution of systems, especially if they are large and sparse.

The problem of serious error can be overcome by means of the fragmentation of the original system into subsystems. The useful idea of division of the initial matrix is known since the fundamental mathematical work [6], wherein the solution of large system is determined by means of resolving of the subsystem with following substitution of the found components instead unknown variables of the initial system. Other idea to protect the solution from various errors leads us to the MGE method. Namely, according to MGE we simultaneously determine solutions of few subsystems, which have common unknowns. Next, solutions are compared with each other, and a value of the component, which is nearly equal for few subsystems and has minimal error in the solution, is chosen as final value for the velocity parameter. Only after such rigorous selection, the obtained value is substituted into the initial system.

There is also another difference from the scheme that has been described in [6], where an overdetermined system is solved by means of an inversion of a square matrix, which has been obtained by multiplying of the initial rectangular matrix and the transpose matrix. In MGE we also build a square (basic) matrix however by means of selecting independent rows (columns) that satisfy Rouche-Capelli theorem. Additionally columns of the square matrix should meet criteria of probability theory that requires their high correlation with the column of observed data [17].

In accordance with MGE the system of (5) can be represented as the next system of subsystems:

$$\begin{cases} \mathbf{A}^1\mathbf{x}^1 = \mathbf{b}^1 \\ \qquad \vdots \\ \mathbf{A}^q\mathbf{x}^q = \mathbf{b}^q \end{cases} \tag{6}$$

The condition giving the final solution is the approximate equality of the components $\mathbf{x}_\mathbf{i}^\mathbf{k}$ and $\mathbf{x}_\mathbf{j}^\mathbf{l}$ that are **i**-th and **j**-th components of the solution for **k**-th and **l**-th subsystems. Because of a sparseness of the system (5) and an uncertain error of the location of regional earthquakes it is reasonable to apply MGE for the hydrocarbon model reconstruction.

Preliminary testing of MGE for the arbitrary model showed that it is effective in retrieving of large-size structure. Below we illustrate two results, which were obtained by application of two different algorithms under the same conditions of inversion experiment. The first is LSQR algorithm [11] that is based on the Lanczos process and it is similar to the conjugate gradients method. Let us note that the LSQR using with a small number of iterations is equivalent to using the Singular Value Decomposition (SVD) technique. Figure 5 (from the left) demonstrates the initial testing model representing a single large-size anomaly that surrounded by an uniform zone. An anomaly and a zone are characterized by opposite signs and by the same absolute value of the velocity perturbation. For example, if the velocity perturbation of anomalous block is equal to +3 % then its value in a zone is equal to −3 % (or vice versa). The geometry of the experiment is the same as in [9]. Figure 5 (from the right) shows the result of the SVD inversion of synthetic data that correspond to this model. One can see that the inverted model does not coincide with the initial one.

The same experiment has been repeated for MGE. Figure 6 illustrates that it is possible to reconstruct this model. Concrete example with numerical description of this model and computational results of both methods can be found in Appendix B of the work [20].

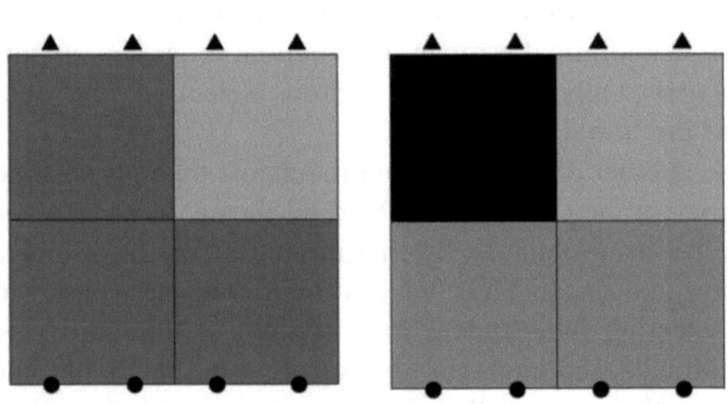

Fig. 5 An illustration of the SVD testing. The initial model (from the *left*) presents the large-size structure, which is a single anomaly (colored in *light gray*) surrounded by uniform zone (colored in *gray*). *Black circles* (*triangles*) denote sources (receivers). The model consists of 4 blocks. The inverted model derived by SVD (from the *right*) does not coincide with the initial one

Fig. 6 An illustration of the MGE testing. The initial model (from the *left*) and the configuration of blocks are the same as has been presented in Fig. 5. *Black circles* (*triangles*) denote 4 sources (receivers). The model consists of 4 blocks. The inverted model derived by MGE (from the *right*) coincides with the initial one

Analyzing of the hydrocarbon model (Figs. 2 and 3) shows that the complex structure of the reservoir mainly consists of the combination of large-size anomalies. Therefore we encourage the MGE application. Owing to MGE we divide our original data into 22 sub-systems, each of which is formed by rays coming to the same receiver from the cluster of events. Farther solutions of all sub-systems should be compared to get stable value to unknown component. Each sub-system is weakly-sparse and does not meet Rouche-Capelli theorem. Therefore we applied the Moore-Penrose pseudo inversion method [10, 13] that is based on SVD. However all inverted models were inadequate comparing with the "true" model.

5 The CSSA Relaxation Scheme

In order to resolve this complicated case we investigated a synthetic field from point of view of a complexity. We have analyzed statistical characteristics of the data vector representing the response of the model for each subsystem. Obviously properties of a medium define this response. We selected few homogeneous vectors, components of which are near equal. This means that travel times of seismic waves through a medium are near equal too. Hence we can guess that the medium has a homogeneous structure. In order to check this assumption one should estimate resolution parameters of the CSSA method that is known as effective technique to reconstruct the structures having separated anomalies [18].

Mathematically, CSSA corresponds to relaxation scheme of the method of steepest descent. We search a minimum of the functional in the sense of least squares:

$$\mathbf{F}(\mathbf{x}) = \|\mathbf{Ax} - \mathbf{b}\|^2 = \sum_{m=1}^{M} \left(\sum_{k=1}^{K} \mathbf{a}_{mk}\mathbf{x}_k - \mathbf{b}_m\right)^2 \rightarrow \min_{\mathbf{x}}\,, \tag{7}$$

where **M** is the total number of seismic rays, **K** is the total number of blocks (cells) of the medium.

The solution is defined in the form of:

$$\mathbf{x}^{(\mathbf{i})} = \mathbf{x}^{(\mathbf{i-1})} + \boldsymbol{\lambda}_{\mathbf{k}}\mathbf{n}_{\mathbf{k}}, \tag{8}$$

where **i** is the iteration number, $\boldsymbol{\lambda}_{\mathbf{k}}$ indicates the value of the velocity parameter in the **k**-th block, $\mathbf{n}_{\mathbf{k}}$ is the unit vector, **k**-th component of which is equal to 1. The solution is searched by applying the next simple condition;

$$\mathbf{F}'(\mathbf{x}) = \mathbf{0} \tag{9}$$

Further $\boldsymbol{\lambda}^*_{\mathbf{k}}$ is determined to meet the condition of (9):

$$\boldsymbol{\lambda}^*_{\mathbf{k}} = \underset{\lambda_{\mathbf{k}}}{\mathbf{argmin}}\ \mathbf{F}_{\mathbf{k}}(\mathbf{x}^{(\mathbf{i})}) \tag{10}$$

Finally we have:

$$\boldsymbol{\lambda}^*_{\mathbf{k}} = -\frac{(\mathbf{A}\mathbf{x}^{(\mathbf{i-1})} - \mathbf{b}, \mathbf{A}\mathbf{n}_{\mathbf{k}})}{(\mathbf{A}\mathbf{n}_{\mathbf{k}}, \mathbf{A}\mathbf{n}_{\mathbf{k}})} \tag{11}$$

The convergence of CSSA has been proved and the degree of convergence has been estimated showing that seismic rays that are uniformly distributed in blocks provide the convergence to the accurate solution in the least square sense [18]. It has been also shown on numerical examples that CSSA easily retrieves a single large-size anomaly that is surrounded by the uniform zone having zero values of velocity perturbation, while this structure can not be reconstructed by means of the SVD algorithm. Figures 7 and 8 illustrate the results of application of two methods under the same conditions of fair match.

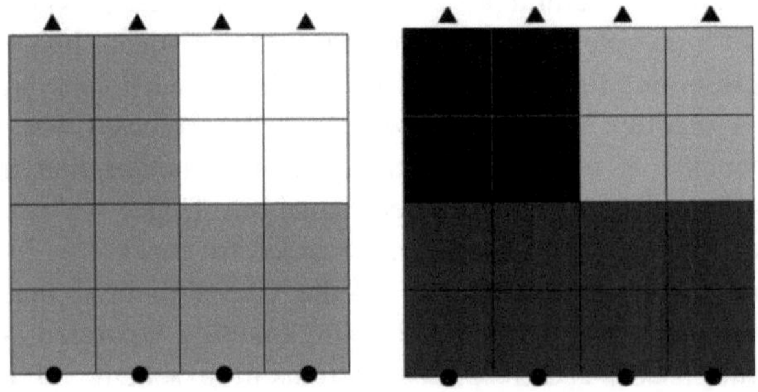

Fig. 7 An illustration of the SVD testing. The initial model (from the *left*) presents the large-size structure, which is a single anomaly (colored in *white*) that surrounded by the zone with zero values of velocity perturbation (colored in *gray*). *Black circles* (*triangles*) denote 4 sources (receivers). The model consists of 16 blocks. The inverted model (from the *right*) derived by SVD does not coincide with the initial one

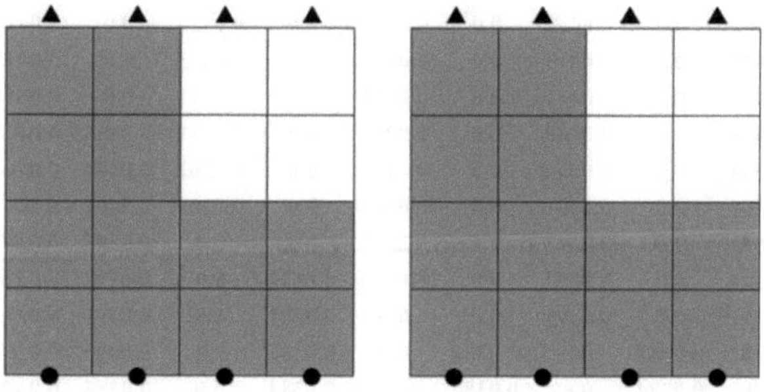

Fig. 8 An illustration of the CSSA testing. The initial model (from the *left*) and the configuration of blocks are the same as in Fig. 7. *Black circles* (*triangles*) denote 4 sources (receivers). The inverted model (from the *right*) derived by CSSA coincides with the initial one

Let us return to the synthetic field of the hydrocarbon model. As said above we can make an assumption that homogenous data vector we have obtained for sub-systems can be a response to a homogeneous structure. By using the formulas (8) and (11)the CSSA solution for the simple case of a homogeneous structure can be estimated as the next:

$$\mathbf{x_k} = \frac{(\mathbf{a_k}, \mathbf{b^{(k)}})}{(\mathbf{a_k}, \mathbf{a_k})} \tag{12}$$

where (,) denotes a scalar product; $\mathbf{a_k}$ is $\mathbf{k}$-th vector-column of sub-matrix $\hat{\mathbf{A}}$; vector $\mathbf{b^{(k)}}$ is equal to component multiplication of vectors $\mathbf{b}$ and $\mathbf{a^{(k)}}$. Here $\mathbf{b}$ be a part of vector $\mathbf{b}^{synth}$, and $\mathbf{a^{(k)}}$ is equal to $\mathbf{a_k}/\sum_{\mathbf{m=1}}^{\mathbf{M}} \mathbf{a_{mk}}$, where $\mathbf{a_{mk}}$ be a row element, $\mathbf{M}$ be the number of elements in $\mathbf{k}$th row-vector.
The CSSA resolution parameters can be determined using the next formula:

$$\mathbf{r_k} = \frac{(\mathbf{a_k}, \mathbf{b^{(k)}})^2}{(\mathbf{a_k}, \mathbf{a_k})(\mathbf{b^{(k)}}, \mathbf{b^{(k)}})} \tag{13}$$

The parameters $\mathbf{r_k}$ estimate the proximity of the CSSA solution to the accurate solution in the least squares sense.

6 Comparison of SVD and CSSA Numerical Outcomes

We will apply CSSA in order to retrieve the homogeneous part of the medium that is beneath the reservoir. Let us consider an example of the sub-system having homogeneous vector $\mathbf{b}$. Below we also will compare the inversion results of two methods SVD and CSSA Let $\hat{\mathbf{A}}$ be the 12×12 matrix:

$$\hat{\mathrm{A}} = \begin{vmatrix} 0.042 & 0.042 & 0.0 & 0.0418 & 0.0 & 0.0418 & 0.0 & 0.0418 & 0.0 & 0.0418 & 0.0418 & 0.0418 \\ 0.0415 & 0.0415 & 0.0 & 0.0415 & 0.0 & 0.0415 & 0.0 & 0.0415 & 0.0 & 0.0415 & 0.0415 & 0.0415 \\ 0.0411 & 0.0411 & 0.0 & 0.0411 & 0.0 & 0.0411 & 0.0 & 0.0411 & 0.0 & 0.0411 & 0.0411 & 0.0411 \\ 0.0408 & 0.0408 & 0.0 & 0.0408 & 0.0 & 0.0051 & 0.0 & 0.0357 & 0.0408 & 0.0408 & 0.0408 & 0.0408 \\ 0.0404 & 0.0404 & 0.0 & 0.0404 & 0.0 & 0.0202 & 0.0 & 0.0202 & 0.0404 & 0.0404 & 0.0404 & 0.0404 \\ 0.0401 & 0.0401 & 0.0 & 0.0401 & 0.0 & 0.0301 & 0.0 & 0.0100 & 0.0401 & 0.0401 & 0.0401 & 0.0401 \\ 0.0397 & 0.0397 & 0.0 & 0.0397 & 0.0 & 0.0397 & 0.0 & 0.0 & 0.0397 & 0.0397 & 0.0397 & 0.0397 \\ 0.0394 & 0.0394 & 0.0 & 0.0394 & 0.0 & 0.0394 & 0.0197 & 0.0 & 0.0197 & 0.0394 & 0.0394 & 0.0394 \\ 0.0391 & 0.0391 & 0.0 & 0.0391 & 0.0 & 0.0 & 0.0391 & 0.0 & 0.0391 & 0.0391 & 0.0391 & 0.0391 \\ 0.0387 & 0.0387 & 0.0387 & 0.0 & 0.0387 & 0.0 & 0.0387 & 0.0 & 0.0097 & 0.0291 & 0.0387 & 0.0387 \\ 0.0384 & 0.0 & 0.0384 & 0.0 & 0.0384 & 0.0 & 0.0384 & 0.0 & 0.0384 & 0.0384 & 0.0384 & 0.0384 \\ 0.0381 & 0.0 & 0.0381 & 0.0 & 0.0381 & 0.0 & 0.0381 & 0.0 & 0.0381 & 0.0381 & 0.0381 & 0.0381 \end{vmatrix}$$

while **b** be the known 12×1column:

$$\mathbf{b} = \begin{vmatrix} -0.0217 \\ -0.0216 \\ -0.0214 \\ -0.0212 \\ -0.0210 \\ -0.0209 \\ -0.0206 \\ -0.0205 \\ -0.0203 \\ -0.0201 \\ -0.0200 \\ -0.0198 \end{vmatrix}$$

First, we have estimated the resolution parameters for both methods. The CSSA parameters can be obtained using formula (13). In accordance with [2] the SVD resolution matrix **R** is equal to:

$$\mathbf{R} = (\mathbf{A}^{\mathrm{T}}\mathbf{A})^{-1}\mathbf{A}^{\mathrm{T}}\mathbf{A}, \tag{14}$$

where $\mathbf{A}^{\mathbf{T}}$ is the transpose of the matrix **A**.

It has been shown in [22] that the diagonal elements of **R** characterize an accuracy of the SVD solution. By analyzing of them and components of the CSSA parameter one can see that the CSSA indicators are more close to the unit values. Thus the resolution gets poorer for SVD (see Table 2) and a researcher should select CSSA in order to reconstruct sub-systems having a homogeneous vector **b.** Next, we inverted synthetic data by using of formula (12) and by the standard MATLAB program. As we expected, the SVD inverted model was inadequate compared with the "true" model, while the CSSA inverted model was preferable (see Table 3).

Figure 9 illustrates a behavior of the error, which is absolute difference between the "true" solution and the solution, which has been obtained for each component of vector **x** in the case of SVD and CSSA in accordance with Table 3.

Table 2 Resolution parameter values

SVD method	0.1671	0.9004	0.3630	0.3522	0.3630	0.8247	0.8247	0.8247	0.8247	0.2212	0.1671	0.1671
CSSA method	1.0175	0.9547	1.0053	1.0009	1.0054	1.0023	1.0064	1.0058	1.0011	0.9306	1.0078	0.9918

Table 3 Velocity perturbations

"True" model	0.065	0.065	0.065	0.065	0.065	0.065	0.065	0.065	0.065	0.065	0.065	0.065
SVD method	0.097	0.052	0.046	0.041	0.046	0.042	0.042	0.042	0.042	0.084	0.087	0.087
CSSA method	0.063	0.062	0.061	0.065	0.062	0.066	0.064	0.068	0.066	0.065	0.068	0.068

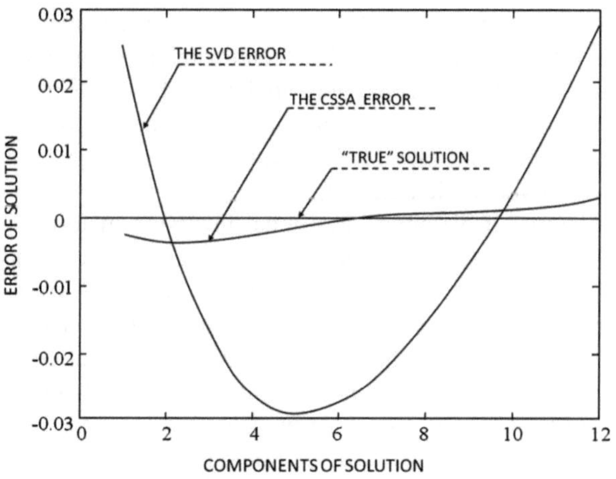

Fig. 9 Errors of each component of the SVD (CSSA) solution are approximated by a curve. The *zero line* corresponds to errors of the "true" model

7 Results of the Hydrocarbon Model Reconstruction

After eliminating of all CSSA stable components from equations of the system of (5) the initial system size is reduced and we should solve a new form of subsystems. Newly born subsystems are over determined. In accordance with the MGE criteria we build basic systems, which are consistent and can be easily inverted by any application that is conventionally used.

We have obtained the next outcomes. Anomalies of the gas reservoir are accurately determined in three blocks that are close to a surface and in the neighborhood with encasing shale beneath the reservoir. Their values are equal to 0.1 Anomalies in two blocks that are located between encasing shale that beneath and above the reservoir have been determined with permissible error. Their values are equal to 0.07.

The block, which is bordered with water, has been retrieved with unacceptable error of reconstruction. Its value is equal to 0.03. Strictly speaking the boundary "gas-water" is fundamental problem in the oil and gas research. The problem can be overcome in future if we will combine efforts of few disciplines, which will include seismic tools together with other geophysical and geochemical information.

8 Discussion and Conclusion

In the present work, we have considered the complex 2-D reservoir model (synthetic example) and have selected the medium parameters that correspond to seismic data from regional events. We have assumed that the number of earthquakes is not big. Thus the tomography experiment has been conducted in the limited conditions, when seismic rays are sparsely distributed and their number is small. Our result is the following. Mathematical methods we have used are able to reliably reconstruct the main part of the initial model, which contains the gas reservoir.

The computational tactic is based on the interaction of two techniques MGE and CSSA. Let us note that seismic velocity can not be properly retrieved if any of these methods is separately applied to the whole data set. Our practice showed that it is impossible to reconstruct the model if the algorithm uses only the MGE technique in spite of the requirement of high resolution of the MGE subsets. Moreover we demonstrated that there is a failure in reconstruction of the large-size structure if we apply traditional inversion methods.

New tactic makes an emphasis on statistical analysis of observation data. We solve the system of linear equations first of all by means of a review of different parts of the vector of constant terms that are seismic observations. In the given work it has been shown that by using statistical characteristic of observations one can to make an assumption about geological property of a medium or in other words about the morphology. This thesis is based on the example of hydrocarbon model. Namely we detected homogeneous data vector. Geophysical sense of these data can be in that seismic waves traveled through the same structure. To check this assumption we applied the CSSA technique that is effective to resolve a large-size structure. By comparing a few inversion results having a high resolution in various directions we found the shale.

Thus we conclude that computational models should not be formally constructed by means of just inverting the initial data sets. The inversion process requires understanding of the physics of a propagation of seismic waves through different kinds of geological structure, interpretation of this process on a language of statistics and only then the choice of calculation tools.

It has been shown that the proposed tactic permitted to retrieve seismic velocities in blocks that form the part of the gas reservoir, which is before the boundary "gas water". Velocities are also determined in encasing shale that are beneath and above of this part of a reservoir. Thus this study confirms that the transmission tomography can be among methods for hydrocarbon exploration in the subsurface layer.

Acknowledgments We thank the organizers of the Symposium. Since the last ICNAAM conference we had a fruitful discussion of subject of complexity. This helped in developing of the computational tactic that is presented in this study. Our thanks go to Reviewers for providing of comments, owing to which the description of consistency between computational and seismic models has been extended.

References

1. K. Aki, W.K.H. Lee, Determination of three-dimensional velocity anomalies under a seismic array using first P-arrival times from local earthquakes: 1. A homogeneous initial model. J. Geophys. Res. **81**(23), 4381–4399 (1976)
2. G.E. Backus, J.F. Gilbert, The resolving power of gross earth data. Geophys. J. R. Astron. Soc. **16**, 169–205 (1968)
3. S. Ito, W. Ellsworth, Y. Iio, Complex fault patterns in Western Nagano, Japan revealed by the double-difference method. Eos, Trans. Am. Geophys. Union. **82**(47), F870 (2001)

4. A. Jakovlev, G. Rümpker, M. Lindenfeld, I. Koulakov, A. Schumann, N. Ochmann, Crustal seismic velocities of the Rwenzori region, east African rift, from local travel-time tomography: evidence for low-velocity anomalies beneath the mountain range. Bull. Seism. Soc. Am. **101**(2), 848–858 (2011)
5. C. Jing, J.J. Carazzone, E.M. Rumpfhuber, R.L. Saltzer, T.A. Dickens, A.A. Mullur, Patent application title: Hydrocarbon Detection with Passive Seismic Data. IPC8 Class: AG01V128FI. USPC Class: 367 73. Class name: Synthetic seismograms and models. Patent application number: 20110255371 (2011)
6. C. Lanczos, *Applied Analysis* (Prentice Hall, Englewood Cliffs, 1956)
7. M.M. Lavrentiev, V.G. Romanov, Three linearized inverse problems for hyperbolic equations. Soviet Math. Dokl. **7**, 1650–1652 (1966)
8. M.M. Lavrentiev, *Some Improperly Posed Problems in Mathematical Physics.* Tracts in Natural Philosophy, vol. 11, (Springer, Berlin, 1967)
9. J.J. Leveque, L. Rivera, G. Wittlinger, On the use of the checker-board test to assess the resolution of tomographic inversions. Geophys. J. Int. **115**, 313–318 (1993) (Blackwell)
10. E.H. Moore, On the reciprocal of the general algebraic matrix. Bull. Am. Math. Soc. **26**, 394–395 (1920)
11. C.C. Paige, M.A. Saunders, LSQR: an algorithm for sparse linear equations and sparse least squares. ACM Trans. Math. Soft. **8**, 43–71, 195–209 (1982)
12. G.L. Pavlis, Appraising earthquake hypocenter location errors: a complete practical approach for single-event locations. Bull. Seism. Soc. Am. **76**, 1699–1717 (1986)
13. R.A. Penrose, Generalized inverse for matrices. Proc. Cambridge Philos. Soc. **51**, 406–413 (1955)
14. H. Ren, G. Goloshubin, Spectra crossplot. Lead. Edge. **1563**, 1562–1566 (2007)
15. J.R.R. Ritter, M. Jordan, U. Christensen, U. Achauer, A mantle plume below the Eifel volcanic fields, Germany. Earth Planet. Sci. Lett. **186**, 7–14 (2001)
16. V.G. Romanov, *Some Inversion Problems for Hyperbolic Equations (in Russian)* (Nauka, Novosibirsk, 1972)
17. A.N. Shiryaev, *Probability* (Springer, New York, 2008)
18. T.A. Smaglichenko, A.V. Nikolaev, S. Horiuchi, A. Hasegawa, The method for consecutive subtraction of selected anomalies: the estimated crustal velocity structure in the 1996 Onikobe (M = 5.9) earthquake area (1996), northeastern Japan. Geophys J. Int. **153**, 627–644 (2003) (Blackwell)
19. T.A. Smaglichenko, Modification of Gaussian elimination for the complex system of seismic observations. Founded by Stephen Wolfram **20**(3), 229–241 (2012) (Complex systems Publications, Inc. USA), http://www.complex-systems.com/index.html
20. T.A. Smaglichenko, H. Shigeki, T. Kaori, A differentiated approach to the seismic tomography problem: method, testing and application to the western Nagano fault area (Japan). Int. J. Appl. Earth Obs. Geoinf. **16**, 27–41 (2012) (Elsevier)
21. V.V. Tetelmin, V.A. Yazev, *Basics of Drilling for Oil and Gas* (Editorial URSS, Moscow, 2009)
22. R.W. Wiggins, The general linear inverse problem: Implication of surface waves and free oscillations for earthstructure. Rev. Geophys. Space Phys. **10**, 251–285 (1972)
23. G.J. Wolfe, I.T. Bjarnason, J.C. VanDecar, S.C. Solomon, Seismic structure of the Iceland mantle plume. Nature **385**(16), 245–247 (1997)

Controlling Complexity

Ivan Zelinka, Petr Saloun, Roman Senkerik and Michal Pavelch

Abstract Complex systems and dynamics are present in many parts of daily life and branches of science. This chapter is continuation of our previous research, that introduced a novelty method of visualization and possible control of complex networks, that are used to visualize dynamics of evolutionary algorithms. Selected evolutionary algorithms are used as an example in order to show how its behavior can be understood as complex network and controlled via conversion into CML system—a model based on mutually joined nonlinear *n* equations. The main aim of this investigation was to show that dynamics of evolutionary algorithms can be converted via complex network to CML system and then controlled. Selected results of conversion of evolutionary dynamics into complex network and consequently to CML as well as controlled CML system are discussed here.

1 Introduction

In this article, we try to merge, at first glance, two completely different areas of research: complex systems with attention to complex networks and evolutionary algorithm dynamics. Complexity can be classified from different point of views

I. Zelinka (✉) · P. Saloun
Department of Computing Science, Faculty of Electrical Engineering and Computing Science, Technical University of Ostrava, Tr. 17. Listopadu 15, Ostrava, Czech Republic
e-mail: ivan.zelinka@vsb.cz

P. Saloun
e-mail: petr.saloun@vsb.cz

R. Senkerik
Faculty of Applied Informatics, Tomas Bata University in Zlin, Nam T.G. Masaryka 5555, 76001 Zlin, Czech Republic
e-mail: senkerik@fai.utb.cz

M. Pavelch
e-mail: pavelch@fai.utb.cz

I. Zelinka et al. (eds.), *How Nature Works*, Emergence, Complexity and Computation 5, DOI: 10.1007/978-3-319-00254-5_11,
© Springer International Publishing Switzerland 2014

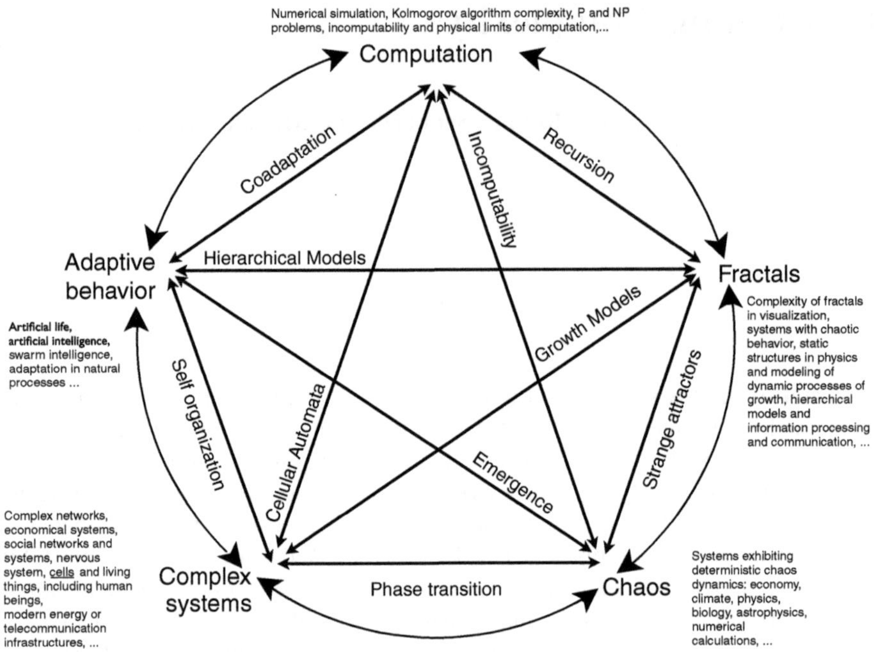

Fig. 1 Different view on complexity

like dynamics, structure, memory and information processing etc. Complex systems create rich set of subclasses, that have mutual relations as shown at Fig. 1.

One of many possible definition of a complex system is like: *A complex system is a system composed of interconnected parts that as a whole exhibit one or more properties (behavior among the possible properties) not obvious from the properties of the individual parts. This characteristic of every system is called emergence.*

or

A systems complexity may be of one of two forms: disorganized complexity and organized complexity. In essence, disorganized complexity is a matter of a very large number of parts, and organized complexity is a matter of the subject system (quite possibly with only a limited number of parts) exhibiting emergent properties.

A complex system is in fact any system featuring a large number of interacting components, whose aggregate activity is non-linear (existence of linear complex systems is also possible) and typically exhibits self-organization under selective pressures. The term complex systems has multiple meaning as mentioned here:

- A field of science studying these systems, see further complex systems.
- A specific kind of systems, that are complex from different point of view.
- A paradigm, that complex systems have to be studied with non-linear dynamics.
- ...

Today a lot of various informal descriptions of complex systems have been put forward, and these may give some insight into their, very often interesting, properties. As mentioned in the special edition of Science journal (Vol. 284. No. 5411, 1999) about complex systems highlighted several of these:

- A complex system is a highly structured system, which shows structure with variations.
- A complex system is one whose evolution is very sensitive to initial conditions or to small perturbations, one in which the number of independent interacting components is large, or one in which there are multiple pathways by which the system can evolve. This is very typical for deterministic chaos system.
- A complex system is one that by design or function or both is difficult to understand and verify.
- A complex system is one in which there are multiple interactions between many different components complex systems are systems in process that constantly evolve and unfold over time.
- ...

Examples of complex systems include natural as well as artificial: ant colonies, human economies, social structures, climate, nervous systems, cells and living things, including human beings, modern energy or telecommunication infrastructures and much more. Many systems of research and technology interest to humans are complex systems. Complex systems are studied by many areas of natural science, mathematics, physics, biology and social science. Fields that specialize in the interdisciplinary study of complex systems include systems theory, complexity theory, systems ecology and mainly cybernetics.

Complex system used in this investigation are complex networks, that are used to model interaction between n independent vertices (agents, neurons, cells,...) and are the main part of this investigation. Typical example are social networks with people in vertices position. In this chapter we are substituting people from social networks by individuals from evolutionary algorithms population and social interaction by offspring creation in the evolution. By this simple idea we get evolutionary complex network.

Large-scale networks, exhibiting complex patterns of interaction amongst vertices exist in both nature and in man-made systems (i.e., communication networks, genetic pathways, ecological or economical networks, social networks, networks of various scientific collaboration, Internet, World Wide Web, power grid etc.). The structure of complex networks thus can be observed in many systems. The word complex networks [1, 2] comes from the fact that they exhibit substantial and non-trivial topological features, with patterns of connection between vertices that are neither purely regular nor purely random. Such features include a heavy tail in the degree distribution, a high clustering coefficient, hierarchical structure, amongst other features. In the case of directed networks, these features also include reciprocity, triad significance profile and other features. Amongst many studies, two well-known and much studied classes of complex networks are the scale-free networks and small-world networks (see examples in

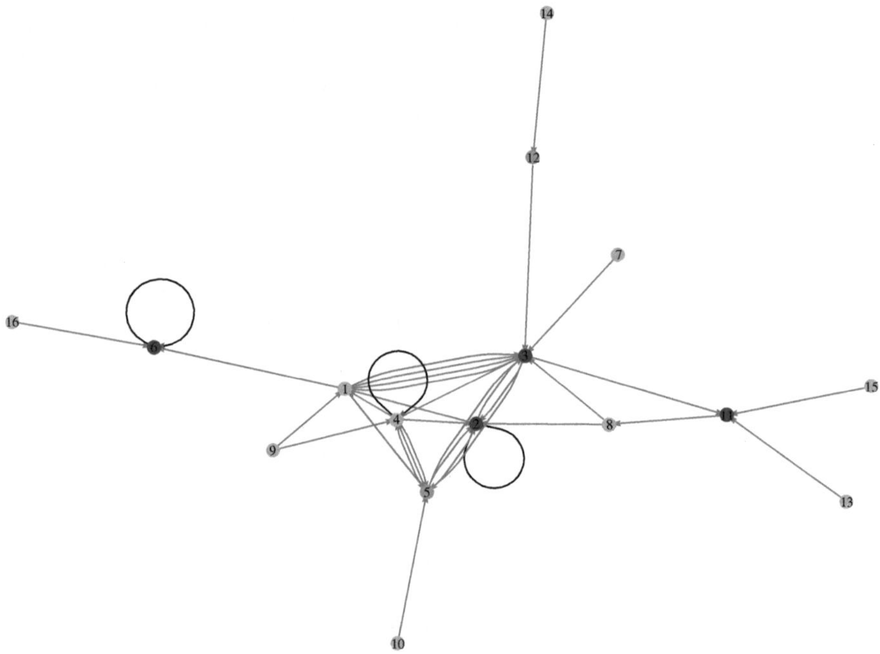

Fig. 2 Example of a small network

Figs. 2 and 3), whose discovery and definition are vitally important in the scope of this research. Specific structural features can be observed in both classes i.e. so called power-law degree distributions for the scale-free networks and short path lengths with high clustering for the small-world networks. Research in the field of complex networks has joined together researchers from many areas, which were outside of this interdisciplinary research in the past like mathematics, physics, biology, chemistry computer science, epidemiology, etc... Evolutionary computation is a sub-discipline of computer science belonging to the bio-inspired computing area. Since the end of the second world war, the main ideas of evolutionary computation has been published [3] and widely introduced to the scientific community [4]. Hence, the golden era of evolutionary techniques began, when Genetic Algorithms (GA) by Holland [4], Evolutionary Strategies (ES), by Schwefel [5] and Rechenberg [6] and Evolutionary Programming (EP) by Fogel [7] had been introduced. All these designs were favored by the forthcoming of more powerful and more easily programmable computers, so that for the first time interesting problems could be tackled and evolutionary computation started to compete with and became a serious alternative to other optimization methods. The main idea of our research is to show in this article that the dynamics of evolutionary algorithms, in general, shows properties of complex networks and evolutionary dynamics can be analyzed and visualized like a complex networks. This article is focused on

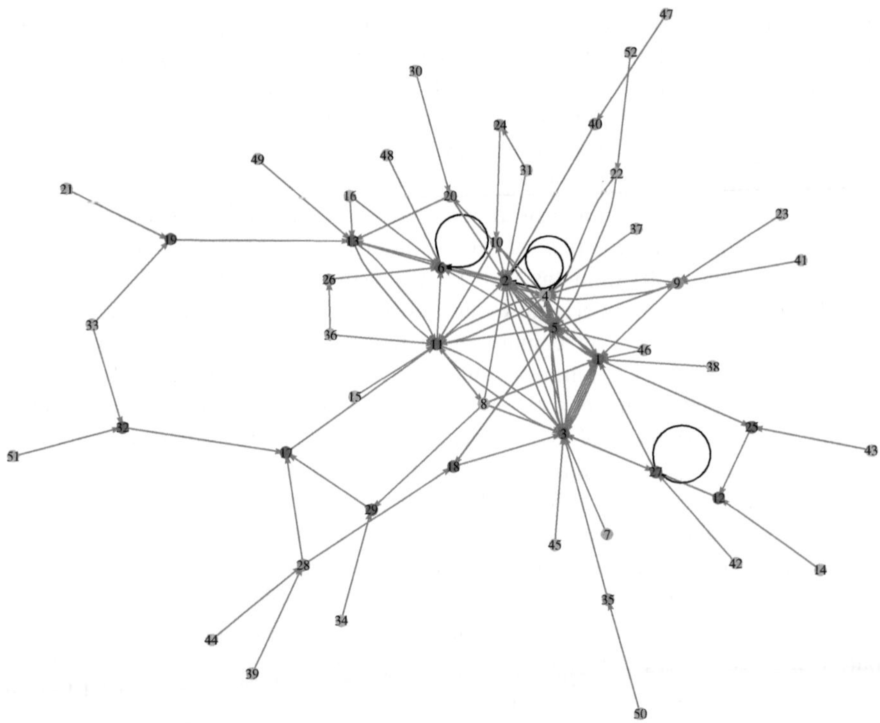

Fig. 3 Example of a more complex network with multiple edges and self-loops

observation and description of complex networks phenomenon in evolutionary dynamics. Possibilities of its use are discussed at the end.

Motivation of this research is quite simple. Evolutionary algorithms are capable of hard problem solving. Its performance depend on algorithm parameter setting that is usually given by heuristic observation. A number of examples on evolutionary algorithms can be easily found like differential evolution, SOMA, simulated annealing, genetic algorithms, etc. Evolutionary algorithms (EA)are used to solve different tasks like for example with chaotic systems that is done for example in [8] where EAs has been used on local optimization of chaos, [9] for chaos control with use of the multi-objective cost function or in [10, 11], where evolutionary algorithms have been studied on chaotic landscapes. Slightly different approach with evolutionary algorithms is presented in [12] where selected algorithms were used to synthesize artificial chaotic systems. In [13, 14] EAs has been successfully used for real-time chaos control and in [15] EAs was used for optimization of chaos control. Other examples of evolutionary algorithms application can be found in [16], which developed statistically robust evolutionary algorithms, alongside research conducted by [17]. Parameters of permanent magnet synchronous motors has been optimized by PSO and experimentally validated on the servomotor. Another research was focused on swarm intelligence, which has been used for IIR filter synthesis, co-evolutionary particle swarm optimization (CoPSO) approach for the design of constrained engineering problems, particularly for pressure vessel, compression spring and welded beam, etc. On the other side, complex networks, widely studied across many branches of science are promising and is a modern interdisciplinary research. Evolutionary algorithms, based on its canonical central dogma (following darwinian ideas) clearly demonstrate intensive interaction amongst individual in the population, which is, in general, one of the important attributes of complex networks (intensive interaction amongst the vertices). The main motivation (as well as a question) is whether it is possible to visualize and simulate underlying dynamics of evolutionary process like complex network. This investigation contain three mutually joined parts, namely introducing a novel approach joining evolutionary dynamics, complex networks and CML systems exhibiting chaotic behavior. The first part will discuss a novel method on how dynamics of evolutionary algorithms can be visualized in the form of complex networks. An analogy between individuals in the populations in an arbitrary evolutionary algorithm and the vertices in a complex network will be discussed as well as the relationship between the communications of individuals in a population and the edges in a complex network. The second part will discuss the possibility of how to visualize the dynamics of a complex network by means of coupled map lattices and to control by means of chaos control techniques. The last part will discuss some possibilities on CML systems control, especially by means of evolutionary algorithms. The spirit of this chapter is to create a closed loop in the following schematic: evolutionary dynamics → complex network → CML system → control CML → control evolutionary dynamics. Reason for this is such that today various techniques for analysis and control of complex networks exists and if complex network structure would be hidden behind EA dynamics, then we believe, that for example above mentioned control techniques could be used to improve dynamics of EAs. All experiments here were designed to analyze and either confirm or reject this idea.

2 Experiment Design

2.1 Selected Algorithms

For the experiments described here, stochastic optimization algorithms, such as DE [18] and Self Organizing Migrating Algorithm (SOMA) [19], have been used. Application of alternative algorithms like GA and Simulated Annealing (SA), ES and/or Swarm Intelligence in general are possible to use too. All experiments have been done on a special server consisting of 16 Apple XServer (2×2 GHz Intel Xeon, 1 GB RAM), each with 4 CPU, so in total 64 CPUs were available for calculations. It is important to note here, that such technology was used to save time due to a large number of calculations, however it must be stated that evolutionary identification described here, is also solvable on a single PC (with longer execution time). For all calculations and data processing, Mathematica version 7.0.1.0 was used. Four versions of SOMA and two versions of DE have been applied for all simulations in this chapter. See details in [20]. Parameters for the optimizing algorithm were set up in such a way as to reach similar value of maximal cost function evaluations for all used versions. Each version of EAs has been applied 50 times in order to get less or more valuable statistical data. Differential Evolution [18] is a population-based optimization method that works on real-number-coded individuals. For each individual $\mathbf{x}_{i,G}$ in the current generation G, DE generates a new trial individual $\mathbf{x}'_{i,G}$ by adding the weighted difference between two randomly selected individuals $\mathbf{x}_{r1,G}$ and $\mathbf{x}_{r2,G}$ to a randomly selected third individual $\mathbf{x}_{r3,G}$. The resulting individual $\mathbf{x}'_{i,G}$ is crossed-over with the original individual $\mathbf{x}_{i,G}$. The fitness of the resulting individual, referred to as a perturbed vector $\mathbf{u}_{i,G+1}$, is then compared with the fitness of $\mathbf{x}_{i,G}$. If the fitness of $\mathbf{u}_{i,G+1}$ is greater than the fitness of $\mathbf{x}_{i,G}$, then $\mathbf{x}_{i,G}$ is replaced with $\mathbf{u}_{i,G+1}$; otherwise, $\mathbf{x}_{i,G}$ remains in the population as $\mathbf{x}_{i,G+1}$. DE is quite robust, fast, and effective, with a global optimization ability. It does not require the objective function be differentiable, and it works well even with noisy, epistatic and time-dependent objective functions. For more about DE see in [18].

SOMA is a stochastic optimization algorithm, modeled based on the social behavior of competitive-cooperating individuals [19]. It was chosen because it has been proved that this algorithm has the ability to converge towards the global optimum [19]. SOMA works on a population of candidate solutions in loops, called migration loops. The population is initialized by being randomly and uniformly distributed over the search space at the beginning of the search. In each loop, the population is evaluated and the solution with the lowest cost value becomes the leader. Apart from the leader, in one migration loop, all individuals will traverse the searched space in the direction of the leader. Mutation, the random perturbation of individuals, is an important operation for evolutionary strategies. It ensures the diversity among all the individuals and it also provides a means to restore lost information in a population. Mutation is different in SOMA as compared with other evolutionary strategies. SOMA uses a parameter called PRT to achieve perturbations. This parameter has the same effect for SOMA as mutation for GA. The novelty of this approach lies in that the PRT vector is created before an individual starts its journey over the search space. The PRT vector defines the final movement of an active individual in the search space. The randomly generated binary perturbation vector controls the permissible dimensions for an individual. If an element of the perturbation vector is set to zero, then the individual is not allowed to change its position in the corresponding dimension. An individual will travel over a certain distance (called the PathLength) towards the leader in a finite number of steps in the defined length. If the PathLength is chosen to be greater than one, then the individual will overshoot the leader. This path is perturbed randomly. More about SOMA, see in [19].

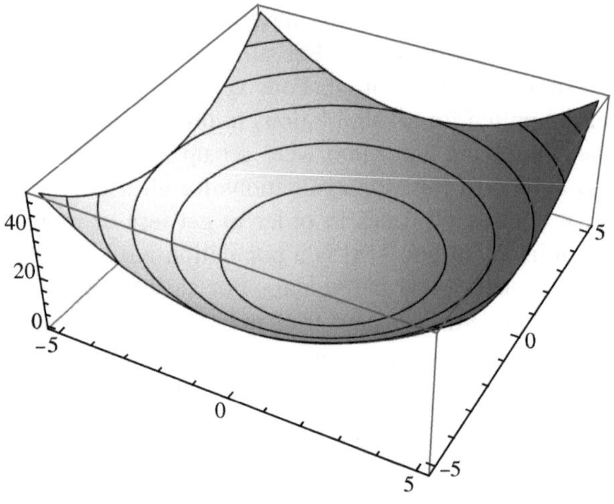

Fig. 4 Selected test functions: 1st DeJong

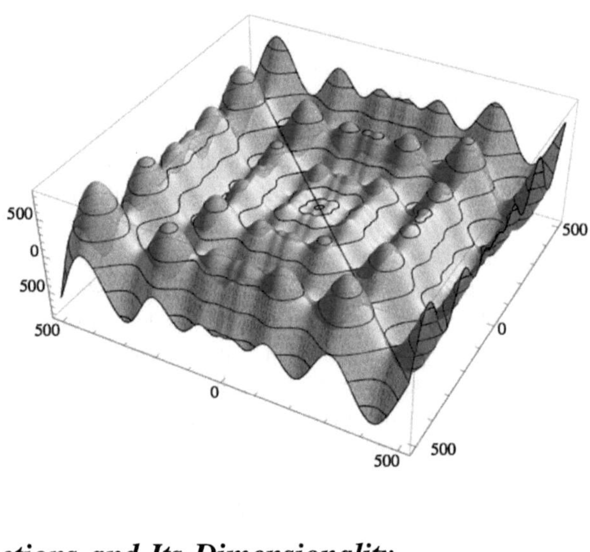

Fig. 5 Selected test functions: Schwefels function

2.2 Selected Test Functions and Its Dimensionality

The test functions applied in this experimentation were selected from the test bed of 17 test functions. In total 16 test function were selected as a representative subset of functions which shows geometrical simplicity and low complexity as well as functions from the opposite side of spectra. Selected functions (see Figs. 4, 5, 6, 7) were: 1st DeJong, Schwefels function, Rastrigins function, Ackleys function amongst the others. Each of them has been used for identification of complex networks dynamics and structure in 5, 10, 20 and 50 dimensions (individual length was 5, 10, 20 and 50). Test functions has been selected due to their various

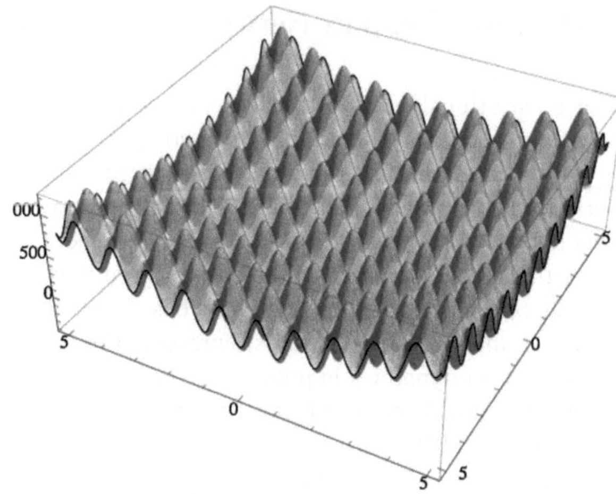

Fig. 6 Selected test functions: Rastrigins function

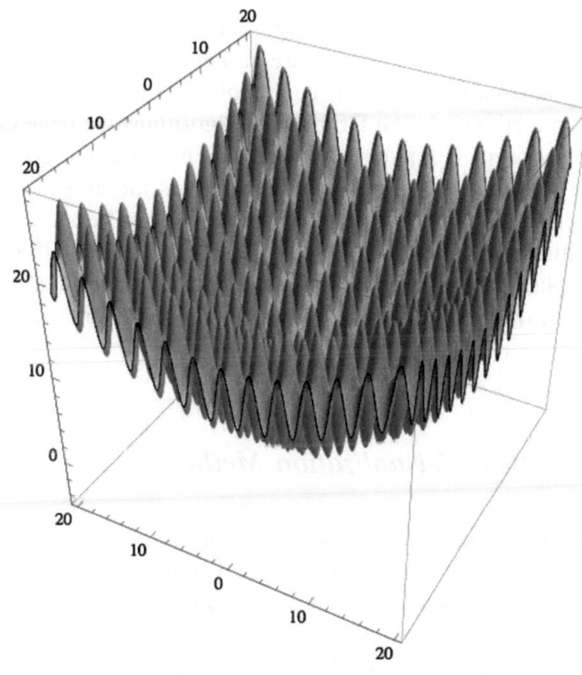

Fig. 7 Selected test functions: Ackleys function

complexity and mainly for the fact that this functions are widely used by researchers working with evolutionary algorithms. Another reason was, that speed of convergence and thus evolutionary dynamics itself, is different for simple functions like 1st DeJong or more complex example Rastrigins function.

3 Data for Complex Network Visualization

The most critical point of this research and related simulations was as to which data and relations should be selected and consequently visualized. Based on investigated algorithms, we believe that there is no universal approach, but rather a personal one, based on the knowledge of algorithm principle. Of course, some conclusions (see Sect. 7 and [21]) can be generalized over a class or family of algorithms. As mentioned in the previous sections, algorithms like DE and SOMA were used. Each class of algorithm is based on a different principle. The main idea was such that each individual is represented by vertex and edges between vertices should reflect dynamics in population, i.e. interactions between individuals (which individual has been used for offspring creation). The SOMA algorithm, as described in [19], consists of a Leader attracting the entire population in each migration loop (equivalent of generation), so in that class of swarm like algorithm, it is clear that the position in the population of activated Leaders shall be recorded like vertex (getting new inputs from remaining vertices—individuals) and used (with remaining part of population) for visualization and statistical data processing. The other case is DE, e.g. DERand1Bin in which each individual is selected in each generation to be a parent. Thus in DE, we have recorded only those individuals-parents, that has been replaced by better offspring (like vertex with added connections). In the DE class of algorithms we have omitted the philosophy that a bad parent is replaced by a better offspring, but accepted philosophical interpretation, that individual (worse parent) is moving to the better position (better offspring). Thus no vertex (individual) has to be either destroyed or replaced in the philosophical point of view. If, for example, DERand1Bin has a parent been replaced by offspring, then it was considered as an activation (new additional links, edges) of vertex-worse parent from three another vertices (randomly selected individuals, see [18]).

3.1 Basic Visualization Methods

Experimental data can be visualized in a few different ways and as an example, a few typical visualizations is depicted here. As mentioned in [21]: *vertices in complex graph are individuals that are activated by other individuals, incrementally from generation to generation. Used visualization is depicted in* Fig. 8, *in which one can see which individual (out of 100) has been activated for offspring creation (in this case selected like Leader in SOMA).* Figure 8 *are sort of auxiliary visualizations, which does not give total view on complex network structure behind evolutionary dynamics. Better visualization that can be used is as in* Fig. 9 *which shows, that interactions between individuals create (at the first glance) structures, which looks like complex networks. However, it has to be said, that we have met results whose visualizations looks like net and resemble complex networks but*

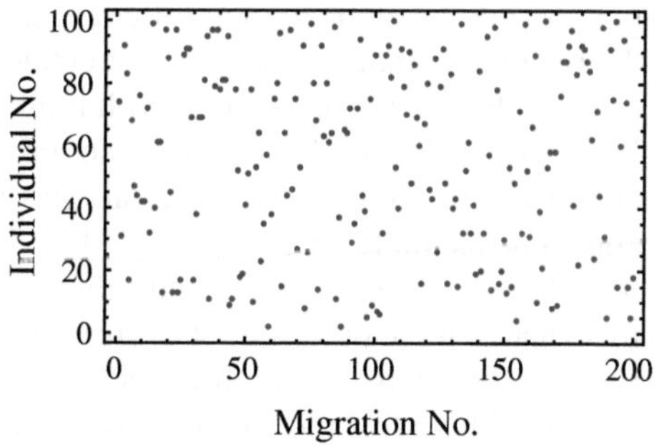

Fig. 8 An example of activated leaders (y axe) with dependance on Migrations (x axe) SOMA. For example 31th individual (y axe) has been select for 4 times during 200 Migrations (x axe)

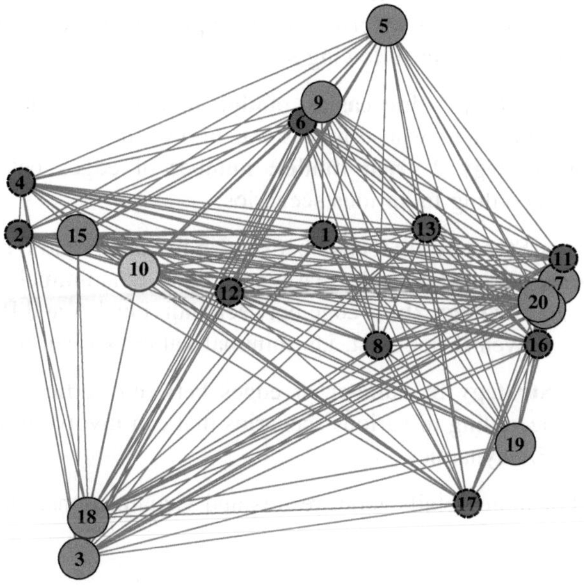

Fig. 9 Complex network example of SOMA dynamics. Vertex (individual) 10 is the most profitable vertex. Visualization of the multiple edges is disabled on all pictures in this chapter for better visualization

after closer complex network characteristics calculations, those networks did not belong to the class of complex networks with small world phenomenon. Meaning of vertices in the above mentioned figures is given by ratio of incoming and outgoing edges and implies that: small vertex (small gray (pink) with dashed edges) has less incoming edges than outgoing. White (middle-sized) vertex is balanced (i.e. has the same incoming number of edges as outgoing) and dark gray (green), the biggest, are vertices with more incoming edges than outgoing. The light gray (yellow) vertex is the most activated individual vertex with the maximum of incoming edges. In EA jargon, small vertex is an individual, which has been used more times for offspring creation rather than as a successful parent and pink vertices reflects the opposite.

In the basic visualization above is proposed that vertices of network are individuals and number of edges between them only increase by integer number. It is also possible to set decrement of edges according to various criteria. For example when parents—individuals does not create better offsprings, then number of edges between them decrease. Each evolutionary algorithm is running under different philosophy—algorithm, so there is no universal recipe how to convert dynamics of studied EA to complex network and has to be developed for each specific algorithm. For example SOMA, see [19], is in each migration loop winner only one individual, so called Leader and other individuals establish connection with this individual. On the other side, DE create for selected parent individual, by means of randomly selected another three individuals (DERand1Bin) new offspring which is understood here like movement of selected parent individual to the new position. If fitness is better, then this selected parent individual get new connections from those randomly selected another three individuals. Decreasing of existing edges is done in similar inverse way. Complex networks, generated from EAs dynamics, can be done in following scenarios:

- **Strategy 1 (increasing mode):** number of established edges increase only for individuals whose fitness was improved.
- **Strategy 2 (real mode):** number of edges between vertices is decreasing according to various scenarios:
  - **Strategy 2a:** if edge between individual A and B exist, then is decreased by 1, if not exist, then decrease is not done, i.e. minimal value that can be reached is 0.
  - **Strategy 2b:** if edge between individual A and B exist, then is decreased by 1 even if not exist, i.e. minimal value that can be negative.
- **Strategy 3:** number of edges between vertices is increasing and decreasing according to various scenarios that can involve fitness and another attributes of EA dynamics.

For more details about results and vizualizations, please see [21].

3.2 Preliminary Results

As reported above, both algorithms, in 10 versions, has been tested on various test function (to reveal its complex networks dynamics) with constant level of test function dimensionality (i.e. individual length) and different number of generations (migrations) in all used algorithms. All data has been processed graphically (see [21], etc.) alongside calculations of basic statistical properties. Emergence of complex network structure behind evolutionary dynamics depend on many factors, however some special versions of used algorithms did not show complex network structure despite the fact that the number of generations was quite large. Some pictures like Figs. 10, 11, 12, 13 and 14 has been generated in order to visualize all important data and relations, see also [21]. All main ideas coming from the results are discussed further. Another part of our experiments has been focused on parallel

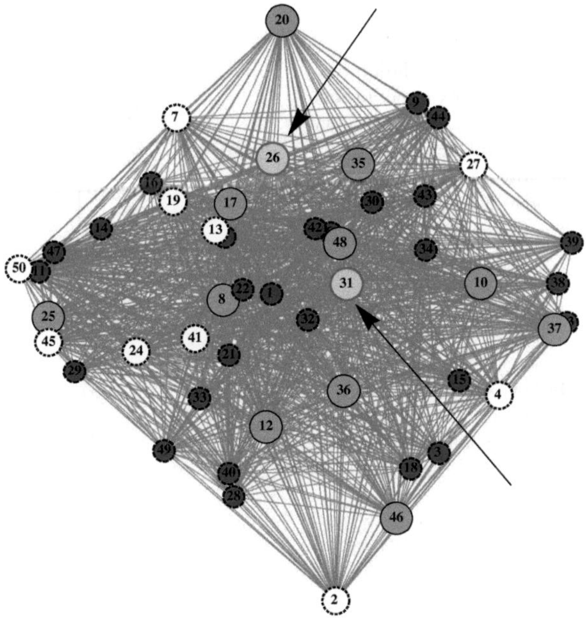

Fig. 10 Complex network of the DELocalToBest with two the most intensively connected vertices (individuals)

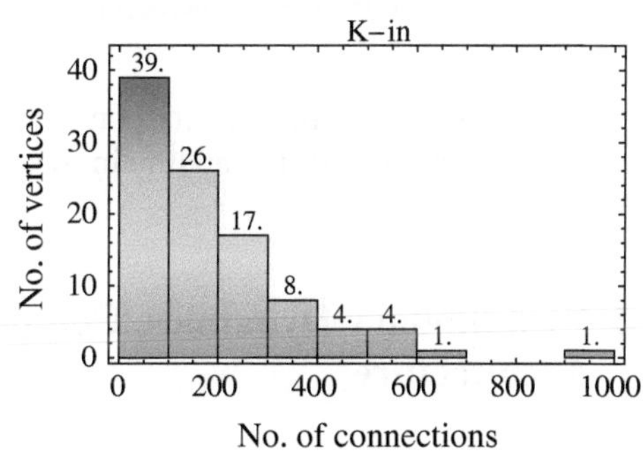

Fig. 11 Complex network of the DELocalToBest with two the most intensively connected vertices (individuals) and its histogram of the vertices connections (note that winning vertex has with almost 900 connections)

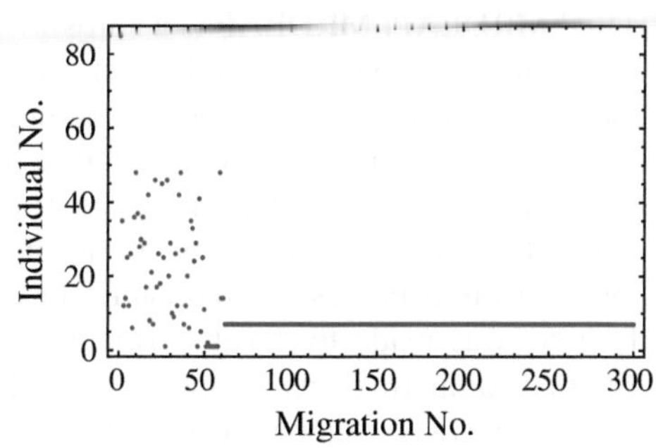

Fig. 12 An example of "rich become to be richer", see also Figs. 13 and 14

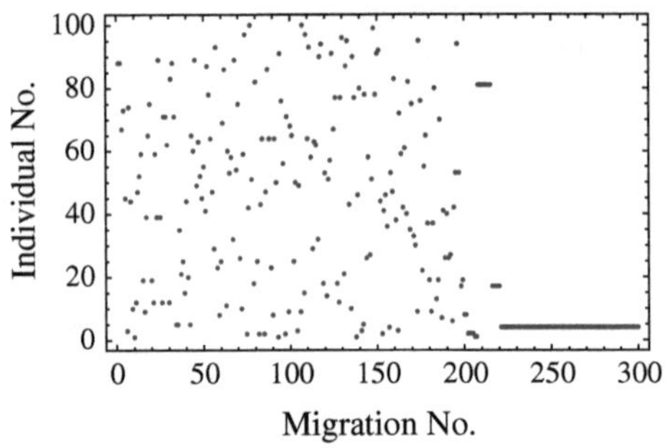

Fig. 13 An example of activated leaders with moment when evolution has found global extreme. In such a moment the best individual is repeatedly selected (see line after 230 migrations) and become to be extremely attractive node of all

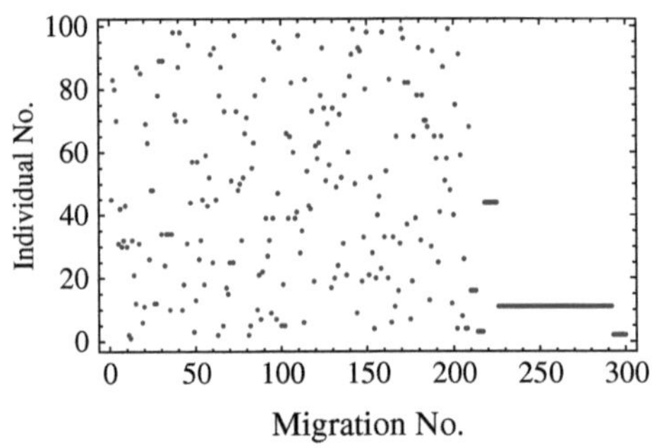

Fig. 14 An example of activated leaders with moment when evolution has found global extreme

evolutionary algorithms and its conversion and visualization as a complex networks. Preliminary results and main idea is reported in the next section.

4 Evolutionary Dynamics, Complex Networks and Parallel Architecture

Simulations presented in this chapter were carried out by a computing platform called MRDEAF. MRDEAF is a software framework running on top of a Hadoop server and incorporates MapReduce programming model in order to distribute compute load across heterogeneous computing cluster. During the development of software for evolution of artificial neural networks we have run into problems regarding computational cost of the software. Evolution of sufficient solution simply took too much time, even after its division into multiple threads on multi core and multi processor computers. In order to achieve acceptable run times, a decision was made to develop version of the software which could exploit distributed computing environment. Instead of single purpose program a flexible framework was developed which allows adaptation of wide variety of evolutionary algorithms. For the ease of development as well as for the previous experiences we chose MapReduce programming model, together with its implementation called Hadoop.

4.1 MapReduce and Hadoop

MapReduce [22] is a programming model developed by Google. Its original purpose was the simplification of development of distributed applications for large scale data processing [22, 23]. The main idea behind MapReduce is inspired by map and reduce functions in functional languages like LISP. The simplest programs require developers to define just two functions (or classes, depending on implementation), which serve as map and reduce parts of resultant program. For data storing and sharing, MapReduce is built with close relations to distributed file systems [24, 25], optimized for large amount of data and shared access. Important thing about data structure is that it needs to be in form of [key, value] pairs, the purpose of keys is explained later. In order to run MapReduce programs the framework servers have to be installed on a number of computers, which are usually of two types: distributed file system servers and actual computation nodes. One server is designated master node, which controls task assignments, monitors active nodes and cooperates all tasks and framework functions. Programs are usually copied to master, which copies them onto slave nodes. Program execution is as follows (also displayed in Fig. 15):

- Program and data to be processed are uploaded to servers.
- Data is divided into M blocks, size of blocks is user controllable. Master must gradually assign these blocks to map nodes.
- Map nodes process blocks and generate intermediate results, also in [key, value] pairs, which are written to local disks or alternatively to distributed file system. Generally only one result is generated for each key. Location of intermediate results is sent to master node.
- Master sorts locations of results according to keys, then sends locations of intermediate results to reduce nodes, so that results with the same key are sent to one reduce node.
- Reduce functions use iterators to remotely read data and process them. Results of produce functions are appended to resultant output file. It is usual for map functions to generate high number of outputs for one key and reduce function generates only one result for each key.
- After all reduce functions are completed program ends.

The simplest example of MapReduce is the word count problem. Problem is defined as follows: count the number of occurrences of words in large number of files. Map functions loads a number of files and generates data pairs, where key is string representation of a word and value is number of its occurrences in an assigned files. Reduce function groups values by words and writes them into output file. MapReduce was developed with emphasis on its usage on commonly available hardware, meaning that hardware and connection errors are likely to occur. To address this shortcomings MapReduce includes mechanisms which deal with failures. The simplest method is the check and restart master periodically checks slave nodes and if a node does not respond it is added to a black list, so no

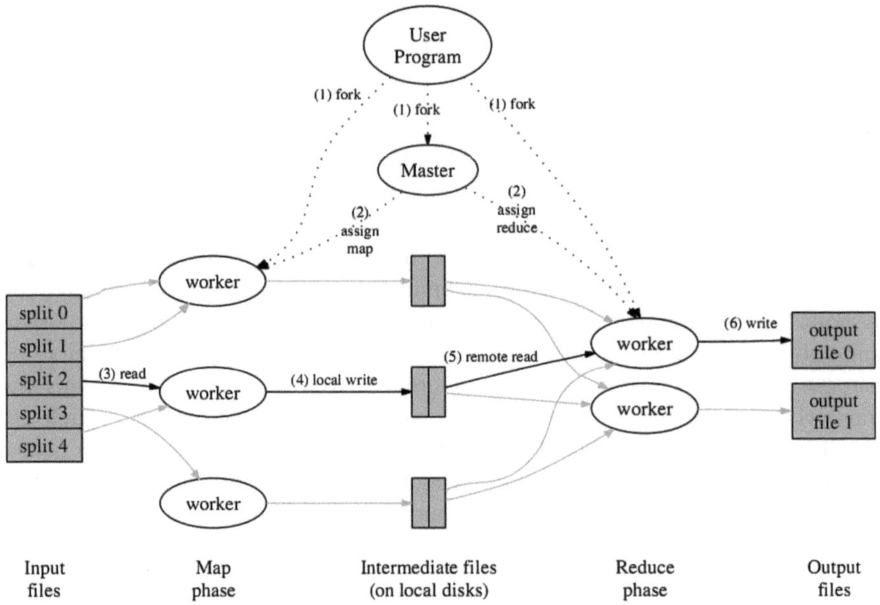

Fig. 15 Overview of MapReduce execution flow [22]

new task are assigned to this node and failed task is reassigned (and restarted) to another node. Status of blacklisted nodes is reevaluated after a period of time. Several implementations of MapReduce exist, possibly the most popular and widely used open implementation is the Hadoop [26] together with Hadoop distributed file system implementation (HDFS) [25].

4.2 MapReduce and Evolutionary Algorithms

There have been several attempts at using MapReduce model together with evolutionary algorithms (EA). EAs are incremental algorithms working in rounds, where output of each round is an input for next round. In contrary, MapReduce is designed to run only once and produce final outputs immediately, this means that basic premises behind either MapReduce model of EAs have to be modified. Probably the first reported usage of MapReduce for improving the execution speed of EAs was made by Jin et al. [27]. They used custom MapReduce implementation, called MRPGA, based on .NET platform, which included additional reduce phase to increase effectiveness of genetic algorithm (GA). Map function was used for fitness function evaluations, where each map function was called for one individual from population. 1st reduce phase chose local optimum individuals and 2nd reduce selected global optimum individual which is emitted as final result. Mutation and crossover are implemented sequentially on master node. Above operations are continuously repeated with previous population as an input until stopping criteria are met. In response to the previous work, Verma et al. presented another approach [28]. Instead of developing their own MapReduce implementation they utilized Hadoop. They identified that Hadoop needs large amounts of time to initialize each map and reduce function, so in order to minimize these overheads, each map function processes more that one individual. Each map function gets data pairs of several individuals, where key represents genome and value represents fitness of individual. Each individual is then evaluated and emitted from map function and also each map function selects and emits the best individuals. To counter partitioning of population into smaller parts, which could lower search space and cause premature convergence, individuals are assigned to reducers randomly. Reduce functions randomly chose a number of individuals and performs crossover and emits new individuals as outputWhole process is repeated until stopping criteria are met. Another work, which uses combination of GA and Hadoop, with approach similar to Verma et al. is [29]. GA was not the only EA which was implemented via the MapReduce model, other EAs include differential evolution [30], self-organizing migration algorithm [31] and particle swarm optimization [32].

4.3 MapReduce Distributed Evolutionary Algorithms Framework

All previous uses of MapReduce with EAs required a MapReduce process to be restarted at least once for each generation. Previous works also indicated that framework overhead significantly decreased performance gain from adding new nodes into the cluster [28, 31]. In order to decrease communication and initialization overhead it was necessary to change the method of distributing the workload across the computing nodes. MRDEAF is an acronym for MapReduce Distributed Evolutionary Algorithms Framework and is developed on top of Hadoop. Its main goal is to provide researchers and developers with an easy solution for development of distributed optimization programs independent on EA used. MRDEAF uses the coarse and fine grained model of parallel EAs (in some source also island models). The whole population is divided into relatively isolated subpopulations, which evolve on their own, and uses migration operator which is used for moving individuals between subpopulations. Migrations are useful for increasing diversity of subpopulations, thus allowing them to explore wider search areas and increasing their chances of finding better solutions. Distinction between coarse and fine grained algorithms is in ratio of subpopulation count versus subpopulation size. Smaller count of bigger subpopulations (coarse grained) is best suitable for distributed architectures as it minimizes communication overheads, while reverse solution is intended for massively parallel computers. Island models introduce 5 additional parameters into EAs:

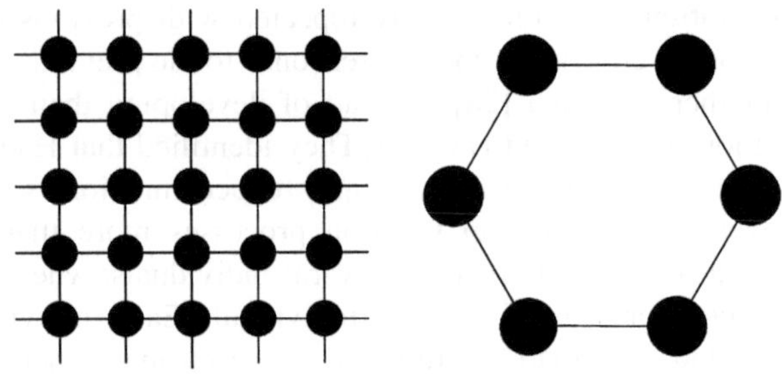

Fig. 16 Fine and coarse grained parallelizations with different topologies [33]

- Topology determines to which subpopulations will individuals migrate. High number of target islands may cause domination of one solution across population and premature convergence (Fig. 16).
- Number of migrating individuals.
- Migration rate is migration occurs too often, algorithm is degraded to global parallelization.
- Method for selection of individuals for migration if only best individuals migrate, they may draw population into local optimum and population will lose diversity.
- Method for replacement of individuals in target population.

Problem with using Hadoop is that for each map or reduce function call new java virtual machine (JVM) is started, this event considerably increases framework overhead. To minimize this overhead it was important to use as little map functions as possible. Each map function in MRDEAF is responsible for evolution of one subpopulation for a given number of generations. This way we are able to lower the number of JVM restarts. Implementation of migration between subpopulations relies on HDFS. The execution of MRDEAF programs can be divided into two separate logical loops: outer loop, or migration round and inner loop or generation. Inner loops are computed on each compute node independently and they represent evolution of a subpopulation. Outer loop represents one MapReduce cycle and encapsulates the evolution of subpopulations, selection of migrating individuals and migration of these individuals between populations and testing of stopping criteria. The compute nodes in Hadoop cluster are not able to communicate directly with each other, therefore all data which should be transferred into next migration round must be stored in HDFS. Hadoop requires its data to be in form of [key, value] pairs. Value is a data structure which holds entire subpopulation, while key is an integer number serving as a subpopulation index. This data structure ensures that all populations with the same population index are assigned to the same compute node. The selection of migrating individuals if performed after the evolution of subpopulation for this round finishes. Genotypes and fitness values of selected individuals are copied into a new data structure of the same class as subpopulation, which has the population index set to the index of subpopulation which should receive these individuals. One new subpopulation is created for each subpopulation which is supposed to receive individuals based on the topology. To distinct migrating individuals from actual subpopulation a boolean value is assigned to each subpopulation which indicates its state. The migration mechanism is flexible enough to handle several destination subpopulations, different selection methods (select best, worst, random, best + random) and numbers of migrating individuals. Prior to a migration round each compute node checks availability of migrating individuals for its corresponding subpopulation. If migrating individuals do exist they are imported into subpopulation using the defined replacement function. Creation of custom topology, selection and replacement functions involves overriding built-in MRDEAF functions. Topology function returns array of integers to identify which subpopulations should receive individuals from current one. Selection and replacement functions work directly with subpopulations, and thus can base their selection on complete state of the subpopulation. As all the data exchange between nodes and between migration rounds is based on files stored in HDFS, the historic states of subpopulations is stored and can be further analyzed. Interesting by-product of this approach is that MRDEAF execution can be started with already stored subpopulations but with different settings, or perhaps completely different EAs. MRDEAF operation is independent of EA used for evolution of subpopulations, therefore it is possible for each subpopulation to have its own variant of EA. The only constraint is that individuals which represent a solution in one subpopulation must be interpreted the same way in all subpopulations. Execution of MRDEAF displayed in Fig. 17 with description in pseudocode in Fig. 18.

4.4 Generating the "Parallel" Networks

MRDEAF implements the functionality to generate networks describing the evolutionary process for several built in algorithms. Each subpopulation maintains records of interactions amongst its individuals and also of migrating individuals. After program finishes, data from all subpopulations is compiled into a single file. We chose the GraphML [34] file format to store networks. It is an XML-based file format for exchange of graphing information. Example of how the data is stored in an GraphML file is in Fig. 19.

Network nodes represent individuals, while edges show interactions between them. Each node in network holds 4 data values:

- Id index of an individual in subpopulation.
- Fitness fitness value of an individual.
- Population index of subpopulation.
- InWeights sum of weights of incoming edges.

Edges contain 3 values:

- Source the starting point of an edge.
- Target the end point of the edge.
- Weight weight value.

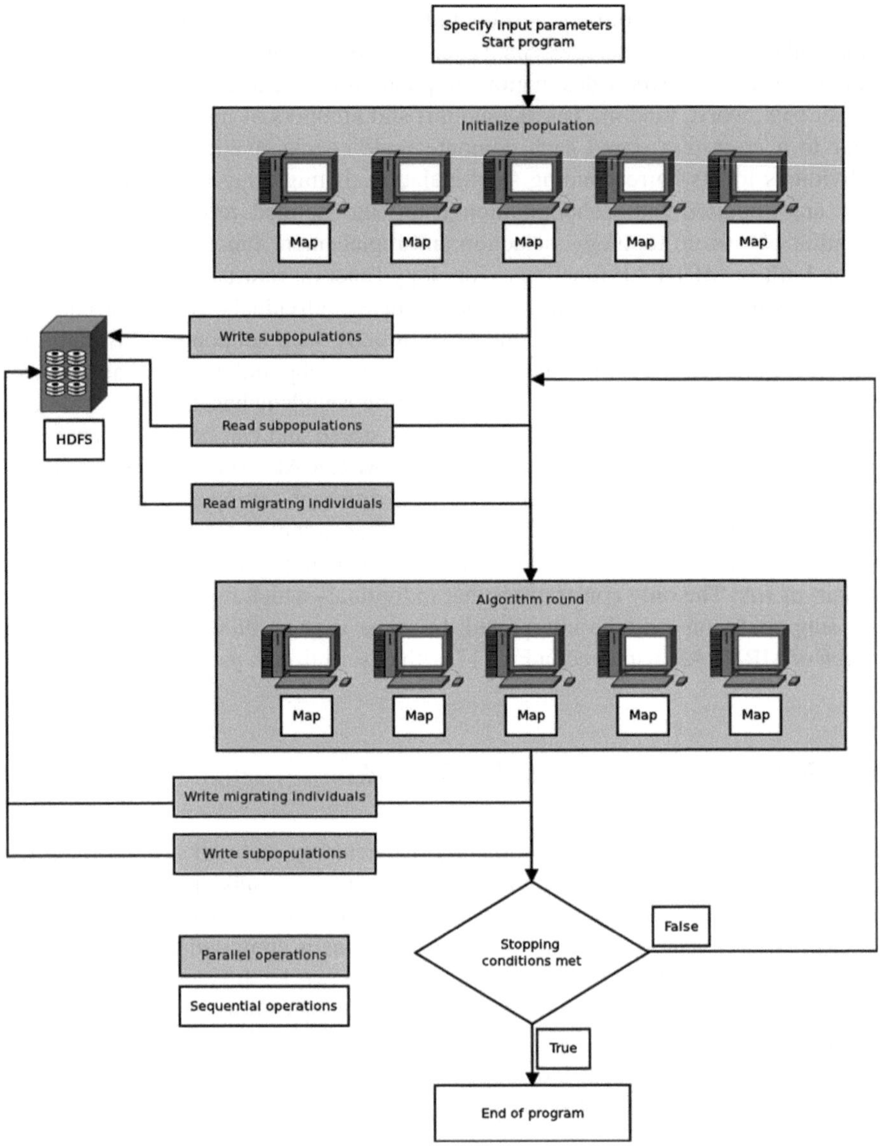

Fig. 17 Overview of MRDEAF execution flow

The process of creating network nodes and connections is dependent on the EA. GraphML format can be further analyzed and visualized. We used open-source software called Gephi [35] for visualization of networks. Three variants of network generating techniques were proposed for SOMA.

```
initialize program

compute in parallel:
        for p in populations:
                create and initialize population
                write population into HDFS
        end for
end parallel

while r < number of rounds:
        compute in parallel:
                for p in populations:
                        read population
                        perform migration
                        for g < number of generations
                                p.generation()
                        end for
                        select individuals for migration
                        write population and migrating individuals into HDFS
                end for
        end parallel
        select best individual from all populations
        test stopping criteria
end while
```

Fig. 18 MRDEAF execution in pseudocode

```
<node id="2">
    <data key="fitness">-3311.938450936671</data>
    <data key="population">0</data>
    <data key="inWeights">45</data>
</node>
<node id="3">
    <data key="fitness">-3312.4873377197027</data>
    <data key="population">0</data>
    <data key="inWeights">47</data>
</node>
<edge source="2" target="0">
    <data key="weight">1</data>
</edge>
<edge source="3" target="0">
    <data key="weight">2</data>
</edge>
```

Fig. 19 Example of GraphML file

- After an individual finishes its movement and its fitness improved, it adds 1 to its connection to the leader.
- The same as previous, except that the connection direction is reversed (1 is added to a connection from leader to individual).
- After one SOMA generation ends, all individuals add 1 to their connection to the leader.

Fig. 20 Network describing evolution of 5 subpopulations

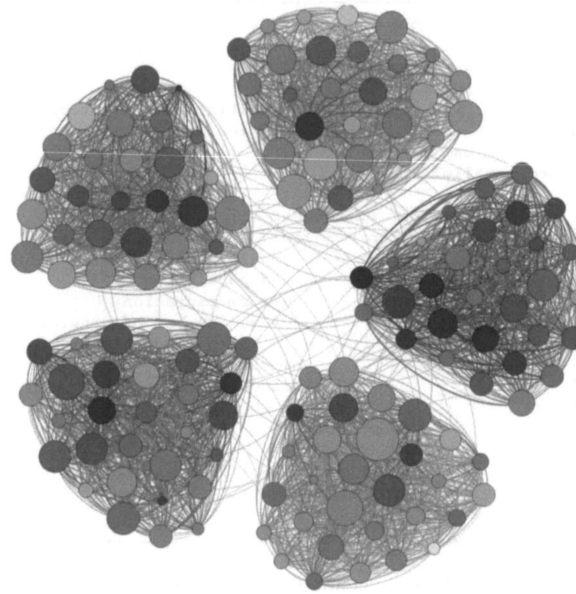

At Fig. 20 is shown network which describes evolution of 5 subpopulations, each running SOMA algorithm. The nodes are clustered according to their subpopulation index. Color of the nodes indicates the fitness value (the darker the better) and node size is a function of inWeights parameter. Edges which connect nodes from different clusters represent migrating individuals.

This is clear demonstration (one of may we got) that parallelization of evolutionary algorithm also create complex network, or better, dynamics of parallelized evolutionary algorithms can be visualized in that way.

5 Complex Networks and CML Systems

In the previous section we have shown, that dynamics of evolutionary algorithms can be, under certain conditions, visualized like a complex network. Following sections will explain CML systems and demonstrate, how can be complex network converted to the CML systems (Coupled Map Lattices, see [36]), for control and analysis (deterministic, chaotic behavior,...) of network (i.e. evolutionary algorithm) dynamics. Complex networks belong to the class of strongly nonlinear systems, which can, in general, generate very wide spectrum of behavior, from deterministic to chaotic one. All kind of behavior can be met in daily engineering activities and can have negative impact on various devices and everyday life. According to our experiences and contemporary research so called deterministic chaos is the most interesting and promising area of behavior analysis and control. Deterministic chaos, discovered by Lorenz [37] is a fairly active area of research in the last few decades. The Lorenz system produces one of the well-known canonical chaotic attractors in a simple three-dimensional autonomous system of ordinary differential equations [37, 38]. For discrete chaos, there is another famous chaotic system, called logistic equation [39]. Logistic equation is based on a predator-prey model showing chaotic behavior. This simple model is widely used in the study of chaos, where other similar models exist (canonical logistic equation [40] and 1D or 2D coupled map lattices [36]). Since then, a large set of nonlinear systems that can produce chaotic behavior have been observed and analyzed. Chaotic systems thus have become a vitally important part of science and engineering in theoretical as well as in practical levels of research. The most interesting and applicable notions are, for example, that chaos control and chaos synchronization are related to secure communication, amongst others. Recently, the study of chaos is focused not only along the traditional trends but also on the understanding and analyzing principles, with the new intention of controlling and utilizing chaos as demonstrated in [41, 42]. The term chaos control was first coined by Ott et al. in [43]. It represents a process in which a control law is derived and used so that the original chaotic behavior can be stabilized on a constant level of output value or a n-periodic cycle. Since the first experiment of chaos control, many control methods have been developed and some are based on the first approach [43], including pole placement [44, 45] and delay feedback [46, 47]. Another research has been done on CML control by [48], special feedback methods for controlling spatiotemporal on-off intermittency has been used there and [48]. This paper introduces a controller (based on discrete-time sliding mode and Lyapunov function) for controlling of spatiotemporal chaos system. Many methods were adapted for the so-called spatiotemporal chaos represented by coupled map lattices (CML). Control laws derived for CML are usually based on existing system structures [36], or by using an external observer [49]. Evolutionary approach for control was also successfully developed, for example in, [50–52]. Many published methods of deterministic chaos control (DCC) were (originally developed for classic DCC) adapted for so called spatiotemporal chaos represented by CML, given by Eq. (1). Models of this kind are based on a set of spatiotemporal (for 1D, Fig. 21, axe x is time). Typical example is CML based on so called logistic equation, [39, 49, 53] which is used to simulate behavior of system which consists of n mutually joined cells (logistic equations) via nonlinear coupling, usually noted like ϵ. Nonlinear coupling is done only and only between nearest cells. Cells that are not direct neighbor of a cell X are not directly influenced by this cell. Mathematical description of CML system is given by Eq. (1), [53]. The function, which is represented by $f(x_n(i))$ is an arbitrary discrete system—in this case study logistic equations has been selected to substitute $f(x_n(i))$, variable is usually set to value that represent nonlinear coupling between systems $f(x_n(i))$. CML description based on (1) in Mathematica software is given in Fig. 21.

$$x_{n+1}(i) = (1-\varepsilon)f(x_n(i)) + \frac{\varepsilon}{2}(f(x_n(i-1)) + f(x_n(i+1))) \tag{1}$$

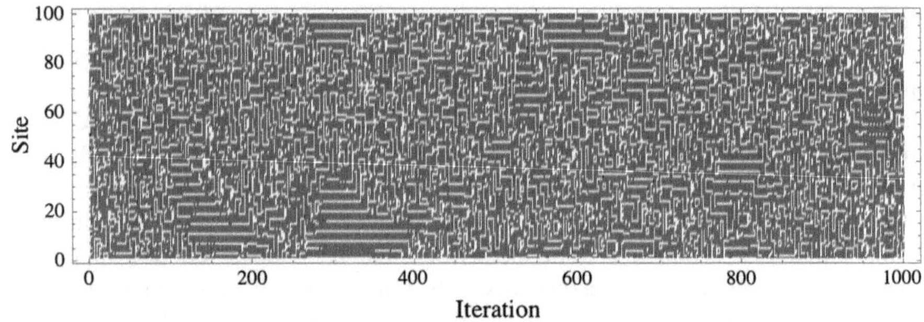

Fig. 21 Typical CML behavior with chaotic and deterministic windows

It is important to say, that CML are main backbone of our approach and this participation.

The term chaos covers a rather broad class of phenomena whose behavior may seem erratic and unpredictable at the first glance. Often, this term is used to denote phenomena, which are of a purely stochastic nature, such as the motion of molecules in a vessel with gas etc. This publication focuses on the deterministic chaos, a phenomenon that—as its name suggests—is not based on the presence of a random, stochastic effects. On the contrary, it is based on the absence of such effects what may seem surprising at the first glance. Broadly used, the term chaos can denote anything that cannot be predicted deterministically (e.g. motion of an individual molecule, numbers in a lottery,...). If, however, the word chaotic is combined with an attribute such as stochastic or deterministic, then a specific type of chaotic phenomena is involved, having their specific laws, mathematical apparatus and a physical origin. Stochastic system (not stochastic chaos) is the appropriate term for a system such as plasma, gas, liquid, which should be studied by using a suitable apparatus of plasma physics, statistical mechanics or hydrodynamics. On the contrary, if a double pendulum, billiard or the similar objects are the subjects of examination; a mathematical apparatus, which is based on classical mathematics and does not exhibit stigmata of statistics, is employed. The mathematical apparatus for the description and study of the systems was not chosen at random; in fact, it is related with the physical nature of the system being studied. Considering the class of systems of deterministic chaos as mentioned above, signs of chaotic behavior are usually conditional on the presence of nonlinearities, either in the system itself (i.e. the system is a nonlinear system) or in links between linear systems [54]. Usually, such nonlinearities are only visible after making up a mathematical model of the system or after analysis of observed data. Simple systems exhibiting deterministic chaos include, for instance, double pendulum, magnetic pendulum, electronic circuit or so called billiard problem through which balls are poured from the same starting position. Note that in CML are observable deterministic—periodical windows. On the x axe are iterations of CML, while on y are sites, mutually joined, see Eq. (1). Chaos is visible here like grainy part while deterministic behavior like periodic parts. Ideas of CML is in this participation used to show, that complex networks behavior can be also visualized and mainly modeled in this way. The possibility to handle with complex network like with CML then allows to use a wide class of CML control methods to control complex networks.

5.1 Complex Networks Dynamics and Its Visualization

Structure and dynamics of complex networks is usually visualized in a classical way that is depicted on Figs. 2, 3. Complex network is depicted like a set of vertices, mutually joined by single and multiple edges. Each edge can be added or cancelled during the evolution of the network, or importance of an edge can be modified by weights associated to the each edge. Adding or canceling of the vertices and modification of the edge weights represents, in fact, dynamics of the network. Network then change its shape, structure and size and as a consequence isolated sub-networks (or its fractions) can be observed. In [55] are reported various techniques how to control and analyze such networks. Our approach is based on well-known CML systems and its analysis and control by means of traditional as well as heuristic methods. Our method of visualization is based on fact that simplest version of CML (i.e. 1D version) is usually depicted like a row of mutually joined sites, where each site is nonlinearly joined with its nearest sites, see Fig. 22.

Our vision of equivalence between CML and complex network is quite simple. Each vertex is equivalent to the site in the CML. Comparing to the standard CML, sites in complex network CML (CNCML) are not joined to the nearest site, but to the sites equal to the complex network vertices. Thus sites in CNCML are not joined symmetrically (i.e. from site X to Y and vice versa) and between different sites is random pattern of connections, which can change in the time, see Fig. 23.

Our experiments of CNCML visualization were based on above described idea. In all cases has been CNCML calculated and visualized in such a way. Different levels of vertices (sites) excitation are depicted by different colors, see Fig. 24. When compared with our previous results, [56, 57] it is clearly visible, that our proposed kind of visualization is usable. It is observable, that CNCML visualization shows complex and obviously nonlinear behavior of tested CNs. On figures are visualized experiments with SOMA algorithm according to Strategy 1 (see Sect. 3.1) for different number of individuals and migrations, see Figs. 25, 26, 27 and 29. On Fig. 28 is final network of SOMA (related CML is in Fig. 29) and in Fig. 30 is depicted history of selected individuals—Leaders of SOMA, that got edges from remaining part of population.

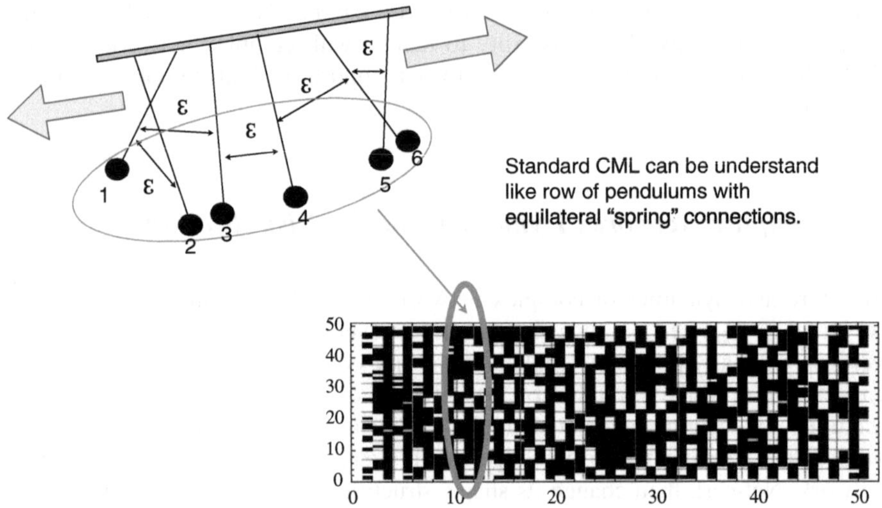

Fig. 22 Typical CML and its mechanical (pendulum-based) interpretation. Parameter ϵ is strength between two neighbor (pendulums, vertices) and in classic CML philosophy is in the interval $[0, 1]$, see [36]

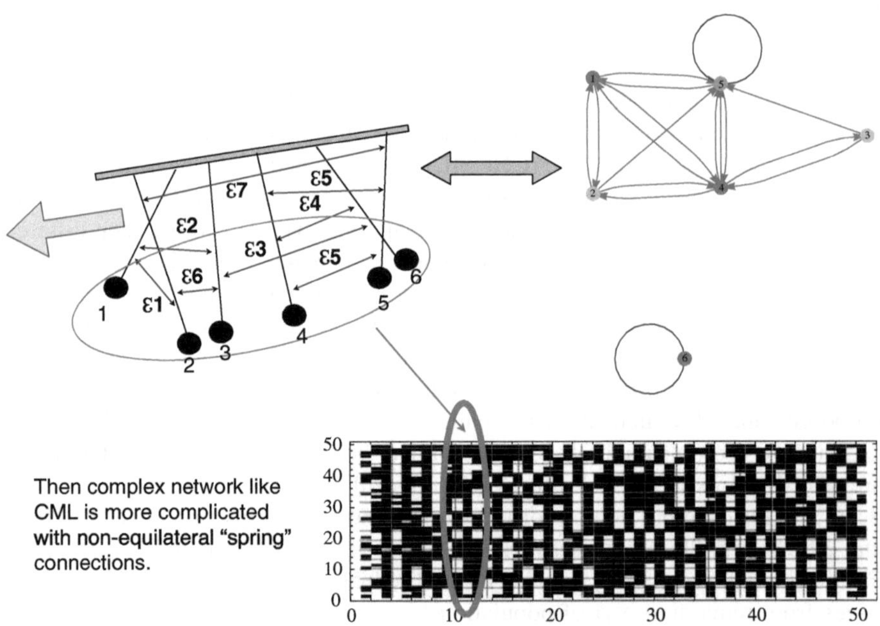

Fig. 23 Complex network like CML, vertices (pining sites in CML) are not connected equilaterally but according to complex network topology. Parameter ϵ is here number of edges (can be also converted to the interval $[0, 1]$) between two neighbor (pendulums, vertices)

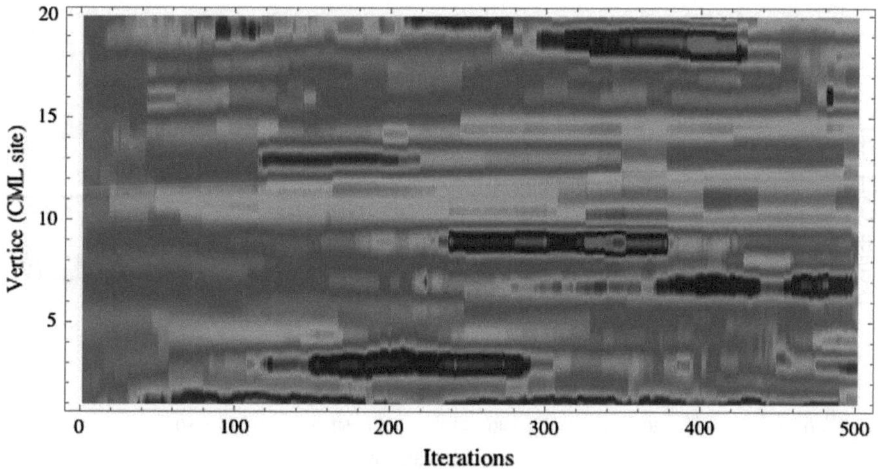

Fig. 24 Zoom of the network with 20 vertices in 500 iterations

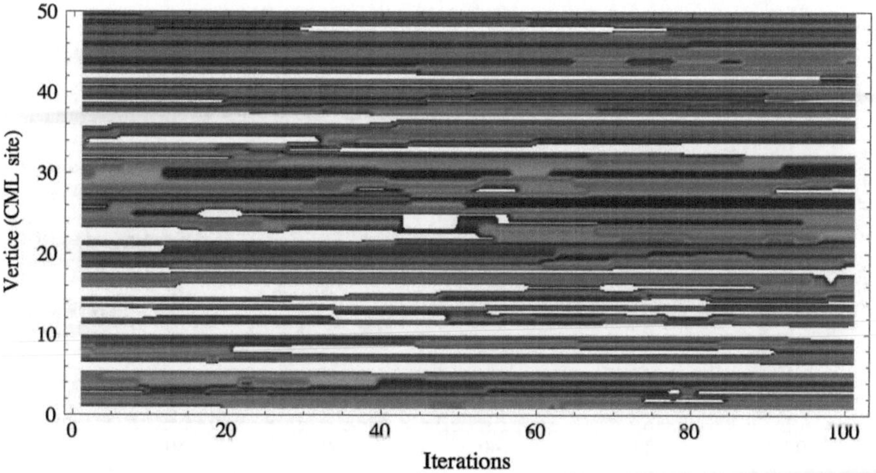

Fig. 25 An example of CML visualization of SOMA dynamics with 50 individuals, according to the Strategy 1 (see Sect. 3.1)

6 CML Control

In the previous parts has been shown, how can be converted dynamics of selected EA to complex network with selected examples and also idea describing conversion from complex network to CML has been introduced too. Here, in the last part, we would like to discuss possibilities about CML control. Because classical control techniques are well known, our attention is focused here on CML control selected examples by means of evolutionary algorithms. Examples mentioned here

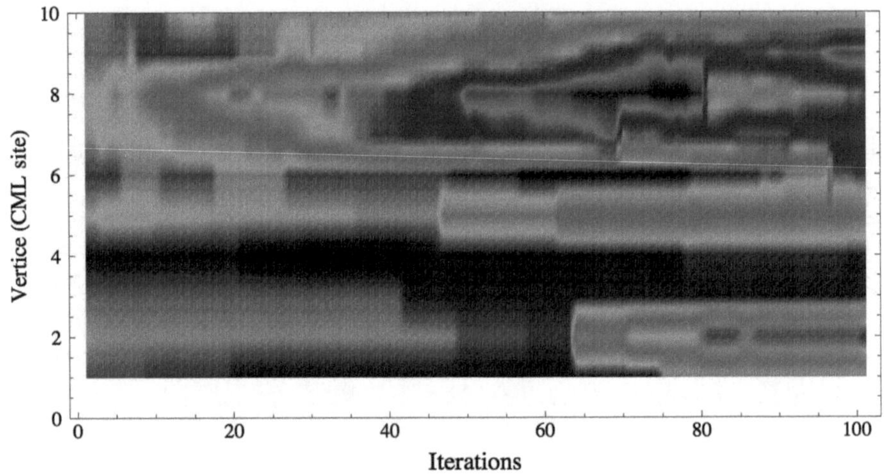

Fig. 26 Another example of CML visualization of SOMA dynamics with 10 individuals, according to the Strategy 1 (see Sect. 3.1)

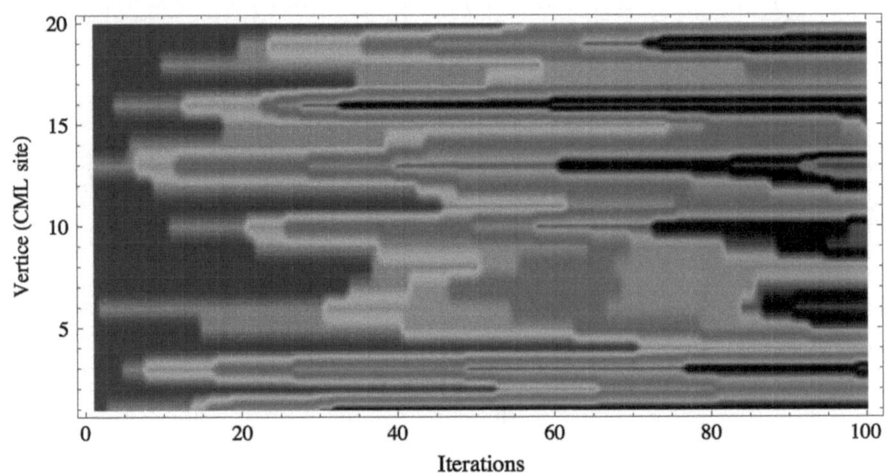

Fig. 27 Another example of CML visualization of SOMA dynamics with 20 individuals, according to the Strategy 1 (see Sect. 3.1)

are results of our research and for its detailed description important references are provided here.

6.1 CML Control—Selected Examples

This part introduces the main ideas of an investigation on deterministic spatio-temporal chaos real-time control by means of selected evolutionary techniques,

Fig. 28 Complex network of SOMA dynamics used for Fig. 27, according to the Strategy 1 (see Sect. 3.1)

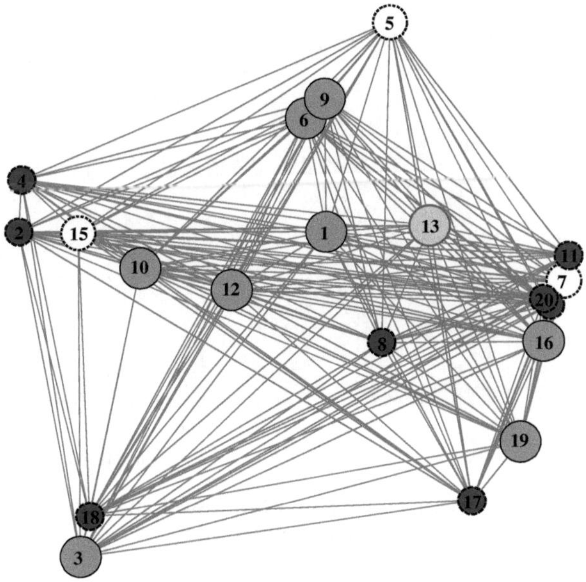

done in our previous research, see for full versions and references [20]. Real-time like behavior is specially defined and simulated with spatiotemporal chaos model based on mutually nonlinearly joined n equations, so called Coupled Map Lattices—CML. In total five evolutionary algorithms has been used for chaos control here: differential evolution, self-organizing migrating algorithm, genetic algorithm, simulated annealing and evolutionary strategies in total of 15 versions. For modeling of spatiotemporal chaos behavior, so called coupled map lattices were used based on logistic equation to generate chaos. The main aim of this investigation was to show that evolutionary algorithms, under certain conditions, are capable of control of CML deterministic chaos, when the cost function is properly defined as well as parameters of selected evolutionary algorithm. Investigation consists of four different case studies with increasing simulation complexity. For all algorithms each simulation was 100 times repeated to show and check robustness of used methods. All data were processed and used in order to get summarizing results and graphs.

6.2 Cost Function

The fitness (cost function) has been calculated according to using the distance between desired CML state and actual CML output, Eq. (2). The minimal value of this cost function, guarantee the best solution, is 0. The aim of all simulations was to find the best solution, i.e. a solution that returns the cost value 0. This cost function was used for reported case studies in [20], the first two case studies

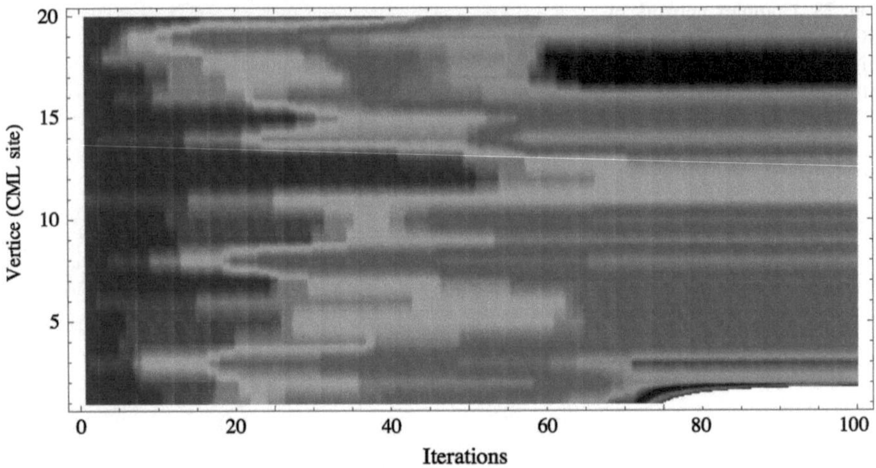

Fig. 29 Another example of CML visualization of SOMA dynamics with 20 individuals, according to the Strategy 1 (see Sect. 3.1)

Fig. 30 Selected Leaders of SOMA, compare Fig. 29 from Iteration (Migration) 140. Growth of the complexity has finished. Since this iteration individual No. 1 is getting all connections, according to the Strategy 1 (see Sect. 3.1)

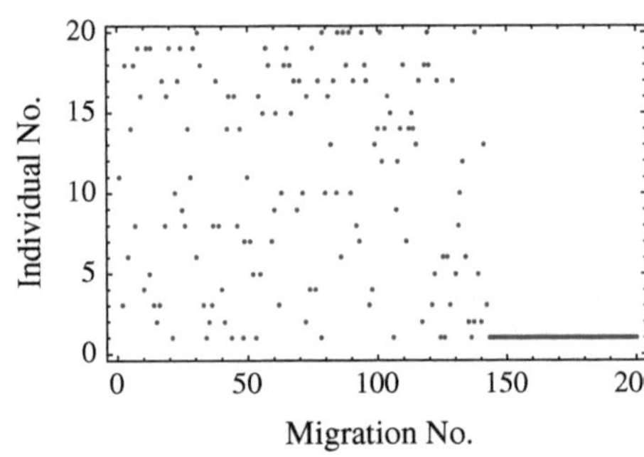

(pinning values setting, pinning sites setting). In the next (last) two case studies cost function (3) was used. It is synthesized from cost function (2) so that two penalty terms are added. The first one, p_1, represents number of used pinning sites in CML. The second one, p_2, is added here to attract attention of evolutionary process on main part of cost function. If this would not be done, then mainly p_1 would be optimized and results should not be acceptable (proved by simulations). Indexes i and j are coordinates of lattice element, i.e. $CML_{i,j}$ is ith site (equation) in jth iteration. For all simulations of T_1S_1 was set stabilized state to $S_1 = 0.75$, and for T_1S_2 to period $S_2 = (0.880129, 0.536537)$, i.e. CML behavior was controlled to this state.

Preliminary knowledge about complexity and variability of used cost function is usually very important. See for example Fig. 32 that shows cost function surface, which is highly erratic and a lot of classical algorithms would fail there. Thus such knowledge can be important when class of optimizing algorithms is selected. A few ideas and examples has been selected here to show complexity and its

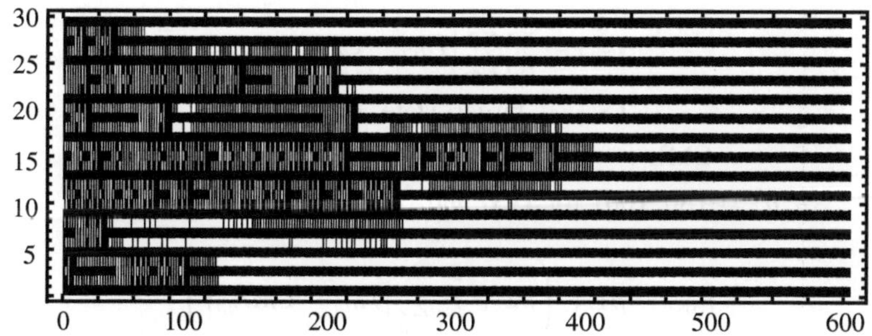

Fig. 31 CML T_1S_2 in configuration 30×600—stabilization after 400 iterations is visible

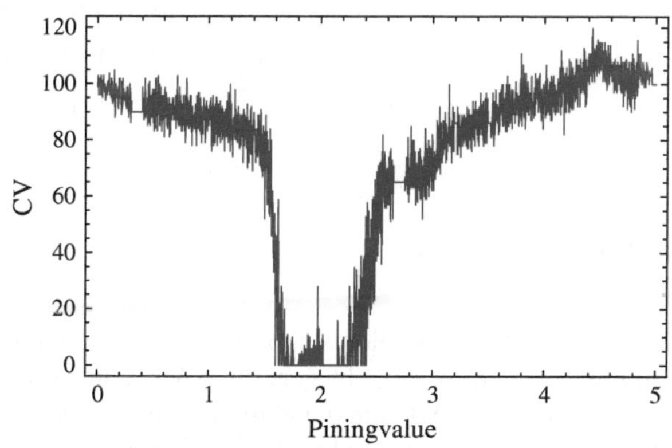

Fig. 32 Landscape of Eq. (2) for T_1S_2 in configuration 30×600. Comparing with landscapes for T_1S_1 is this much more complex, see also [20]

dependance on chaotic system parameter setting. How complex can be such a cost function is clearly visible from Fig. 31, for more see [20]. It is clearly visible, that cost function is partly chaotic and for certain pining value global minimum representing stabilization is accessible. Intensity of chaos of such a graphical representation is caused by fact that calculations are based on chaotic system. If average value (over many of such runs) would be calculated, then we would get smooth graphs however, because our simulations were running on single run, not over a lot of them, are Fig. 32 real representation of landscape of our cost function. Over such or similar landscape were running our simulations. It is also important to note, that for each simulation of CML can exact shape and its intensity of chaos slightly be different from previous one, due to the sensitivity of initial conditions. Another more complex visualizations of landscape of Eq. (2) are depicted in [20]. It is clear that complexity of used cost function is of a high level, despite simple mathematical description. Also suitable stabilizing combination of control parameters depend on number of CML iterations and configuration ($T_1S_{1,2}$) of stabilized state.

$$f_{\cos t} = \sum_{i=1}^{30} \sum_{j=a}^{b} \left| TS_{i,j} - CML_{i,j} \right|^2$$
$$\begin{aligned} &\mathrm{TS_{i,j}} - \text{target state of CML} \\ &\mathrm{CML_{i,j}} - \text{actual state of controlled CML} \\ &\{a,b\} = \{80,100\} \text{ for } \mathrm{T_1S_1} \text{ and } \{\mathrm{a,b}\} = \{580,600\} \text{ for } \mathrm{T_1S_2} \end{aligned} \tag{2}$$

$$f_{\cos t} = p1 + \left(p2 + \sum_{i=1}^{30} \sum_{j=a}^{b} \left| TS_{i,j} - CML_{i,j} \right| \right)^2$$
$$\begin{aligned} &\mathrm{TS_{i,j}} - \text{target state of CML} \\ &\mathrm{CML_{i,j}} - \text{actual state of controlled CML} \\ &p1 - \text{number of actually selected pinning sites} \\ &p2 - 100, \text{ heuristically set weight constant} \\ &\{a,b\} = \{80,100\} \text{ for } \mathrm{T_1S_1} \text{ and } \{\mathrm{a,b}\} = \{580,600\} \text{ for } \mathrm{T_1S_2} \end{aligned} \tag{3}$$

6.3 Case Studies

The class of CML problems chosen for this comparative study was based on case studies reported in [36]. In general, CML control means setting of such pinning sites (controlled CML sites) and their pining values (control values) so that system stabilizes itself on expected spatiotemporal pattern. CML as an object of study was chosen because it shows chaotic behavior and its level of complexity can be quite rich.

All simulations designed and reported here are based on previous simulations, like [58–60]. In the previous simulations has been found that EAs are capable to control CML chaos. Some of them were modified (cost functions was redefined) to increase speed (i.e. number of cost function evaluations) of simulations. To highlight impact of proposed changes in this chapter, we have used all five evolutionary algorithms to control CML, size of 10 inputs, see Figs. 33 and 34.

Comparing to simulations described further, this simulation was defined so that 20 unknown parameters has been estimated. The reason, why exactly 20, is simple. CML size was 10 and EAs estimated which pining site (10 inputs of CML) shall be used and what pining value (10 control signals) will be applied to each input. Thus evolutionary search has run in the 20 dimensional solution space. Based on informations in [36] and previous experiences [58, 59] cost function (3) has been used and empirically has been discovered that cost value ≤ 5 should guarantee stabilized CML (at least in our realization in *Mathematica* code).

We have found, that when EAs stop above cost value ≤ 5.1, then CML is stabilized in almost all cases between 300–600 iterations. To safely stabilize CML

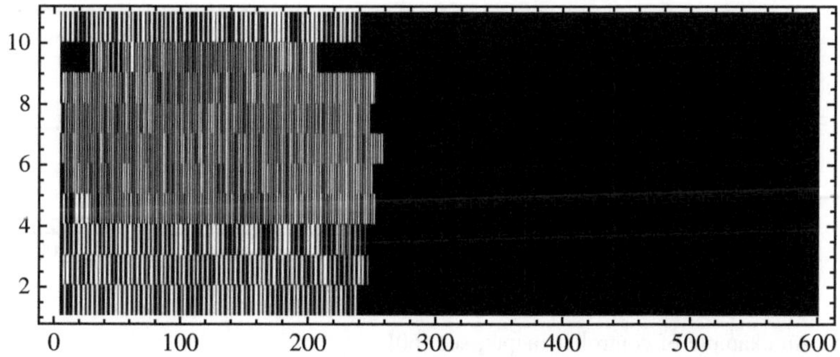

Fig. 33 An example from 1500 simulated CML behavior. Stabilization to T_1S_1 has been reached after 250 iterations

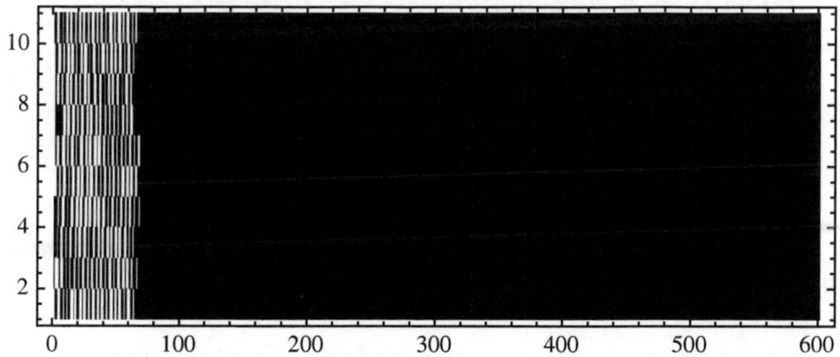

Fig. 34 Another example from 1500 simulated CML behavior. Stabilization to T_1S_1 has been reached before 100 iterations

before 100 iterations, it is enough when EAs stop below this level, like for example cost value ≤ 5.00001. Thus it is quite critical, what stopping cost value is selected.

In [20] are summarized cost function needed to find stabilizing combination of all 20 parameters. It is visible, that number of cost function evaluations needed to reach solution was 3964 in the average. All those results are valid for CML of 10 inputs only, however, more often one can use CML with 50, 100 or more inputs and in such a case search algorithms would search in really high dimensional space. Expected cost function evaluations would be much more higher.

To improve performance and speed of simulations, two modifications are suggested here. Number of used pining sites is omitted, only period of used pining site is estimated (i.e. **only one variable instead of *n* variables**) which means that only each *n*th site is used to apply pining value. Pining value is estimated in the same manner. Only one value is estimated and then applied to each *n*th pining site. In such a case, problem of generally *n* dimensional problem (*n* can be 20, 50, 100,...) is reduced only to search in the 2D solution space.

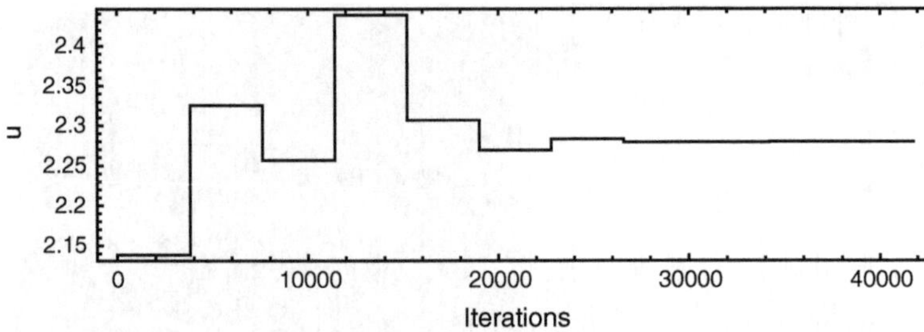

Fig. 35 An example of controller output, see [60]

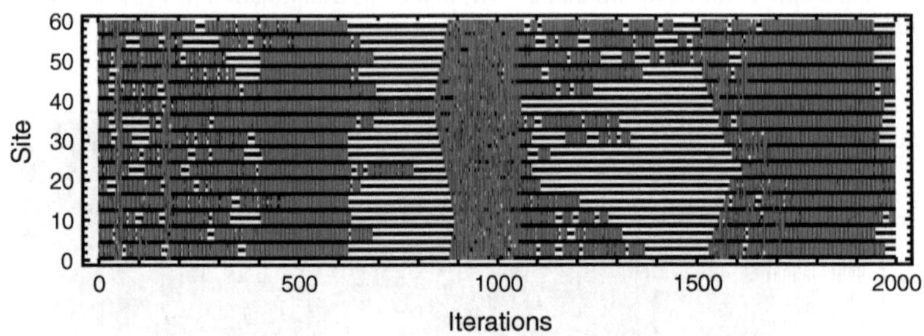

Fig. 36 Partial stabilization of realtime CML in T_1S_2 pattern—pattern is temporarily stabilized, see [60]

Selected modifications has improved performance of selected algorithms, as reported in following section. Investigation consists of four parts in increasing order from complexity point of view and was based on paper [61]. The first one is focused on pining values estimation for *a priori* given pinning sites. In the second one pinning sites with *a priori* given pinning values were estimated by EAs. The third simulation was enlargement of the previous simulation—EAs were used to find minimal number of pinning sites and the fourth simulation was focused on mutual estimation of pinning sites and values, i.e. EA was searching for the minimal number of pinning sites and optimal (as much as possible) pinning values. All simulations were based on the same model and 100 times repeated for each EA with new initial conditions for each simulation. Simulations were done for two basic CML stabilized configuration—T_1S_1 (CML is stabilized on period Time $= 1$ and Space $= 1$, i.e. after stabilization is CML as reported in [20], in total $4 \times 2 \times 1500 = 12000$ independent simulations ($4\times$ $T_1S_{1,2}$, 15 algorithms, each for 100 independent runs) of spatiotemporal CML were carried out.

Next logical step is to control CML in real time, that is the main aim of this process, because it shall allow to us to control complex networks dynamics as well as dynamics of used evolutionary algorithm. An example of real time control of CML by means of evolutionary algorithms is reported in [59] and extended modified study in [60]. Capability of EAs on such a black-box processes control has been demonstrated there. Comparing to non realtime CML was in [59]

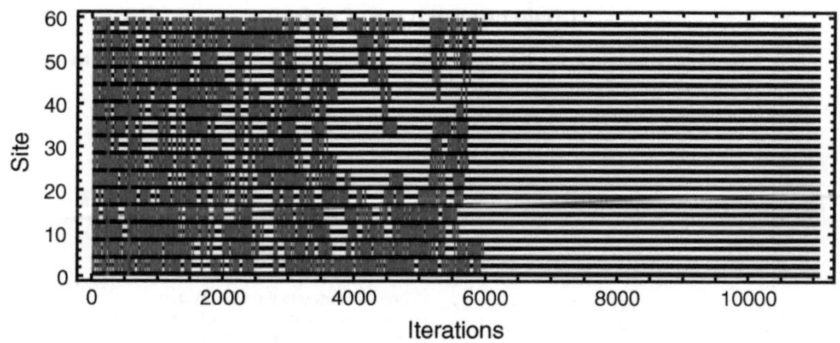

Fig. 37 Succesfull stabilization of realtime CML in T_1S_2 pattern—pattern is permanently stabilized, see [60]

simulated realtime in such a way that each configuration (individual) has been applied after n iterations without possibility to start from initial conditions again. For demonstrative purposes are three figures depicted here. Figure 35 show typical of controller output developed during evolution. Each change of its value is related to another, better individual, whose parameter (or one of them) was estimated controlled output. Figures 36 and 37 show process of T_1S_2 stabilization. For more exact information, it is recommended to see [60].

6.4 Mutual Intersections and Open Problems

Based on previous informations, is can be stated that a numerous unsolved research items is still open. Some of them are for example:

- Use control techniques to control CML and EAs dynamics, as mentioned above.
- Convert to the complex net and CML another, till now not used, evolutionary algorithms.
- Use our proposed philosophy evolutionary dynamics → complex network → CML system → control CML to verify-measure presence of chaos in used EA and/or in complex network.
- Study relations between chaotic regimes in EA dynamics and structure of related complex network.
- Study relations between previous item and fractal dimensions.
- ...

All this problems, symbolically depicted at Fig. 38, are now in process of research of our group.

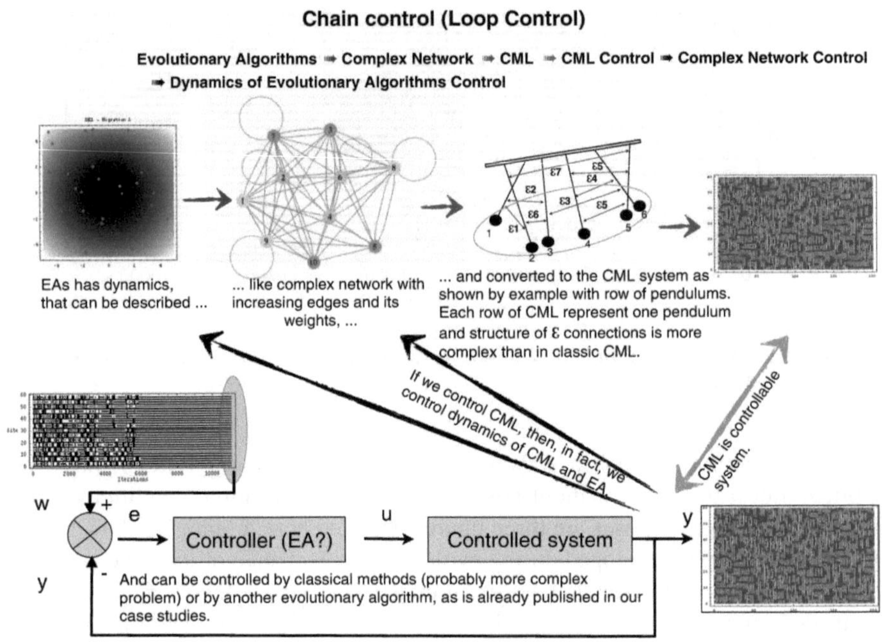

Fig. 38 Proposed scheme of complex networks and evolutionary dynamic control

7 Conclusion

The main motivation of research reported in this chapter is whether it is possible to visualize and simulate underlying dynamics of an evolutionary process as a complex network. Based on preliminary results (based only on 2 algorithms in 10 versions and 16 test function out of 17) it can be stated that number of generations has such impact that occurrence of the complex network structure (CNS) sensitively depends on the number of generations. If the number of generations was small, then no CNS was established. This effect can be easily understand so that low number of generations means that EAs has no time, long enough, to establish CNS. During our experiments has been observed that the moment of CNS establishing depend on cost function dimension, population size, used algorithm and cost function. Very generally, EAs searching for global extreme is on the beginning quite random-like and when domain of global extreme is discovered, then CNS is quite quickly established.

Impact of dimensionality on CNS forming has been observed when the dimension of the cost function was big and number of generations was too low, the selected EA was not able to finish successfully the global extreme search not all connections had been properly established. Thus if high dimensional cost functions are used, then number of generations has to be selected so that at least domain of the global extreme is found. On the other side, if number of generations (or

Migrations in the case of the SOMA algorithm) if very big, then it is possible observe effect that rich become to be richer, i.e. one vertex (individual) become to be winner repeatedly. This moment usually means that global extreme has been found and further searching is not necessary.

Population size: Comparing to results reported in [20] CNS forming was observed usually from population size of 10 and more individuals for dimensions 5 and more. Again, it is parameter, which does not influent CNS forming alone, but in the combination with another parameters.

Used algorithm: CNS forming has also been clearly observed with algorithms, that are more or less based on swarm philosophy or partly associated with it. For example DERand1Bin did not show any CNS formatting (in principle each individual is selected to be parent), see [21], while in the case of the DELocalToBest (Fig. 10) in which the best solution in the population play an important role, CNS has been observed, as well as in the SOMA strategies, see [21]. The conclusion reached is that CNS formatting is more likely observable with swarm like algorithms rather than randomly driven algorithms. We think that this is quite logical and close to the idea of preferred linking in the complex networks modeling social behavior (citation networks, etc).

Parallelization: Part of our research is also focused on parallel evolutionary techniques and analysis of its dynamics as of complex networks. Simple implementation based on MRDEAF and visualized at Fig. 20. It is shown network which describes evolution of 5 subpopulations of SOMA algorithm. The nodes are clustered according to their subpopulation index. Color of the nodes indicates the fitness value (the darker the better) and node size is a function of inWeights parameter. Edges which connect nodes from different clusters represent migrating individuals. It is clear demonstration that also dynamics of parallel population based algorithms can be visualized like complex networks.

As noted in the Sect. 6, [20, 36] it is possible to control CML by means of classical as well as evolutionary methods. So the last step of our proposed investigation *evolutionary dynamics* $\rightarrow$ *complex network* $\rightarrow$ *CML system* $\rightarrow$ *control CML* $\rightarrow$ *control evolutionary dynamics* is **control evolutionary dynamics**, see Fig. 38. This shall be possible benefit of this approach. Based on numerically demonstrated fact (no mathematical proof has been made) that EAs dynamics can be visualized like complex networks we believe that there is new research area for study of EAs dynamics and its possible control via techniques of complex network control [55].

Acknowledgments The following two grants are acknowledged for the financial support provided for this research: Grant Agency of the Czech Republic—GACR 13-08195S, by the Development of human resources in research and development of latest soft computing methods and their application in practice project, reg. no. CZ.1.07/2.3.00/20.0072 funded by Operational Programme Education for Competitiveness, co-financed by ESF and state budget of the Czech Republic.

References

1. S.N. Dorogovtsev, J.F.F. Mendes, Evolution of networks. Adv. Phys. **51**, 1079 (2002)
2. S. Boccaletti et al., Complex networks: structure and dynamics. Phys. Rep. **424**, 175–308 (2006)
3. A. Turing, Intelligent Machinery, Unpublished Report for National Physical Laboratory. in *Machine Intelligence*, vol. 7, ed. by D. Michie (1969) [A.M. Turing (ed.), The Collected Works, vol. 3 (Ince D. North-Holland, Amsterdam, 1992)]
4. J. Holland, *Adaptation in Natural and Artificial Systems* (University of Michigan Press, Ann Arbor, 1975)
5. H. Schwefel, Numerische Optimierung von Computer-Modellen, PhD thesis (1974), Reprinted by Birkhauser (1977)
6. I. Rechenberg, Evolutionsstrategie–Optimierung technischer Systeme nach Prinzipien der biologischen Evolution, PhD thesis (1971), Printed in Fromman-Holzboog (1973)
7. D.B. Fogel, Unearthinga fossil from the history of evolutionary computation. Fundamenta Informaticae **35**(1–4), 116 (1998)
8. H. Richter, K.J. Reinschke, Optimization of local control of chaos by an evolutionary algorithm. Physica D **144**, 309–334 (2000)
9. H. Richter, An evolutionary algorithm for controlling chaos: the use of multi-objective fitness functions, in *Parallel Problem Solving from Nature-PPSN VII*, ed. by M. Guervos, J.J. Panagiotis, A. Beyer, F.H.G. Villacanas, J.L. Schwefel, H.P. Schwefel, Lecture Notes in Computer Science, vol. 2439 (Springer, Berlin, 2002), pp. 308–317
10. H. Richter, in *Evolutionary Optimization in Spatio-temporal Fitness Landscapes*, Lecture Notes in Computer Science. NUMB 4193 (Springer, 2006), pp. 1–10, ISSN 0302–9743
11. H. Richter, A study of dynamic severity in chaotic fitness landscapes, evolutionary computation. IEEE Congr. **3**(2–5), 2824–2831 (2005)
12. I. Zelinka, G. Chen, S. Celikovsky, Chaos Synthesis by means of evolutionary algorithms. Int. J. Bifurcat. Chaos **18**(4), 911–942 (2008)
13. I. Zelinka, Real-time deterministic chaos control by means of selected evolutionary algorithms. Eng. Appl. Artif. Intell. (2008). doi:10.1016/j.engappai.2008.07.008
14. I. Zelinka, Investigation on realtime deterministic chaos control by means of evolutionary algorithms. in *1st IFAC Conference on Analysis and Control of Chaotic Systems* (Reims, France, 2006)
15. R. Senkerik, I. Zelinka, E. Navratil, Optimization of feedback control of chaos by evolutionary algorithms. in *1st IFAC Conference on Analysis and Control of Chaotic Systems* (Reims, France, 2006)
16. Y. Dashora et al., Improved and generalized learning strategies for dynamically fast and statistically robust evolutionary algorithms. Eng. Appl. Artif. Intell. (2007). doi:10.1016/j.engappai.2007.06.005
17. L. Li, L. Wenxin, A.C. David, Particle swarm optimization-based parameter identification applied to permanent magnet synchronous motors. Eng. Appl. Artif. Intell. (2007). doi:10.1016/j.engappai.2007.10.002
18. K. Price, *An Introduction to Differential Evolution, New Ideas in Optimization*, ed. by D. Corne, M. Dorigo, F. Glover (McGraw-Hill, London, UK, 1999), pp. 79–108
19. I. Zelinka, SOMA self organizing migrating algorithm Chapter 7. in *New Optimization Techniques in Engineering*, ed. by B.V. Babu, G. Onwubolu (Springer, 2004), p. 33. ISBN 3-540-20167X
20. I. Zelinka, S. Celikovsky, H. Richter, G. Chen (eds.), *Evolutionary Algorithms and Chaotic Systems* (Springer, Germany, 2010), p. 550s
21. I. Zelinka, D. Davendra, M. Chadli, R. Senkerik, T.T. Dao, L. Skanderova, in *Evolutionary Dynamics and Complex Networks*, ed. by I. Zelinka, V. Snasel, A. Ajith. Handbook of Optimization (Springer, Germany, 2012), p. 1100s

22. J. Dean, S. Ghemawat, MapReduce: simplified data processing on large clusters. Commun. ACM **51**(1), 107113 (2008)
23. J. Dean, S. Ghemawat, MapReduce: a flexible data processing tool. Commun. ACM **53**(1), 7277 (2010)
24. S. Ghemawat, H. Gobioff, S.T. Leung, The Google file system. ACM SIGOPS Oper. Syst. Rev. **37**, 2943 (2003)
25. D. Borthakur, The Hadoop distributed file system: architecture and design. Hadoop Project Website (2007)
26. A. Bialecki, M. Cafarella, D. Cutting, O. O'Malley, Hadoop: a framework for running applications on large clusters built of commodity hardware (2005). http://lucene.apache.org/Hadoop
27. C. Jin, C. Vecchiola, R. Buyya, MRPGA: an extension of mapreduce for parallelizing genetic algorithms. in *Fourth IEEE International Conference on eScience*, 2008, pp. 214–221
28. A. Verma, X. Llora, D.E. Goldberg, R.H. Campbell, Scaling genetic algorithms using mapreduce. in *Ninth International Conference on Intelligent Systems Design and Applications ISDA'09*, 2009, pp. 13–18
29. D. Logoftu, D. Dumitrescu, Parallel evolutionary approach of compaction problem using mapreduce. in *Parallel Problem Solving from Nature PPSN XI* (Springer, Berlin, 2011), pp. 361–370
30. C. Zhou, Fast parallelization of differential evolution algorithm using mapreduce. in *Proceedings of the 12th Annual Conference on Genetic and, Evolutionary Computation*, 2010, pp. 1113–1114
31. M. Pavlech, Distributed SOMA algorithm in mapreduce framework. in *Proceedings of the International Masaryk Conference*, 2010, p. 1380
32. A.W. McNabb, C.K. Monson, K.D. Seppi, Parallel pso using mapreduce. in *IEEE Congress on Evolutionary Computation CEC 2007*, 2007, p. 714
33. E. Cant-Paz, A survey of parallel genetic algorithms, Calculateurs paralleles, reseaux et systems repartis (Citeseer) 10(2), 141–171 (1998)
34. The GraphML File Format, http://graphml.graphdrawing.org/
35. The Open Graph Viz Platform, http://gephi.org/
36. H. Schuster, *Handbook of Chaos Control* (Wiley/Wiley-Interscience, New York, 2002)
37. E. Lorenz, Deterministic nonperiodic flow. J. Atmos. Sci. **20**(2), 130141 (1963)
38. I. Stewart, The Lorenz attractor exists. Nature **406**, 948949 (2000)
39. R. May, Simple mathematical model with very complicated dynamics. Nature **261**, 4567 (1976)
40. R. Gilmore, M. Lefranc, *The Topology of Chaos: Alice in Stretch and Squeezeland* (Wiley, New York, 2002)
41. G. Chen, X. Dong, *From Chaos to Order: Methodologies, Perspectives and Applications* (World Scientific, Singapore, 1998)
42. X. Wang, G. Chen, Chaotification via arbitrarily small feedback controls: theory, method, and applications. Int. J. Bifur. Chaos **10**, 549570 (2000)
43. E. Ott, C. Grebogi, J. Yorke, Controllingchaos. Phys. Rev. Lett. **64**, 11961199 (1990)
44. C. Grebogi, Y.C. Lai, Controlling Chaos, in *Handbook of Chaos Control*, ed. by H. Schuster (Wiley, New York, 1999)
45. Y. Zou, X. Luo, G. Chen, Pole placement method of controlling chaos in DC-DC buck converters. Chin. Phys. **15**, 1719–1724 (2006)
46. W. Just, in *Principles of Time Delayed Feedback Control*, ed. by H. Schuster, Handbook of Chaos Control (Wiley, New York, 1999)
47. W. Just, H. Benner, E. Reibold, Theoretical and experimental aspects of chaos control by time-delayed feedback. Chaos **13**, 259–266 (2003)
48. M. Deilami, C. Rahmani, M. Motlagh, *Control of Spatio-temporal on-off Intermittency in Random Driving Diffusively Coupled Map Lattices* (Chaos, Solitons Fractals, 2007)
49. G. Chen, *Controlling Chaos and Bifurcations in Engineering Systems* (CRC Press, Boca Raton, 2000)

50. H. Richter, K. Reinschke, Optimization of local control of chaos by an evolutionary algorithm. Physica D **144**, 309–334 (2000)
51. H. Richter, An evolutionary algorithm for controlling chaos: the use of multi-objective fitness functions. in *PPSN*, ed. by J.J.M. Guervos, P.A. Adamidis, H.G. Beyer, J.-L. Fernandezsps Villacanas, H.P. Schwefel, LNCS. vol. 2439 (Springer, Heidelberg, 2002), pp. 308–317
52. I. Zelinka, Investigation on real-time deterministic chaos control by means of evolutionary algorithms. in *Proceedings of First IFAC Conference on Analysis and Control of Chaotic Systems* (Reims, France, 2006), pp. 211–217
53. R. Hilborn, *Chaos and Nonlinear Dynamics* (Oxford University Press, Oxford, 1994)
54. Q. He, L. Wang, An effective co-evolutionary particle swarm optimization for constrained engineering design problems. Eng. Appl. Artif. Intell. **20**(1), 8999 (2007)
55. S.P. Meyn, *Control Techniques for Complex Networks* (Cambridge University Press, Cambridge, 2007)
56. I. Zelinka, D. Davendra, V. Snasel, R. Jasek, R. Senkerik, Z. Oplatkova, Preliminary Investigation on Relations Between Complex Networks and Evolutionary Algorithms Dynamics. in *CISIM* (Poland, 2010)
57. I. Zelinka, D. Davendra, R. Enkek, Do evolutionary algorithm dynamics create complex network structures? Complex Syst. 20(2), 127–140 (2011). ISSN 0891–2513
58. I. Zelinka, *Investigation on Evolutionary Deterministic Chaos Control* (IFAC, Prague, 2005)
59. I. Zelinka, Investigation on Evolutionary Deterministic Chaos Control–Extended Study. in *19th International Conference on Simulation and Modeling (ECMS 2005)*, Riga, Latvia, 1–4 June 2005
60. I. Zelinka, Real-time deterministic chaos control by means of selected evolutionary algorithms. Eng. Appl. Artif. Intell. doi:10.1016/j.engappai.2008.07.008
61. G. Hu, F. Xie, J. Xiao, J. Yang, Z. Qu, Control of Patterns and Spatiotemporal Chaos and its Application. in *Handbook of Chaos Control*, ed. by H.G. Schuster (Wiley, New York, 1999)
62. M. Molga, C. Smutnicky, Test functions for optimization needs (2005) www.zsd.ict.pwr.wroc.pl/files/docs/functions.pdf

Influence of Chaotic Dynamics on the Performance of Differential Evolution Algorithm

Roman Senkerik, Donald Davendra, Ivan Zelinka and Zuzana Oplatkova

Abstract This paper outlines the extended investigations on the concept of a chaos driven Differential Evolution. The focus of this paper is the embedding of chaotic systems in the form of chaos number generator for Differential Evolution. The chaotic systems of interest are the discrete dissipative systems. Three chaotic systems were selected as possible chaos number generators for Differential Evolution. Repeated simulations were performed on the set of six basic benchmark functions. Finally, the obtained results are compared with canonical Differential Evolution.

1 Introduction

During the recent years, usage of new intelligent systems in engineering, technology, modeling, computing and simulations has attracted the attention of researchers worldwide. The most current methods are mostly based on soft computing, which is a discipline tightly bound to computers, representing a set of methods of special algorithms, belonging to the artificial intelligence paradigm. The most popular of these methods are neural networks, evolutionary algorithms, fuzzy logic, and genetic programming. Presently, evolutionary algorithms are known as a powerful set of tools for almost any difficult and complex optimization problem. Ant Colony (ACO), Genetic Algorithms (GA), Differential Evolution (DE), Particle Swarm Optimization (PSO) and Self Organizing Migration Algorithm (SOMA) are some of the most potent heuristics available.

R. Senkerik (✉) · Z. Oplatkova
Faculty of Applied Informatics, Tomas Bata University in Zlin,
Nam T.G. Masaryka 5555 76001 Zlin, Czech Republic
e-mail: senkerik@fai.utb.cz

D. Davendra · I. Zelinka
Faculty of Electrical Engineering and Computer Science,
Technical University of Ostrava, 17. listopadu 15 70833
Ostrava-Poruba, Czech Republic

I. Zelinka et al. (eds.), *How Nature Works*, Emergence,
Complexity and Computation 5, DOI: 10.1007/978-3-319-00254-5_12,
© Springer International Publishing Switzerland 2014

Recent studies have shown that Differential Evolution (DE) [1] has been used for a number of optimization tasks, [2], [3] has explored DE for combinatorial problems, [4] has hybridized DE whereas [5]–[7] has developed self-adaptive DE variants.

This paper is aimed at investigating the chaos driven DE. Although a several of papers have been recently focused on the connection of DE and chaotic dynamics either in the form of hybridizing of DE with chaotic searching algorithm [8] or in the form of chaotic mutation factor and dynamically changing weighting and crossover factor in self-adaptive chaos differential evolution (SACDE) [9], the focus of this paper is the embedding of chaotic systems in the form of chaos number generator for DE and its comparison with the canonical DE.

This research is an extension and continuation of the previous initial application based experiment with chaos driven DE [10].

The primary aim of this work is not to develop a new type of random number generator, which should pass many statistical tests, but to try to use and test the implementation of natural chaotic dynamics into evolutionary algorithm as a chaotic random number generator.

The chaotic systems of interest are discrete dissipative systems. Three different chaotic systems were selected as the chaos number generator for DE.

Firstly, Differential Evolution is explained. The next sections are focused on the description of used chaotic systems and benchmark test functions. Results and conclusion follow afterwards.

This work presents the extension and summarization of initial results presented in [11] and [12].

2 Differential Evolution

DE is a population-based optimization method that works on real-number-coded individuals [1]. A schematic is given in Fig. 1.

There are essentially five sections to the code depicted in Fig. 1. Section 1 describes the input to the heuristic. D is the size of the problem, G_{max} is the maximum number of generations, NP is the total number of solutions, F is the scaling factor of the solution and CR is the factor for crossover. F and CR together make the internal tuning parameters for the heuristic.

The Section 2 (See Fig. 1) outlines the initialization of the heuristic. Each solution $x_{i,j,G=0}$ is created randomly between the two bounds $x^{(lo)}$ and $x^{(hi)}$. The parameter j represents the index to the values within the solution and i indexes the solutions within the population. So, to illustrate, $x_{4,2,0}$ represents the fourth value of the second solution at the initial generation.

After initialization, the population is subjected to repeated iterations in section 3.

Section 4 describes the conversion routines of DE. Initially, three random numbers r_1, r_2, r_3 are selected, unique to each other and to the current indexed solution i in the population in 4.1. Henceforth, a new index j_{rand} is selected in the

1. Input: $D, G_{\max}, NP \geq 4, F \in (0,1+), CR \in [0,1]$, and initial bounds: $\vec{x}^{(lo)}, \vec{x}^{(hi)}$.

$$\text{2. Initialize: } \begin{cases} \forall i \leq NP \wedge \forall j \leq D : x_{i,j,G=0} = x_j^{(lo)} + rand_j[0,1] \bullet (x_j^{(hi)} - x_j^{(lo)}) \\ i = \{1,2,...,NP\},\ j = \{1,2,...,D\},\ G = 0,\ rand_j[0,1] \in [0,1] \end{cases}$$

3. While $G < G_{\max}$

$\forall i \leq NP$:

4. Mutate and recombine:

4.1 $r_1, r_2, r_3 \in \{1,2,....,NP\}$, randomly selected, except: $r_1 \neq r_2 \neq r_3 \neq i$

4.2 $j_{rand} \in \{1,2,...,D\}$, randomly selected once each i

$$\text{4.3 } \forall j \leq D, u_{j,i,G+1} = \begin{cases} x_{j,r_3,G} + F \cdot (x_{j,r_1,G} - x_{j,r_2,G}) \\ \qquad \text{if } (rand_j[0,1] < CR \vee j = j_{rand}) \\ x_{j,i,G} \quad \text{otherwise} \end{cases}$$

5. Select

$$\vec{x}_{i,G+1} = \begin{cases} \vec{u}_{i,G+1} & \text{if } f(\vec{u}_{i,G+1}) \leq f(\vec{x}_{i,G}) \\ \vec{x}_{i,G} & \text{otherwise} \end{cases}$$

$G = G + 1$

Fig. 1 DE Schematic

solution. j_{rand} points to the value being modified in the solution as given in 4.2. In 4.3, two solutions, $x_{j,r1,G}$ and $x_{j,r2,G}$ are selected through the index r_1 and r_2 and their values subtracted. This value is then multiplied by F, the predefined scaling factor. This is added to the value indexed by r_3.

However, this solution is not arbitrarily accepted in the solution. A new random number is generated, and if this random number is less than the value of CR, then the new value replaces the old value in the current solution. The fitness of the resulting solution, referred to as a perturbed vector $u_{j,i,G}$., is then compared with the fitness of $x_{j,i,G}$. If the fitness of $u_{j,i,G}$ is greater than the fitness of $x_{j,i,G}$., then $x_{j,i,G}$. is replaced with $u_{j,i,G}$; otherwise, $x_{j,i,G}$. remains in the population as $x_{j,i,G+1}$. Hence the competition is only between the new *child* solution and its *parent* solution.

DE is quite robust, fast, and effective, with global optimization ability. It does not require the objective function to be differentiable, and it works well even with noisy and time-dependent objective functions. Description of the used DERand1Bin strategy is presented in Table 1 together with the description and comparison of the three other most common and used DE strategies. These strategies differ in the way of calculating the perturbed vector $u_{j,i,G}$. Please refer to [1, 13] for the detailed complete description of all other strategies.

Table 1 Description of selected DE Strategies

Strategy	Formulation
DERand1Bin	$u_{j,i,G+1} = x_{j,r1,G} + F \cdot \left(x_{j,r2,G} - x_{j,r3,G}\right)$
DERand2Bin	$u_{j,i,G+1} = x_{j,r5,G} + F \cdot \left(x_{j,r1,G} - x_{j,r2,G} - x_{j,r3,G} - x_{j,r4,G}\right)$
DEBest2Bin	$u_{j,i,G+1} = x_{j,Best,G} + F \cdot \left(x_{j,r1,G} - x_{j,r2,G} - x_{j,r3,G} - x_{j,r4,G}\right)$
DELocalToBest	$u_{j,i,G+1} = x_{j,i,G} + F_{rand}\left(x_{j,Best,G} - x_{j,i,G}\right) + F \cdot \left(x_{j,r1,G} - x_{j,r2,G}\right)$

3 Chaotic Maps

This section contains the description of three discrete chaotic maps used as the random generator for DE. Iterations of the chaotic maps were used for the generation of real numbers in the process of crossover based on the user defined *CR* value and for the generation of the integer values used for selection of solutions (individuals).

3.1 Dissipative Standard Map

The Dissipative Standard map is a two-dimensional chaotic map. The parameters used in this work are $b = 0.1$ and $k = 8.8$ as suggested in [14]. The Dissipative standard map is given in Fig. 2. The map equations are given in Eqs. 1 and 2.

$$X_{n+1} = X_n + Y_{n+1} (\bmod 2\pi) \tag{1}$$

$$Y_{n+1} = bY_n + k \sin X_n (\bmod 2\pi) \tag{2}$$

3.2 Arnold's Cat Map

The Arnold's Cat map is a simple two dimensional discrete system that stretches and folds points (x, y) to $(x + y, x + 2y)$ mod 1 in phase space. The map equations are given in Eqs. 3 and 4. This map uses parameter $k = 2.0$ as suggested in [14]. The $x{-}y$ plot of this map is depicted in Fig. 2.

$$X_{n+1} = X_n + Y_n (\mathrm{mod} 1) \tag{3}$$

$$Y_{n+1} = X_n + kY_n (\mathrm{mod} 1) \tag{4}$$

3.3 Sinai Map

The Sinai map is a simple two-dimensional discrete system similar to the Arnold's Cat map. The map equations are given in Eqs. 5 and 6. The parameter used in this work is $\delta = 0.1$ as also suggested in [14]. The Sinai map is depicted in Fig. 2.

$$X_{n+1} = X_n + Y_n + \delta \cos 2\pi Y_n (\mathrm{mod} 1) \tag{5}$$

$$Y_{n+1} = X_n + 2Y_n (\mathrm{mod} 1) \tag{6}$$

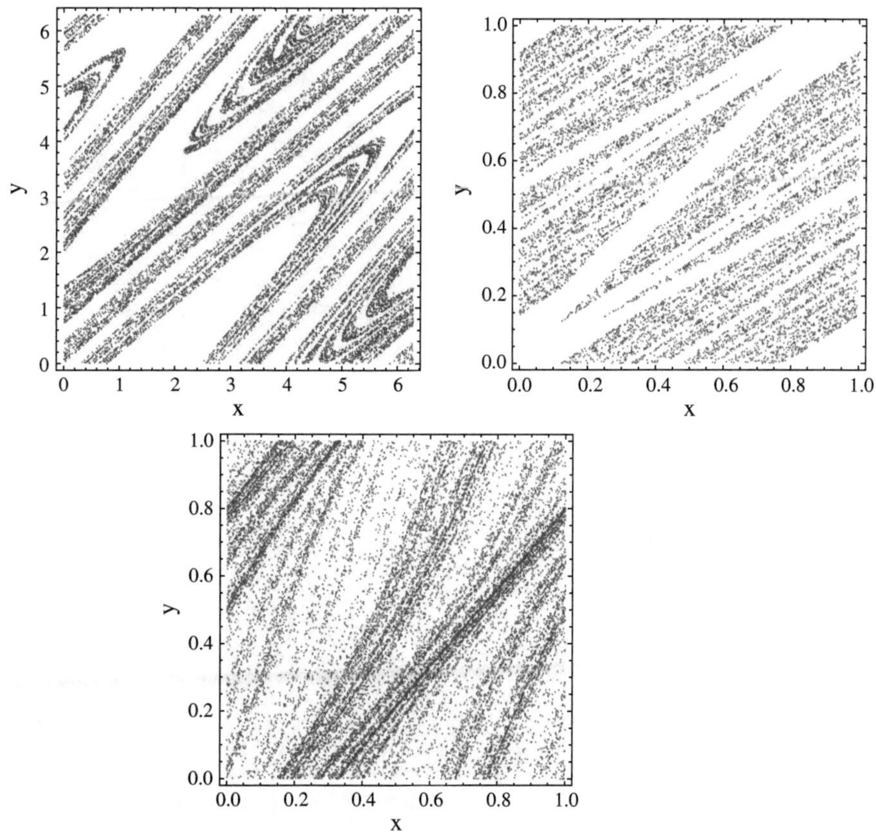

Fig. 2 Dissipative standard map (*upper left*), Arnold's Cat Map (*upper right*), Sinai map (*below*)

4 Benchmark Functions

For the purpose of evolutionary algorithms performance comparison within this research, the following six basic test functions were selected: Schwefel's function (7), Ackley I (8), Ackley II (9), Rastrigin's (10), Stretched V Sine (11) and Griewangk's function (12). The 3D diagrams for $D = 2$ are depicted in Figs. 3, 4, 5, 6, 7 and 8.

$$f(x) = \sum_{i=1}^{D} -x_i \sin\left(\sqrt{|x_i|}\right) \tag{7}$$

$$f(x) = \sum_{i=1}^{D-1} \left(\frac{1}{e^5}\sqrt{(x_i^2 + x_{i+1}^2)} + 3(\cos(2x_i) + \sin(2x_{i+1}))\right) \tag{8}$$

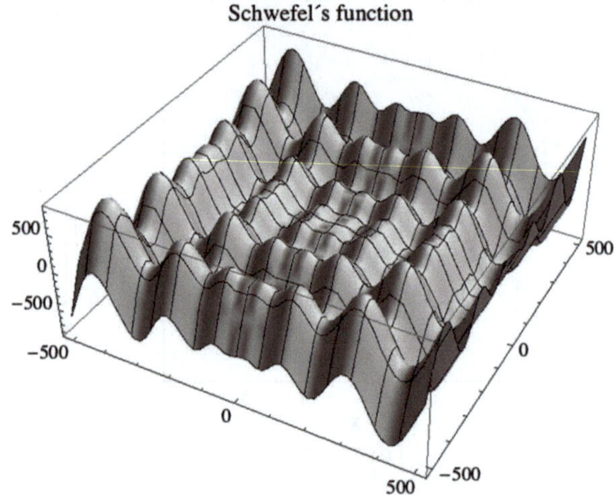

Fig. 3 Schwefel's benchmark function, $D = 2$

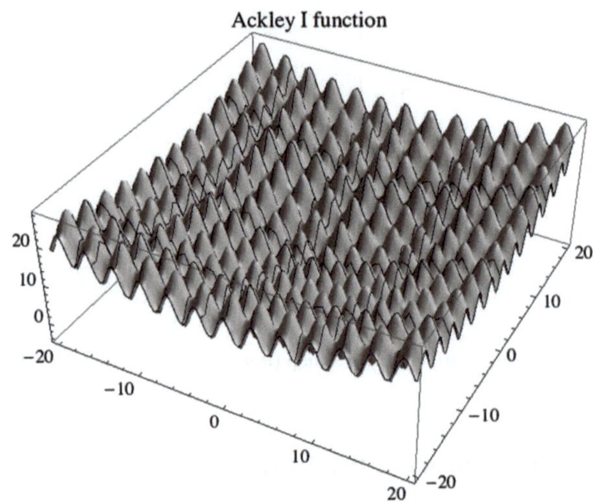

Fig. 4 Ackley I function, $D = 2$

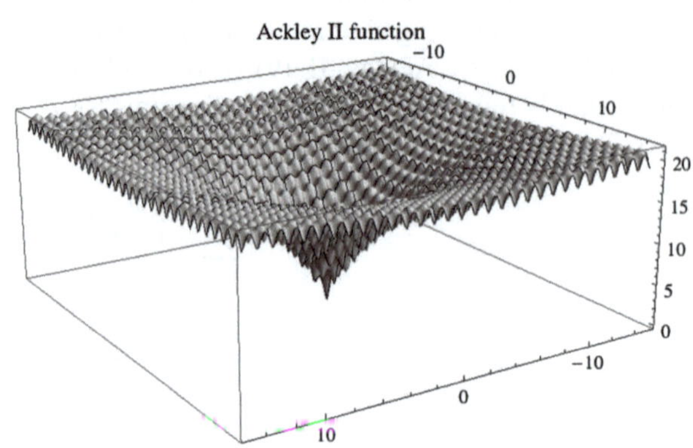

Fig. 5 Ackley II function, $D = 2$

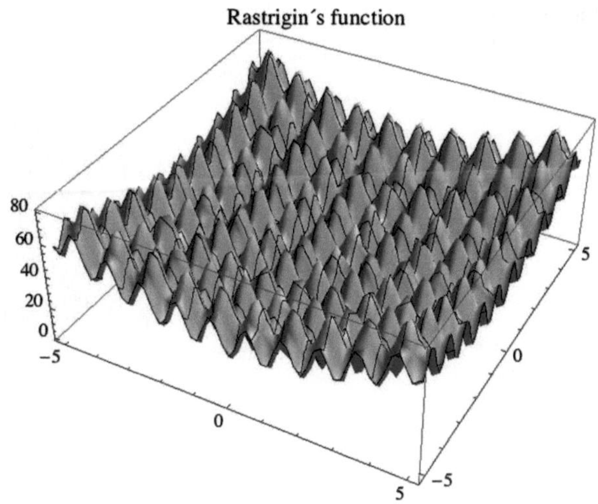

Fig. 6 Rastrigin's function, $D = 2$

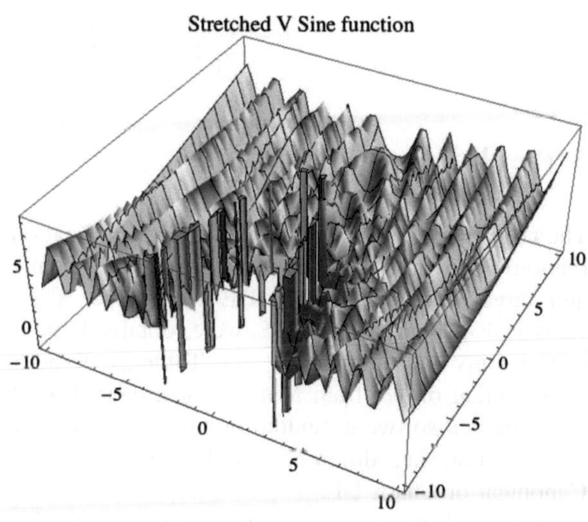

Fig. 7 Stretched V Sine function, $D = 2$

$$\sum_{i=1}^{D-1}\left(20 + e - \frac{20}{e^{0,2\sqrt{\frac{\left(x_i^2+x_{i+1}^2\right)}{2}}}} - e^{0,5(\cos(2\pi x_i)+\cos(2\pi x_{i+1}))}\right) \tag{9}$$

$$f(x) = 10D + \sum_{i=1}^{D} x_i^2 - 10\cos(2\pi x_i) \tag{10}$$

$$\sum_{i=1}^{D-1}\left(\sqrt[4]{\left(x_i^2 + x_{i+1}^2\right)}\sin\left(50\sqrt[10]{\left(x_i^2 + x_{i+1}^2\right)}\right)^2 + 1\right) \tag{11}$$

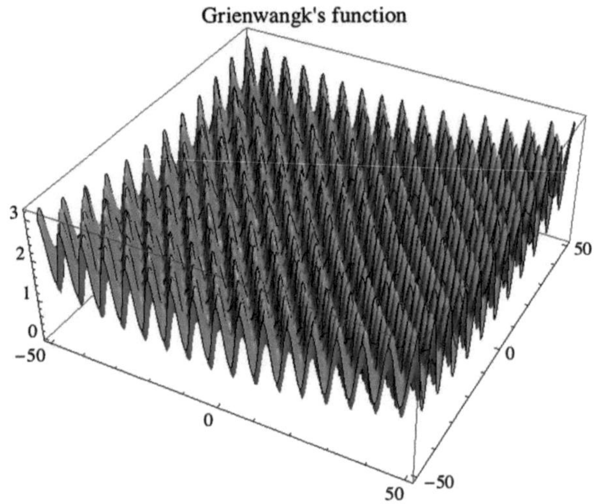

Fig. 8 Griewangk's function, $D = 2$

$$1 + \sum_{i=1}^{D} \frac{x_i^2}{4000} - \prod_{i=1}^{D} \cos\left(\frac{x_i}{\sqrt{i}}\right) \tag{12}$$

5 Results

The novelty of this approach represents the utilization of discrete chaotic maps as a random generator for DE. In this paper, the canonical DE strategy DERand1Bin and three versions of ChaosDE were used. The parameter settings for both canonical DE and ChaosDE were obtained analytically based on numerous experiments and simulations (see Table 2). Experiments were performed in an environment of Wolfram Mathematica, canonical DE therefore used the built-in Mathematica software random number generator. All experiments used different initialization, i.e. different initial population was generated in each run of Canonical or Chaos DE.

Within this research, two experiments were performed. The first one utilizes the maximum number of generations fixed at 200 generations. This allowed the possibility to analyze the progress of DE within a limited number of generations

Table 2 Parameter set up for canonical DE

Parameter	Value—Experiment 1	Value—Experiment 2
Popsize	10D	10D
F	0.8	0.8
Cr	0.8	0.8
Dimensions	2–8	10
Generations	200	∞
Max Cost Function Evaluations (CFE)	4000–16000	∞

and cost function evaluations. In the second case, the number of generations was unlimited. The main observed parameters were the total number of cost function evaluations and the time in seconds required for finding of the global minimum of the used test functions. The stopping criterion was the finding of the real global minimum.

The results of the first experiment are shown in Tables 3, 5, 7, 9, 11 and 13, which represent the average deviations from the known global minimum for 20 repeated runs of the different variants of utilized DE.

Tables 4, 6, 8, 10, 12 and 14 contain the results for the second experiment. These tables show the average time in seconds and number of CFE (Cost Function Evaluations) required for the finding of global minimum for 20 repeated runs of the evolutionary algorithms. The bold values depict the best value.

5.1 Results for the Schwefel's Test Function

Table 3 Average difference from Global minimum for the Schwefel's Function: 2D–8D

Dim.	Known Value	Canonical DE	ChaosDE Dissipative	ChaosDE Arnold Cat	ChaosDE Sinai
2	−837,966	**2,2545.10^{-4}**	**2,2545.10^{-4}**	**2,2545.10^{-4}**	**2,2545.10^{-4}**
4	−1675,932	7,1499.10^{-4}	**5,0084.10^{-4}**	3,4091.10^{-3}	2,0311.10^{-3}
6	−2513,898	228,818	**158,723**	315,082	306,371
8	−3351,864	**800,731**	805,248	872,714	822,857

Table 4 Average CFE and evaluation time for the Schwefel's Function: 10D

	Canonical DE	ChaosDE Dissipative	ChaosDE Arnold Cat	ChaosDE Sinai
CFE	304400	**167100**	340800	263700
Time (s)	162.63	**124.62**	249.58	193.77

5.2 Results for the Ackley I Test Function

Table 5 Average difference from Global minimum for the Ackley I function: 2D–8D

Dim.	Known Value	Canonical DE	ChaosDE Dissipative	ChaosDE Arnold Cat	ChaosDE Sinai
2	−4,5901	**1,6341.10^{-6}**	**1,6341.10^{-6}**	**1,6341.10^{-6}**	**1,6341.10^{-6}**
4	−10,46137	2,39.10^{-3}	**2,32.10^{-3}**	0,01275	0,07054
6	−16,29877	4,3451	**3,31813**	4,68185	4,5361
8	−22,13611	14,1343	**11,7813**	14,5216	14,6120

Table 6 Average CFE and evaluation time for the Ackley I function: 10D

	Canonical DE	ChaosDE Dissipative	ChaosDE Arnold Cat	ChaosDE Sinai
CFE	377400	**222500**	331300	474100
Time (s)	237.598	**172.520**	210.188	371.515

5.3 Results for the Ackley II Test Function

Table 7 Average difference from Global minimum for the Ackley II function: 2D–8D

Dim.	Known Value	Canonical DE	ChaosDE Dissipative	ChaosDE Arnold Cat	ChaosDE Sinai
2	0	$1.8527.10^{-13}$	$2.5419.10^{-13}$	$1.1954.10^{-13}$	$\mathbf{3.7836.10^{-14}}$
4	0	$1.6623.10^{-5}$	$\mathbf{1.219.10^{-5}}$	$1.6262.10^{-5}$	$1.7792.10^{-5}$
6	0	0.01606	**0.01248**	0.02064	0.01578
8	0	0.7009	**0.37949**	0.82246	0.74808

Table 8 Average CFE and evaluation time for the Ackley II function: 10D

	Canonical DE	ChaosDE Dissipative	ChaosDE Arnold Cat	ChaosDE Sinai
CFE	226000	**205700**	244300	242100
Time (s)	174.81	**146.93**	200.45	200.36

5.4 Results for the Rastrigin's Test Function

Table 9 Average difference from Global minimum for the Rastrigin's function: 2D–8D

Dim.	Known Value	Canonical DE	ChaosDE Dissipative	ChaosDE Arnold Cat	ChaosDE Sinai
2	0	**0**	**0**	**0**	**0**
4	0	1,31435	**0,66387**	1,81004	1,83477
6	0	10,8232	**10,4494**	11,9057	11,7926
8	0	24,4106	**21,6622**	23,6621	24,578

Table 10 Average CFE and evaluation time for the Rastrigin's function: 10D

	Canonical DE	ChaosDE Dissipative	ChaosDE Arnold Cat	ChaosDE Sinai
CFE	1997800	**1022100**	2089600	2414600
Time (s)	1686.921	**1123.790**	1720.638	2160.638

5.5 Results for the Stretched V Sine Test Function

Table 11 Average difference from Global minimum for the Stretched V Sine function: 2D–8D

Dim.	Known Value	Canonical DE	ChaosDE Dissipative	ChaosDE Arnold Cat	ChaosDE Sinai
2	0	**$6.3633.10^{-4}$**	$1.538.10^{-3}$	$2.225.10^{-3}$	$9.09.10^{-4}$
4	0	0.43779	**0.39181**	0.45610	0.50088
6	0	3.00447	**2.7542**	2.97886	2.81253
8	0	6.83747	**6.51784**	7.40881	7.28129

Table 12 Average CFE and evaluation time for the Stretched V Sine function: 10D

	Canonical DE	ChaosDE Dissipative	ChaosDE Arnold Cat	ChaosDE Sinai
CFE	1781900	**1387600**	1850600	1868300
Time (s)	1294.516	**1026.029**	1403.32	1432.738

5.6 Results for the Griewangk's Test Function

Table 13 Average difference from Global minimum for the Griewangk's function: 2D–8D

Dim.	Known Value	Canonical DE	ChaosDE Dissipative	ChaosDE Arnold Cat	ChaosDE Sinai
2	0	$6.86.10^{-3}$	**$5.78.10^{-3}$**	$6.84.10^{-3}$	$8.03.10^{-3}$
4	0	0.112184	**0.0925748**	0.126363	0.115032
6	0	0.335629	**0.307843**	0.338102	0.337148
8	0	0.595801	**0.505621**	0.570298	0.528804

Table 14 Average CFE and evaluation time for the Griewangk's function: 10D

	Canonical DE	ChaosDE Dissipative	ChaosDE Arnold Cat	ChaosDE Sinai
CFE	327400	**242500**	334600	333100
Time (s)	217.838	**192.542**	240.192	338.513

6 Brief Analysis of the Results

The results in Tables 3, 4, 5, 6, 7, 8, 9, 10, 11, 12, 13, 14 show that using the Dissipative standard map as a random generator has actually improved the performance of DE. The performance of DE significantly improved in both experiments for the limited number of generations (2D–8D) and for unlimited simulation (10D). The results achieved in the first experiment for the other two chaotic

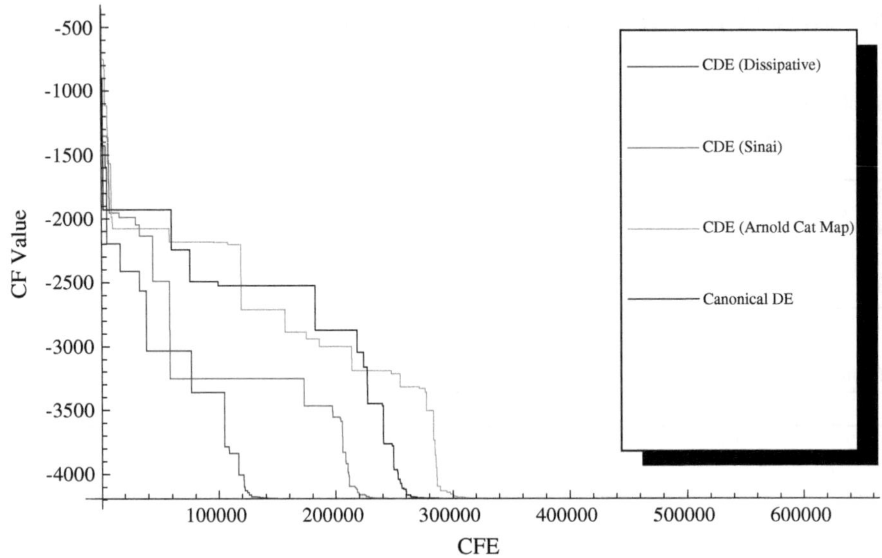

Fig. 9 Example of the time evolution of the cost function values for Schwefel's function and the second experiment—10D

systems are comparable with the results of canonical DE. The Fig. 9 shows the evolution of the cost function value in time (iterations). Time (iterations) are converted here to a number of cost function evaluations, where in each iteration the cost function is evaluated exactly as many times as there are individuals (solutions) in the population.

Based on data presented in Tables 3, 4, 5, 6, 7, 8, 9, 10, 11, 12, 13, 14 and an illustrative example of the time evolution of the cost function values depicted in Fig. 9, the results can be summarized as follows:

Using the Dissipative standard map has significantly improved performance of DE for both experiments.

Using the Sinai map or Arnold's cat map in specific cases helped to improve the behavior of the DE for higher dimension problems; however, for lower dimension problems it is comparable to the performance of canonical DE.

7 Conclusion

In this paper chaos driven DE were tested and compared with canonical DE strategy.

Based on obtained results, it may be claimed, that the developed ChaosDE driven by means of the chaotic Dissipative standard map gives better results than other compared heuristics. Furthermore, all obtained results point to the fact that they are very sensitive to the selection of chaotic system being used as a random generator. Any change in the selection of chaotic system or its parameter adjustment can cause radical improvement of DE performance, however on the downside it can cause a worsening of observed parameters and subsequently the behavior of the evolutionary algorithms as well.

Since this was an initial study, future plans include experiments with benchmark functions in higher dimensions, testing of different chaotic systems and obtaining a large number of results to perform statistical tests.

Furthermore chaotic systems have additional parameters, which can by tuned. This issue opens up the possibility of examining the impact of these parameters to generation of random numbers, and thus influence on the results obtained using differential evolution. One of possible approach in this issue is to use meta-evolution.

Acknowledgments This work was supported by Grant Agency of the Czech Republic GACR P103/13/08195S; by the project Development of human resources in research and development of latest soft computing methods and their application in practice, reg. no. CZ.1.07/2.3.00/20.0072 funded by Operational Program Education for Competitiveness, co- financed by ESF and state budget of the Czech Republic; and by European Regional Development Fund under the project CEBIA-Tech No. CZ.1.05/2.1.00/03.0089.

References

1. K. Price, in *An Introduction to Differential Evolution*, ed. by D. Corne, M. Dorigo, F. Glover. New Ideas in Optimization (McGraw-Hill, London, 1999), pp. 79–108. ISBN 007-709506-5
2. M.F. Tasgetiren, P.N. Suganthan, Q.K. Pan, An ensemble of discrete differential evolution algorithms for solving the generalized traveling salesman problem. Appl. Math. Comput. **215**(9), 3356–3368 (2010)
3. G. Onwubolu, D. Davendra (eds.), *Differential Evolution: A Handbook for Permutation-Based Combinatorial Optimization* (Springer, Germany, 2009)
4. S. Das, A. Konar, U.K. Chakraborty, A. Abraham, Differential evolution with a neighborhood based mutation operator: a comparative study. IEEE Trans. Evolut. Comput. **13**(3), 526–553 (2009)
5. A.K. Qin, V.L. Huang, P.N. Suganthan, Differential evolution algorithm with strategy adaptation for global numerical optimization. IEEE Trans. Evolut. Comput. **13**(2), 398–417 (2009)
6. J. Zhang, A.C. Sanderson, JADE: Self-adaptive differential evolution with fast and reliable convergence performance. in *Proceedings of IEEE Congress on Evolutionary* (Computation, Singapore, 2007), pp. 2251–2258
7. J. Zhang, A.C. Sanderson, Self-adaptive multiobjective differential evolution with direction information provided by archived inferior solutions. in *Proceedings of IEEE World Congress on Evolutionary* (Computation, Hong Kong, 2008), pp. 2801–2810
8. W. Liang, L. Zhang, M. Wang, The chaos differential evolution optimization algorithm and its application to support vector regression machine. J. Softw. **6**(7), 1297–1304 (2011)
9. G. Zhenyu, C. Bo, Z. Min, C. Binggang, in *Self-Adaptive Chaos Differential Evolution*, Lecture Notes in Computer Science. vol. 4221 (2006), pp. 972–975
10. D. Davendra, I. Zelinka, R. Senkerik, Chaos driven evolutionary algorithms for the task of PID control. Comput. Math. Appl. **60**(4), 1088–1104 (2010). ISSN 0898-1221

11. R. Senkerik, D. Davendra, I. Zelinka, M. Pluhacek, Z. Oplatkova, An investigation on the differential evolution driven by selected discrete chaotic systems. in *Proceedings of the 18th International Conference on Soft Computing, MENDEL* (2012), pp. 157–162
12. R. Senkerik, D. Davendra, I. Zelinka, M. Pluhacek, Z. Oplatkova, An investigation on the chaos driven differential evolution: an initial study. in *Proceedings of the Fifth International Conference on Bioinspired Optimization Methods and Their Applications, BIOMA* (2012), pp. 185–194
13. Price, K., Storn, R., *Differential Evolution Homepage*, 2001, [Online]. Available: http://www.icsi.berkeley.edu/~storn/code.html
14. J.C. Sprott, *Chaos and Time-Series Analysis* (Oxford University Press, 2003)

MIX
Papier aus verantwortungsvollen Quellen
Paper from responsible sources
FSC® C105338

If you have any concerns about our products,
you can contact us on
ProductSafety@springernature.com

In case Publisher is established outside the EU,
the EU authorized representative is:
Springer Nature Customer Service Center GmbH
Europaplatz 3, 69115 Heidelberg, Germany

Printed by Libri Plureos GmbH
in Hamburg, Germany